I. M. Ward / P. D. Coates / M. M. Dumoulin

Solid Phase Processing of Polymers

Polymer Processing Society

Progress in Polymer Processing

Series Editor: L. A. Utracki

A. I. Isayev
Modeling of Polymer Processing

L. A. Utracki
Two-Phase Polymer Systems

A. Singh / J. Silverman
Radiation Processing of Polymers

Series Editor: W. E. Baker

I. Manas-Zloczower / Z. Tadmor
Mixing and Compounding of Polymers

T. Kanai / G. A. Campbell
Film Processing

R. S. Davé / A. C. Loos
Processing of Composites

Series Editor: K. S. Hyun

I. M. Ward / P. D. Coates / M. M. Dumoulin
Solid Phase Processing of Polymers

I. M. Ward / P. D. Coates / M. M. Dumoulin (Editors)

Solid Phase Processing of Polymers

With Contributions from

A. Ajji, D. C. Bassett, C. W. M. Bastiaansen,
W. Bessey, P. D. Coates, K. C. Cole, M. M. Dumoulin,
M. J. N. Jacobs, M. Jaffe, P. J. Lemstra, J. E. McIntyre,
S. Osawa, T. Peijs, R. S. Porter, J. Sweeny, A. K. Taraiya,
J.-F. Tassin, I. M. Ward.

HANSER

Hanser Publishers, Munich
Hanser Gardner Publications, Inc., Cincinnati

The Editors:
I. M. Ward, IRC in Polymer Science & Technology, Dept of Physics & Astronomy, University of Leeds, Leeds LS2 9JT, UK
P. D. Coates, IRC in Polymer Science & Technology, Dept of Mechanical & Manufacturing Engineering, University of Bradford, Bradford BD7 1 DP, UK
M. M. Dumoulin, Industrial Materials Institute, National Research Council Canada
75, Boulevard de Mortagne, Boucherville, Québec, Canada J4B 6Y4

Distributed in the USA and in Canada by
Hanser Gardner Publications, Inc.
6915 Valley Avenue, Cincinnati, Ohio 45244-3029, USA
Fax: (513) 527-8977 or 1-800-950-8977
Internet: http://www.Hansergardner.com

Distributed in all other countries by
Carl Hanser Verlag
Postfach 86 04 20, 81631 München, Germany
Fax: +49 (89) 98 12 64

The use of the general discriptive names, trademarks, etc., in this publication, even if the former are not especially identified, is not to be taken as a sign that such names, as understood by the Trade Marks and Merchandise Marks Act, may accordingly be used freely by anyone.

While the advice and information in this book are believed to be true and accurate at the date of going to press, neither the authors nor the editors nor the publisher can accept any legal responsibility for any errors or omissions that may be made. The publisher makes no warranty, express or implied, with respect to the material contained herein.

Library of Congress Cataloging-in-Publication Data
Solid phase processing of polymers / edited by I. M. Ward, P. D. Coates, and M. M. Dumoulin
with contributions from A. Ajii ... [et al].
    p. cm.
Includes bibliographical references and index.
ISBN 1-56990-307-7 (hardcover)
1. Plastics-Forming. I. Ward, I. M. (Ian Macmillan), 1928- II. Coates, P.D. (Phil D.) III. Dumoulin, M. M.

TP1150.S665 2000
668.4'12-dc21   00-039660

Die Deutsche Bibliothek – Einheitsaufnahme
Solid phase processing of polymers / ed. by I. M. Ward ... With contributions from A. Ajii ... -
Munich : Hanser; Cincinnati : Hanser Gardner, 2000
ISBN 3-446-19622-6

© Carl Hanser Verlag, Munich 2000
Printed and bound in Germany by Kösel, Kempten

Dr. A. **Misra**
IIT – New Delhi
INDIA

Dr. Tadamoto **Sakai**
Japan Steel Works
JAPAN

Dr. Christine **Strohm**
Hanser Gardner Publications
U.S.A.

Prof. Zehev **Tadmor**
Technion
ISRAEL

Dr. Hideroh **Takahashi**
Toyota Central Research and Development
Laboratories Inc.
JAPAN

Dr. Leszek A. **Utracki**
National Research Council of Canada
CANADA

Dr. George **Vassilatos**
E. I. Du Pont Co.
U.S.A.

Prof. John **Vlachopoulos**
McMaster University
CANADA

Prof. I. M. **Ward**
The University of Leeds
UNITED KINGDOM

Prof. James L. **White**
University of Akron
U.S.A.

Prof. Xi **Xu**
Chengdu University of Science and
Technology
CHINA

Foreword

Since the Second World War, the industry based on polymeric materials has developed rapidly and spread widely. The polymerization of new polymeric species advanced rapidly during the sixties and the seventies, providing a wide range of properties. A plethora of specialty polymers have followed as well, many with particularly unique characteristics. This evolution has been invigorated by the implementation of metallocene catalyst technology. The end-use of these materials has depended on the development of new techniques and methods for forming, depositing, or locating these materials in advantageous ways, which are usually quite different from those used by the metal or glass fabricating industries. The importance of this activity, „Polymer Processing", is frequently underestimated when reflecting on the growth and success of the industry.

Polymer processes such as extrusion, injection molding, thermoforming, and casting provide parts and products with specific shapes and sizes. Furthermore, they must control, beneficially, many of the unusual and complex properties of these unique materials. Because of their high molecular weights and, in many cases, tendency to crystallize, polymer processes are called upon to control the nature and extent of orientation and crystallization, which in turn have a substantial influence on the final performance of the products made. In some cases, these processes involve synthesizing polymers within a classical polymer processing operation, such as reactive extrusion. Pultrusion and reaction injection molding both synthesize the polymer and form a finished product or part all in one step, evidence of the maturing of the industry. For these reasons, successful polymer process researchers and engineers must have a broad knowledge of fundamental principles and engineering solutions.

Some polymer processes have flourished in large industrial units, synthetic fiber spinning for example. However the bulk of the processes are rooted in small- and medium sized entrepreneurial enterprises in both developed and new developing countries. Their energy and ingenuity have sustained growth to this point but clearly the future will belong to those who progressively adapt new scientific knowledge and engineering principles, which can be applied to the industry. Mathematical modeling, online process control and product monitoring, and characterization based on the latest scientific techniques will be important tools in keeping these organizations competitive in the future.

The Polymer Processing Society was started in Akron, Ohio in 1985 with the aim of providing a focus, on an international scale, for the development, discussion, and dissemination of new and improved polymer processing technology. The Society facilitates this by sponsoring several conferences annually and by publishing the journal International Polymer Processing, and the volume series Progress in Polymer Processing. This series of texts is dedicated to the goal of bringing together the expertise of accomplished academic and industrial professionals. The volumes have a multi-authored format, which provides a broad picture of the volume topic viewed from the perspective of contributors from around the world. To accomplish these goals, we need the thoughtful insight and effort of our authors and volume editors, the critical overview of our Editorial Board, and the efficient production of our Publisher.

This volume deals with the solids phase processing of polymers for preparing polymer products, which attain the special mechanical properties due to orientation in the forms of fibres, films and solid sections including rods, sheets and tubes by special fabrication techniques. These processes have developed into what is arguably the single largest outlet for enhancing the properties of synthetic polymers. They are dependent on the best achievements in polymer design to provide the appropriate shear and extensional viscosity for successful processing. These design achievements have also produced the mechanical and optical properties so important in applications. Therefore, most important in this volume are the extensive discussions on the developments of molecular orientation and structural changes and how these lead to improvements in properties especially mechanical properties. This volume includes numerous contributions, industrial and academic, from Europe as well as Asia and North America and, as such, forms a very useful contribution to the solid forming industries. This volume was initiated with Dr. Warren Baker, my predecessor and became the first volume I had a pleasure to be associated.

Midland, Michigan,   *Kun Sup Hyun*
U.S.A.   *Series Editor*

May 2000

Contents

SOLID PHASE PROCESSING OF POLYMERS
Author details

A. Ajji and M. M. Dumoulin, Industrial Materials Institute, National Research Council Canada, 75, Boul. de Mortagne, Boucherville, Québec, Canada J4B 6Y4

D. C. Bassett, J J Thomson Physical Laboratory, University of Reading, Whiteknights, Reading, RG6 6AF, UK

C. W. M. Bastiaansen, Dutch Polymer Institute, Eindhoven University of Technology, Den Dolech 2, P O Box 513, 5600 MB Eindhoven, The Netherlands

W. Bessey, SpaceNet Products Inc, 7159 Windyrush Road, Charlotte NC 28226, USA

P. D. Coates, IRC in Polymer Science & Technology, Dept of Mechanical & Medical Engineering, University of Bradford, Bradford BD7 1DP, UK

K. C. Cole and A. Ajji, Industrial Materials Institute, National Research Council Canada, 75, Boul. de Mortagne, Boucherville, Québec, Canada J4B 6Y4

M. J. N. Jacobs, DSM High-Performance Fibers, Eisterweg 3, 6422 PN Heerlen, The Netherlands

M. Jaffe, The New Jersey Center for Biomaterials, Medical Device Concept Laboratory, 111 Lock Street, Newark NJ 07013, USA

P. J. Lemstra, Dutch Polymer Institute, Eindhoven University of Technology, Den Dolech 2, P O Box 513, 5600 MB Eindhoven, The Netherlands

J. E. McIntyre, School of Textile Industries, University of Leeds, Leeds LS2 9JT, UK

S. Osawa, Department of Materials Science and Engineering, Kanazawa Institute of Technology , Nonoichi, Ishikawa 921-8501, JAPAN

T. Peijs, Dutch Polymer Institute, Eindhoven University of Technology, Den Dolech 2, P O Box 513, 5600 MB Eindhoven, The Netherlands
and
Department of Materials, Queen Mary and Westfield College, University of London, Mile End Road, London E1 4NS, UK

R. S. Porter, Department of Polymer Science and Engineering, University of Massachusetts, Amherst MA 01003-4530, USA

J. Sweeney, IRC in Polymer Science & Technology, Dept of Mechanical & Medical Engineering, University of Bradford, Bradford BD7 1DP, UK

A. K. Taraiya, IRC in Polymer Science & Technology, Dept of Physics & Astronomy, University of Leeds, Leeds LS2 9JT, UK

J.-F Tassin, Chimie et Physique des Matériaux Polymères, UMR CNRS 6515, Université du Maine, Avenue Olivier Messiaen, 72085 Le Mans Cedex, France

I. M. Ward, IRC in Polymer Science & Technology, Dept of Physics & Astronomy, University of Leeds, Leeds LS2 9JT, UK

Preface

This book is intended to be a comprehensive up to date account of the solid phase processing of polymers with particular emphasis on the production of oriented polymers, in the form of fibres, films and solid sections including rods, sheets and tubes.   There are detailed discussions of how such oriented materials can be produced by a very wide range of techniques including tensile drawing and die drawing, ram extrusion and hydrostatic extrusion. There are also extensive discussions of the development of molecular orientation and structural changes and how these lead to improvements in properties, especially mechanical properties, where the last thirty years have seen spectacular achievements. We consider that this book should form a useful bridge between polymer engineering and polymer physics and this catholicity of interests is very strongly reflected in many of the chapters.

The editors have been fortunate in being able to draw on a vast range of expertise from their many colleagues around the world, especially in Europe and North America. It is very sad to report that Professor Roger Porter, who made seminal contributions to this field, is no longer with us and we are very grateful that he was so determined to complete his chapter.

In conclusion, we would like to thank the Polymer Processing Society and in particular Warren Baker, for the invitation to produce this book and hope that it will serve both as a valuable aid-memoire to the experienced polymer scientist/technologist and a useful introduction to those entering a very challenging and exciting area.

1 Introduction

I M Ward* and P D Coates+
IRC in Polymer Science & Technology
*University of Leeds and +University of Bradford, UK

The last forty years have seen considerable advances in polymer processing, one key area of progress being the production of highly oriented polymers. Processing is receiving increasing emphasis for several reasons. First, it has been recognised that the properties of polymers do not depend only on their chemical composition, but also on the structure which is imparted during processing. Secondly, there is the challenge that by optimising processing operations, polymers can be used for significant load bearing applications, with properties equivalent to or in some cases better than competitive materials. Finally, the combination of improved modelling and increasing computer power should make possible higher degrees of process control.

The contents of this text book reflect these several themes in some detail. Several chapters are devoted to the development of morphology and orientation in polymer processing, which are especially important for high modulus and high strength polymers. These considerations lead naturally to the outstanding commercial developments of aramid and polyethylene fibres by solution and gel spinning, and the gradual emergence of similar products by melt phase and solid phase processing, all of which are discussed.

In the case of the fibre developments, the traditional emphasis has been on new chemistry and on very inventive structural considerations such as those relating to the formation of liquid crystalline phases or gel structures. For the solid phase processing of oriented polymers, the emphasis has been on polymer engineering, to set up useful constitutive relationships to describe solid phase deformation and to provide quantitative analyses of the mechanics of the processes.

1.1 Key Scientific Issues

From the viewpoint of basic scientific considerations there are several principal issues.

1.1.1 The chemical structure of the polymer and its degrees of regularity

Such factors determine whether the polymer can crystallise, whether it can form a liquid crystalline phase and whether it can form high strength and high modulus products for which a high degree of anisotropy is required. Molecular weight and molecular weight distribution are also of key importance in determining both processability and high strain mechanical properties, tensile strength and toughness.

The contribution of chemistry has been twofold:

(1)   New preparative methods for "conventional" polymers, via new catalysts or new cost effective methods for intermediates. Key recent examples here are the development of the metallocene catalysts for polyolefines and the new intermediates for polyesters i.e. dicarboxynaphthoic acid for polyethylene naphthalate and 3 ethylene glycol for polytrimethylene terephthalate. To obtain optimum advantage from the new polymers produced requires the careful optimisation of polymer processing procedures.

(2)   Polymers of very novel chemical composition, such as the polyaramids or the rigid chain bisthiazoles. Commercial pressures have reduced the industrial thrust for really new polymers because of difficulty of making a satisfactory financial return on "niche" products. The recent ACZ0-NOBEL M5 polymer [1] (polypyrido bisimidazole, PIPD) is a notable exception.

The design of new polymers has been given a degree of stimulus by recent developments in the molecular modelling of polymer properties, due to both academic research and the establishment of major software packages by commercial organisations for example Molecular Simulations Incorporated and Oxford Materials.

Molecular modelling is most effective in predicting the structure of crystalline polymers (and perhaps liquid crystalline polymers) and their elastic properties. The results of such calculations enable the polymer technologist to set realistic targets for optimal polymer processing. Table 1.1 illustrates this point by comparing the elastic constants for an ideal polyethylene fibre (the crystal elastic constants suitably averaged to fibre symmetry) with those determined experimentally for several processing methods, fibre melt spinning and hot compaction, die-drawing, and a unidirectional fibre composite [2]. It can be seen that all these production methods are quite successful and only the axial stiffness is very significantly less than that predicted theoretically. As shown in Chapter 5, gel-spinning provides a method of closing the gap still further between practical achievement and theory.

TABLE 1.1   Comparison of stiffness constants (in GPa) for hot compacted sheet of melt-spun PE fibre, die-drawn PE sheet, melt-spun PE fibre/Epoxy composite (55% fibre volume fraction), and theoretical values for uniaxially oriented sheet [3]

Process	C_{33}	C_{11}	C_{13}	C_{12}	C_{44}
Compacted Fibre	62.3	7.16	5.09	4.15	1.63
Die drawn sheet	66.0	6.90	4.40	3.90	1.60
PE/epoxy composite	54.8	7.62	5.89	4.39	1.71
Theoretical values for uniaxially oriented sheet	290	9.15	5.15	3.95	2.86

In the case of strength and toughness, the gap between ideal theoretical properties and those achieved is somewhat greater, but here again theoretical guidelines based on the chemical structure can be valuable. One example is given by the compressive strengths of oriented fibres. There are good theoretical grounds for believing that there is a correlation between compression strength and shear modulus, and the latter can be estimated by molecular modelling. Figure 1.1 shows this correlation, based primarily on results presented by Deteresa and Ferris [4], but adding recent data for the Akzo-Nobel PIPD fibre [5].

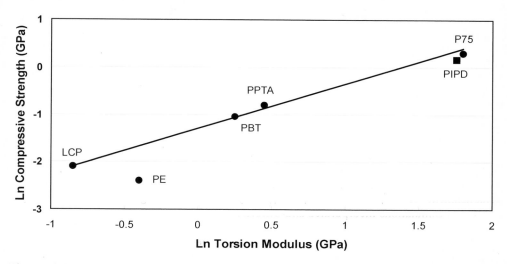

Figure 1.1   Compressive strength (GPa) versus torsional modulus (GPa) for high performance fibres   [LCP Vectran, PE Polyethylene, PPTA poly(paraphenylene terephthalamide),   PBT poly(p phenylene benzobisthiazole), P75 pitch based graphite, PIPD polypyridobisimidazole]

1.1.2   The effect of plastic deformation, the concept of the true stress-true strain curve

It has been long recognised that molecular orientation can produce very great enhancement of the mechanical properties of polymers, and processes for producing oriented polymers are an integral part of the commercialisation of fibres and films. There are two routes to producing highly oriented polymers, which differ in principle.

1.1.2.1 Starting with an amorphous or semi-crystalline polymer and deforming to very high degrees of stretch

This is the conventional tensile drawing route followed for the production of nylon and polyester fibres and for most film processing. It is discussed in detail in Chapters 4 and 6 of this book for fibres and films respectively. The structural changes which occur are of crucial significance and these are discussed in Chapter 2.

The key conceptual idea in this route is the stretching of a molecular network in which the junction points of the network are formed by physical entanglements. Even if crystallisation occurs, the molecular network places overall constraints on any subsequent deformation

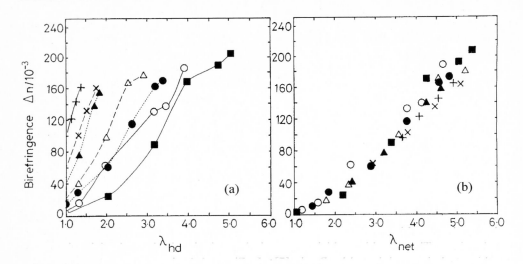

Figure 1.2 Birefringence vs (a) actual hot draw ratio λ_{hd} and (b) network draw ratio λ_{net} for pin-drawn yarns and their spun yarn precursors ■ A   O B   ● C   △ D   ▲ E X F   + H   (A to H is increasing spun yarn orientation due to increased wind-up speed).

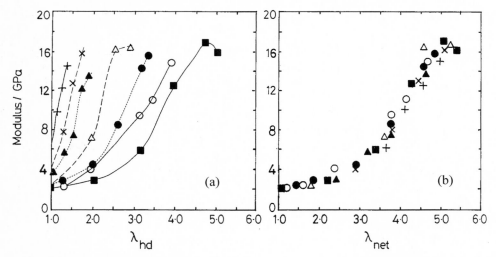

Figure 1.3 Initial modulus vs (a) actual hot draw ratio λ_{hd} and (b) network draw ratio λ_{net} for spin-drawn yarns and their precursors (symbols as in Figure 1.2)

which can occur as the polymer cools down to ambient temperature during the fibre or film extrusion process and in subsequent tensile drawing processes.   The development of molecular orientation and hence enhancement of physical properties relates to the deformation of the molecular network.   It will always be recognised that for fibre and film processing, properties such as modulus and strength could usually be correlated with draw ratio, but it is important to appreciate that the true correlation is with the total network draw ratio, including the total deformation which has occurred from the point at which the network forms. This issue has been clearly demonstrated for the production of polyester (polyethylene terephthalate) PET fibres.

A series of PET fibres was produced by a route where molecular orientation was introduced at spinning by varying the wind-up speed and further orientation was introduced by hot drawing at a temperature above the glass-transition where the polymer stretches homogeneously [6]. The properties of the final oriented fibres were then determined, principally birefringence and modulus, strength and extension to break. As shown in Figures 1.2a and 1.3a, the birefringence and modulus increase with draw ratio for each starting spun yarn of different orientation, but there is no general correlation with the hot draw ratio.

As Figures 1.2b and 1.3b show, the key correlation is with the network draw ratio which embraces the stretching of the network which is imposed during spinning as well as during hot drawing. The stretching which occurs during spinning is determined by the curve matching technique originated by Brody [7]. In Figure 1.4a the room temperature tensile stress-strain curves of all the spun yarns are shown and in Figure 1.4b they are superimposed by horizontal shifts on the log strain axis. When these shifts are used to

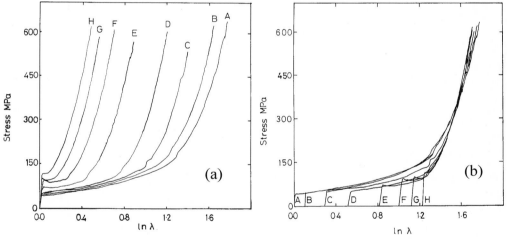

Figure 1.4 (a) True stress-strain curves for spun yarns: A to H is increasing spun yarn orientation. (b) Matching of true stress-strain curves for spun yarns.

determine the network extension in the threadline, the total network draw ratio can be calculated, and the excellent correlations between properties and network draw ratio shown in Figures 1.2b and 1.3b are obtained.

It is important to note that the curve-matching method does not require that the room temperature deformation is homogeneous. During the intermediate strain region, drawing can occur through a neck, but the curve-matching depends on the final upturn of the stress-strain curve where the deformation is homogeneous.

The concept of a true-stress true-strain curve is important for understanding and modelling the plastic deformation and has been used to provide a base-line for the analysis of the mechanics of solid phase deformation processes [8], including hydrostatic extrusion and die-drawing. In these analyses it is assumed that the flow stress depends only on the total plastic deformation and an appropriate measure of the current strain rate. To avoid the problem of neck formation and the consequent inhomogeneous deformation, the flow stress for intermediate degrees of deformation may have to be determined by redrawing oriented samples prepared by hydrostatic extrusion, where the homogeneous deformation is imposed by extrusion through a reducing conical die. For uniaxial deformation it is thus possible to produce a three dimensional description - a flow stress diagram, as shown in Figure 1.5. This diagram defines the flow stress behaviour of a specific grade of polyethylene (Rigidex 50) at a specific temperature (90°C). Each solid phase deformation process then involves taking the polymer on a defined route across this flow stress/strain/strain rate surface, illustrated for tensile drawing and hydrostatic extrusion in Figure 1.6. It can be seen that in tensile drawing, where a neck is formed, the high strain rates in the neck give rise to a peak in the flow stress. In hydrostatic extrusion, through a reducing conical die, on the other hand, the strain and strain-rate rise monotonically as the polymer is deformed and the flow stress also rises monotonically to a maximum value at the die exit. These ideas are developed and discussed in detail in Chapter 9.

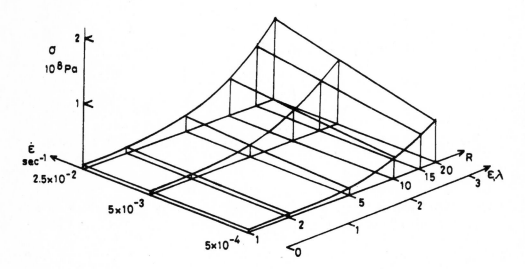

Figure 1.5 True stress-strain-strain rate surfaces for Rigidex 50 LPE at 100°C

Recent research has focused on the possible extension of the true-stress true-strain curve concept to a multidimensional situation, in particular the biaxial drawing of tubes, or the biaxial stretching of films. These are discussed in Chapters 10 and 11 respectively. The simplest approach is to define the stress and strain in terms of equivalent stress and equivalent strain. A rather more sophisticated approach is to undertake the analysis in terms of the octahedral shear stress and the octahedral shear strain rate obtained by transforming the strain rate tensor onto the axes defined by the octahedral stresses.

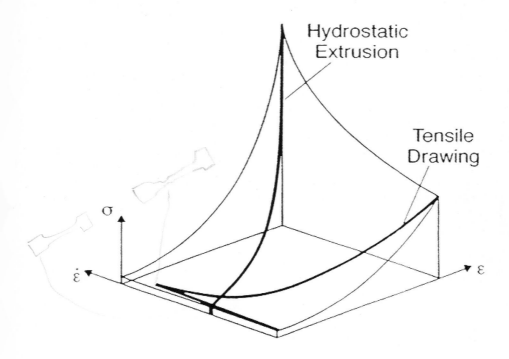

Figure 1.6 Process flow stress paths for hydrostatic extrusion and tensile drawing.

1.1.2.2 Development of orientation in Liquid Crystalline Polymers and Chain-Extended Polyethylene

The processing route for liquid crystalline polymers is different in kind from the deformation to high strains which is required for conventional crystalline and amorphous polymers. In the liquid crystalline polymer highly oriented domains are formed in solution (lytropic liquid crystalline polymers) or in the melt (thermotropic liquid crystalline polymers). The alignment of the domains takes place primarily in the solution or melt spinning process, and any further perfection is produced by annealing the structure under tension in the solid phase. This is discussed in detail in Chapter 5. The hydrostatic extrusion of chain-extended polyethylene is somewhat analogous, in that the isotropic aggregate of highly aligned chain-extended crystalline domains is aligned by extrusion through a conical die (Chapter 9).

1.1.3   Structural considerations: molecular understanding of plastic deformation

There are several aspects to developing a molecular understanding of solid phase deformation and some simple semi-phenomenological ideas which help to develop this understanding.

One key issue is the nature of the yield or flow process and the associated issues of temperature and strain rate dependence. At the comparatively high temperature where polymers are processed in the solid phase, it appears that it is appropriate to adopt the Eyring formulation of a thermally activated rate process. The plastic strain-rate

$$\dot{\varepsilon} = \dot{\varepsilon}_o \, \exp{-\left(\frac{\Delta H - \sigma v}{kT} \right)}$$

where $\dot{\varepsilon}_o$ is the pre-exponential constant, ΔH and v are the activation energy and

activation volume of the process. In the simplest formulation, this activated rate process can be incorporated into the network stretching model as a viscosity stress acting in parallel with the network stress as shown in Figure 1.7. The flow stress σ then has two terms and we can write[8]

$$d\sigma = \left(\frac{\partial \sigma}{\partial \varepsilon} \right)_{\dot{\varepsilon}} d\varepsilon + \left(\frac{\partial \sigma}{\partial \dot{\varepsilon}} \right)_{\varepsilon} d\dot{\varepsilon}$$

where $\left(\dfrac{\partial \sigma}{\partial \varepsilon} \right)_{\dot{\varepsilon}}$ represents the strain hardening spring E in Figure 1.7,

and $\left(\dfrac{\partial \sigma}{\partial \dot{\varepsilon}} \right)_{\varepsilon}$ represents the strain rate sensitivity dashpot η.

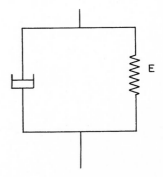

Figure 1.7   Simple mechanical model incorporating viscosity stress acting in parallel with network stress.

Although there is considerable debate regarding the validity of parallel or series formulations as a starting point for a more sophisticated analysis of plastic deformation in polymers, this approach is being generally followed in one form or another and is discussed in detail in Chapter 10. Another important general point is that recent research [9] shows that the viscosity term can be related to key deformation processes, such as interlamellar shear and c-shear in polyethylene, as the isotropic structure deforms due to imposed plastic deformation. In general terms these studies explain why it is usually necessary to undertake the solid phase deformation at comparatively high temperatures, often close to the melting point so that there is adequate mobility at a molecular level.

A further key development is the use of molecular network models to relate the growth of molecular orientation on plastic deformation to the degrees of stretch, either uniaxial or biaxial. It has been shown that simple schemes based on stretching rubber networks can give good first-order predictions for the molecular orientation of both crystalline and amorphous polymers. Generally the approach is to follow the Kuhn and Grün model where the actual molecular network is replaced by an equivalent network of identical chains each containing freely-jointed rigid links [10]. Figure 1.8 shows the predicted and measured development of orientation in PET, following such a scheme [11] (called the affine deformation scheme). At high draw ratios the chains are fully extended and the deformation pattern changes from a rubber network model [12] to the rotation of fully oriented units (the Kratky floating rod model or pseudo-affine deformation scheme). Present developments include the use of molecular modelling procedures to give more accurate prediction and greater physical insight.

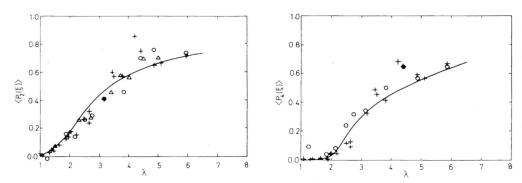

Figure 1.8. Development of molecular orientation as a function of draw ratio $<P_2(\xi)>$ and $<P_4(\xi)>$ are the second and fourth Legendre functions describing molecular orientation, theoretical predictions —, experimental results from refractive index measurements Δ, florescence measurements +, and Raman measurements o.

A final issue is the change in morphology which occurs when a crystalline polymer is transformed from its initial spherulite texture to a final fibrillar texture on plastic deformation. This is dealt with in Chapter 2.

References

1.   D.J. Sikkema, Polymer, **39**, 5981 (1998).
2.   I.M. Ward & P.J. Hine, Polym. Eng. Sci., **37**, 1809 (1997).
3.   D.J. Lacks & G.C. Rutledge, J. Phys. Chem., **98**, 1222 (1994).
4.   S.J. Deteresa and R.J. Farris, Mater. Res. Soc. Proc., **134**, 375 (1989).
5.   P.J. Hine, B. Brew & I.M. Ward, Composites Science & Technology **59**, 1109 (1999).
6.   S.D. Long & I.M. Ward, J. Appl. Polym. Sci., **42**, 1911 (1991).
7.   H. Brody, J. Macromol. Sci. Phys., **B22**, 19 (1983).
8.   P.D. Coates, I.M. Ward, J. Mater. Sci., **15,** 2897 (1980).
9.   N.W. Brooks, A.P. Unwin, R.A. Duckett & I.M. Ward, J. Polymer Sci., Polymer Physics **35**, 545 (1997).
10.   W. Kuhn and F. Grün, Kolloid Z., **101**, 248 (1942).
11.   J.H. Nobbs, D.I. Bower & I.M. Ward, J. Polymer Sci., Polymer Physics Edn., **17,** 259 (1979).
12.   O. Kratky, Kolloid Z., **64**, 213 (1933).

2   Deformation Mechanisms and Morphology of Crystalline Polymers

D C Bassett
University of Reading
Reading, UK

2.1   Introduction

Polymeric materials are especially valuable for their exceptional toughness and plasticity, properties which are, nevertheless, continually sought to be combined with enhanced stiffness and strength appropriate to the molecular chain. To this end it is important to have a framework of understanding within which to address such matters. One can note at the outset that their toughness derives from molecular length and the capability this bestows of being able both to blunt incipient cracks by spreading and so reducing the local tensile stress and of having to break the strong covalent bonds which constitute the backbone of organic polymers. Falling temperature reduces the former accordingly, leading to the onset of brittleness, while crystallization tends to stiffen and harden polymers. In a crystalline polymer yield is linked to that of the crystal lamellae but, if crystallinity is too high, it can bring easier failure in the form of cleavage and fibrillation. To approach these matters it is necessary to decide whether it is adequate to treat the material as a continuum, homogeneous at the molecular or lamellar levels, or whether larger features of the microstructure must be taken into account on the principle that it is failure of the weakest internal link which causes the whole specimen to founder and macroscopic properties to fall well below those of a defect-free material or the crystal lattice.

The former approach includes the molecular and free volume theories applied to the deformation of amorphous polymers [1]. Allied to this, in that physical texture is not specifically considered, is the study of yield criteria for which, once a suitable combination of stresses is reached, yield is predicted to ensue in a homogeneous specimen containing an adequate distribution of defects, such as dislocations in the crystalline component. These are mature subjects on which an extensive literature exists [2, 3, 4]. On the other hand, the involvement of microstructure in yield is less well-explored, in large measure because of the lack of suitable microscopies with which to view the complex morphologies. Nevertheless, significant recent advances have been made and it is with these that this article is predominantly concerned.

2.2   Macroscopic Phenomena

The subject of the involvement of physical microstructure in the deformation and yield of polymers may be approached from two sides, phenomenological or textural. The tensile test, probably the most widely used measurement of the mechanical properties of a polymer, illustrates the former. A specimen deformed at constant extension rate will show load-extension behaviour whose character changes with temperature. At one extreme are brittle specimens extending elastically to break at small strains of a few per cent as for cold poly(methyl methacrylate) or a Kevlar fibre, while one or more distinct maxima preceding a region of draw at constant load before strain hardening, is a familiar pattern for e.g. cold-drawn polyethylene [4]. This is a similar pattern to yield drop and easy glide in metals for which more or less elaborate explanations are given in terms of dislocation movement [2]. In a polymer such as polyethylene, in which interlamellar regions are rubbery at room temperature, yield drops are certainly associated with combinations of twinning, martensitic transformation to a second crystal phase and slip [5,6] but there is not the detailed understanding attained in metals.

Understanding is gained by comparison of the deformed product with the starting material. X-rays show, in wide angle diffraction, WAXS, the development of molecular orientation, most readily evaluated for crystalline samples, plus, at small angles, SAXS, the size and direction of long-range electron density variations (crystallites). For the latter at least, one needs a microstructural model on which to base an evaluation. The discovery of chainfolded lamellae has been very influential in this regard providing the basis of most subsequent interpretations. It has proved more difficult to incorporate higher order structures, notably spherulites, into models of deformation: clarification of important aspects is still required but there is significant progress in this area.

The complementary approach of deforming model systems, beginning with individual solution-grown lamellae, has revealed something of the complexity of behaviour but comparable information is not generally obtainable for melt-crystallized samples because of the difficulties of microscopic examination, especially the use of diffraction techniques on selected areas to determine crystallography and orientation. On the other hand it is now possible to obtain good quality topographic data, at least to lamellar resolution. Especially when applied to well-defined systems this has enabled useful gains in understanding deformation mechanisms to be made. The interaction of morphology with deformation is the principal theme of this chapter.

2.3   Cracks, Crazing and Brittleness

One of the simplest concepts is of tensile failure in relation to flaws in the material emanating either from fabrication or from fatigue in use [2]. Cracks, and crazes which may be their precursors [4], develop normal to tensile stresses. Inspection of a brittle amorphous material will show them appearing first where the corresponding tensile stresses are maximum [7] but cracks and crazes are not confined to such materials. Polypropylene, for example, undergoes crazing in both amorphous and spherulitic conditions, with crazes forming normal to the applied tensile stress at low and high temperatures but deviating up to 15° for spherulitic material just above the glass transition temperature [8]. In this last case crazes always pass through the centre of spherulites and never through their boundaries, showing *inter alia,* that spherulitic boundaries are not inherently weak. This finding is relevant to suggestions that there would be high concentrations in these boundaries of segregated species viz. poorly tactic and shorter molecules, the latter of

De Dengi lecture :-

Include solid state meaning - die chemistry
                          ↓
                    provide a way of

enlarging material properties.

                    Sweeny book ?
                    Glen...?

Magic Toffles - hallucinogenic.

which would certainly weaken boundaries. Weakness due to segregation is an unusual phenomenon in contemporary materials.

Two more recent examples discussed below of crazing in ductile crystalline materials are the drawing of polytetrafluoroethylene, PTFE, and in rubber-toughened polypropylene. Crazes parallel to the draw direction are also the probable origin of lateral textures in highly drawn polyethylene and polypropylene.

The technology of manufacturing PTFE for textile apparel uses fast tensile drawing typically at 300 °C; specimens drawn at slow speeds are liable to fail [9]. Under optimum conditions the material crazes uniformly normal to the applied stress developing a porous texture on the micron scale (Figure 2.1). The controlling factors are the morphology of the undrawn material, prepared from aggregates of as-polymerized particles, which affects where the crazes start and thereby the scale of the porosity together with the lifetime of the transient entanglements in the system. It is essential for a sample to maintain its integrity that it can sustain the applied stress; in polymers this requires covalent connectivity across the specimen. The application of stress tends to unravel those entanglements which exist giving a lifetime within which drawing must operate [10]. Unpublished measurements from this laboratory give lifetimes in PTFE extending from seconds to minutes which are independent of drawing speed; longer lifetimes correlate with better products.

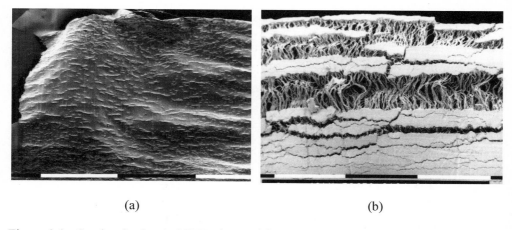

(a)   (b)

Figure 2.1   Crazing in drawn PTFE observed by SEM; a) low magnification with small crazes lying normal to the (vertical) draw direction, bar = 1 mm; b) higher magnification of a different sample revealing the fibrous substructure which is able to transmit stress, bar = 0.1 mm.

Such behaviour emphasizes that the existence of a network of covalent molecules extending through samples underlies polymeric mechanical behaviour although it is difficult to quantify. Molecular networks provide a background, in space and time, within which other mechanisms operate and in providing tie molecules which are essential to the integrity of crystalline polymers.

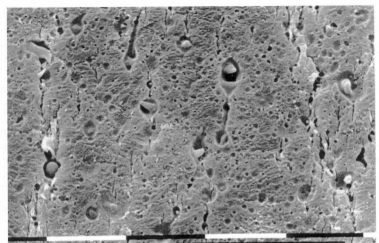

Figure 2.2   Crazes in rubber-toughened polypropylene lying normal to the (horizontal) tensile stress and emanating from the waist of rubber particles. SEM image; bar = 10 μm.

The second example referred to, rubber toughening, is widely used in improving impact properties [11]. The different moduli of rubber particles and their matrix – which may be glassy as in polystyrene or ductile as in polypropylene – means that a common applied strain creates differential stresses at their mutual interface, tending to initiate crazes - which are energy-absorbing - at the waist of the rubber particle, perpendicular to the applied stress. Figure 2.2 shows this phenomenon in a sample subjected to fast tensile drawing in experiments designed to simulate impact behaviour but in a more controlled way.

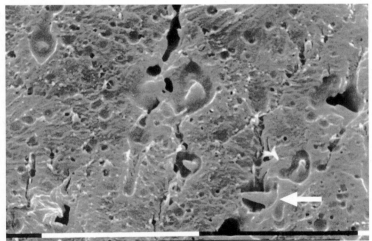

Figure 2.3   Higher magnification of the sample of Figure 2.2 showing permanent deformation in the draw direction of the polyethylene core (arrowed) of rubber-toughened polypropylene when it penetrates through the rubber into the polypropylene matrix. SEM image; bar = μm.

The same samples, whose particles were of ethylene/propylene rubber surrounding a polyethylene core, within a polypropylene matrix are liable to show permanent deformation of the polyethylene core (Figure 2.3). This occurs when a core penetrates the rubber through to the polypropylene so that its applied strain is that of the polypropylene matrix and is not reduced by the surrounding rubber. For sufficiently large strain the polyethylene will acquire a permanent set while the rubber will retract when the stress is removed. Such retraction is due to rubber elasticity, a polymeric property arising from molecular length and the many associated degenerate states of equal energy, omnipresent in deformed polymers. Its effects will, however, be less in lower molecular weight materials, because their ability to form entangled networks is less. Accordingly, lower molecular weight polymers are less able to blunt cracks and so are liable to fail in tension sooner than their higher molecular weight counterparts.

2.4   Segregation-Induced Brittleness

An example where segregation of short molecules does give tensile failure is in the relaxation of drawn pressure-annealed polyethylene. The first reports of experiments designed to increase the lamellar thickness of drawn commercial grade polyethylene, (BP Rigidex 2) using high pressure annealing, were that drawing gave way to brittleness when the annealing temperature reached that of the orthorhombic transition [12]. In the range immediately prior to this temperature behaviour conformed to a series crystalline/amorphous arrangement with the Young's modulus of the latter equal to 3 kbar. The brittleness could be removed with the shorter molecules if these were extracted from the starting polymer beforehand and specimens made from the modified polymer. Subsequent study [13] revealed a characteristic pattern in the polarizing microscope of a lattice of lenticular flaws composed of shorter molecules and increasing in size with the quantity of molecules less than ~10,000 molecular weight. Electron microscopy showed that the chain axis had buckled around the flaws as the sample had shortened during annealing. The overall process demonstrates that shorter molecules are mobile within specimens, allowing them to segregate and facilitating retraction and buckling of the sample. At the same time their very shortness renders them unable to form a sufficiently entangled molecular network to withstand tensile stresses resulting in failure of the sample. It is a corollary that the absence of tensile failure at low strains in contemporary commercial polyolefines implies that that there has been no significant segregation of shorter molecules.

These examples by no means exhaust the ways in which failure occurs normal to a tensile stress. Further examples will be given later concerning transverse stresses in drawing and relaxation of oriented samples.

In contrast to failure normal to the applied stress, failure in planes near 45° to the tensile stress is likely to be by shear, these being where shear stress is maximum. The precise predictions differ according to which particular yield criterion [2, 4] is chosen. Unlike crazes which result from tensile stress, shear bands [14] may occur in both tension and compression. They lie near to the planes of maximum shear stress but may be deflected away from this by frictional forces. For shear bands as for crazes, it is often sufficient to treat samples as homogeneous with no specific knowledge of physical microstructure generally required.

2.5   Cold Drawing

Bunn and Alcock [15] refer to cold drawing, through a neck, as a common property of all high polymers. Brown [16] and Horsley & Nancarrow [17] found necking in polyethylene stretched at room temperature (and as low as -55 °C) but that no neck was produced at 96 °C. Necking itself is a geometrical consequence of instability when the sample can no longer sustain the applied stress and may be described in terms of the Considère construction [2, 4]. Specifically, its occurrence requires that there be a tangent to the true stress/engineering strain curve from the point (-1,0) on the strain axis and cold-drawing that there be a second tangent for greater extensions when there has been sufficient strain hardening. Although the ultimate result of drawing is an oriented product with the **c** axis aligned along the draw direction this is not adopted smoothly but exhibits intermediate alignment of certain crystal planes both during drawing and subsequently on heat relaxation. Related phenomena were explained by Frank, Keller and O'Connor [18] in terms of crystal plasticity within an enveloping rubber-elastic matrix, capable of providing restoring stress and accommodating the crystals' change of shape, with no more specific textural model. The relevant elements invoked were mechanical twinning, on {110} and {310} planes of the orthorhombic lattice, stress-induced (martensitic) transformation to the monoclinic phase (on irrational planes) and dislocation glide transverse to the chain axis.

The discovery of chainfolded lamellar crystals [19] not only gave new insight into the mechanics of deformation but provided model systems in which to observe their details. On the one hand the reduced electron density at lamellar fold surfaces will give rise to SAXS and lamellae have been used as the basis of interpretation of such scattering ever since. On the other hand, chainfolding will reduce covalent continuity and with it stiffness and strength. The presence of the folds themselves will interfere with possible deformation modes by hindering the movement of those dislocations gliding on planes intersecting them and add to toughness by dissipating tensile stresses [20]. A pertinent and confirmatory observation [19] was that when hollow pyramidal crystals were sedimented on a substrate, cracks parallel to the presumed fold planes were clean while those in other directions, which would cross fold planes, exhibited pulled threads. Moreover, because fold planes differ in different sectors of lamellae dislocation glide into adjacent sectors will be inhibited. Breaking of covalent bonds does occur as testified by the formation of free radicals [21] but being a high energy process it will be relatively disfavoured. Experiments drawing lamellae on an extensible substrate, most commonly poly(ethyleneterephthalate), Mylar, and recording diffraction patterns of the deformed products for different orientations to the tensile axis. were able to identify several different mechanisms, including stress-induced formation of the monoclinic phase by martensitic transformation [22, 23].

The relevance of such model experiments to melt-crystallized polymers was initially open to debate especially in the continuing absence of secure morphological knowledge. For example, whether regular folding existed in melt-crystallized systems was much debated at least until the Faraday Discussion of 1979 [24].   Nevertheless Bowden, Young and colleagues, using polyethylene with close to single crystal orientation, prepared by combinations of deformation and subsequent annealing, showed that the pole figures of deformed polyethylene were as expected on the basis of dislocation glide, twinning and martensitic transformation to the monoclinic cell even if the WAXS data were too broadened to allow precise identification of the particular mode [25].   The principles involved are, first, that dislocation glide will, in uniaxial tension, rotate the glide direction and the slip plane towards the tensile axis but from opposite sides while in uniaxial compression both slip plane and slip direction tend to become normal to the stress.   Second, twinning and martensitic transformations both of which occur suddenly can be identified from the relative lattice orientations before and after the change.

Bowden, Young, Ritchie and Rider [26] also provided detailed evidence showing common deformation of polyethylene by so-called chain slip, i.e. parallel to **c**, the chain axis and specifically in {010} the main fold planes in melt-crystallized lamellae.   This is denoted <001>{010} slip which they also inferred to have a lower critical resolved shear stress than transverse slip <100>{010} in the same planes.   Although **c**[001] is the smallest Burgers vector in the unit cell, in view of the strong anisotropy of the system it is not certain, *a priori,* that this will be favoured [18].   Experimentally, however, it is commonly found not least in direct imaging of lamellae in drawn systems.   More recent experiments in compression [27] have concluded that <001>{100} and <010>{100} slip processes are the most important deformation mechanisms, with interlamellar sliding operative only at low strains (to 1.8 compression ratio) while {110} twinning occurred only at very high ratios ~12.

The assumption underlying such interpretations of the scattering experiments is that a sample is equivalent to a collection of single crystals.   The long molecular nature of the polymer is left implicit in the properties of the matrix and explicitly recognized only in the relaxation after stress is removed whose effects typically occur within a few hours.   Only quite recently with the availability of simultaneous WAXS and SAXS cameras on high intensity synchrotron sources has it become possible to examine deformed sample over short times before significant relaxation has occurred.   These revealed that, in polyethylene of 311,000 weight average molecular weight, the monoclinic phase disappeared within 1 h. of removing the applied stress but, in other respects, conformed to previous understanding [6].   In other circumstances, especially involving UHMW polymer, the monoclinic phase can remain for long periods and be detected in post-drawing examination [18].

2.6   Morphological Factors

Major attempts to combine morphology with scattering data after deformation were made by Keller and by Peterlin with their respective co-workers.   The former had a particular interest in how spherulites transformed under tension [28], the latter was more concerned with the development of oriented fibres.   In banded spherulites the circular rings (in section) first became elliptical but could recover elastically before entering a range of inhomogeneous plastic behaviour.   According to the preparation and drawing conditions there could be yielding between spherulites or within them. In some circumstances isolated caps of more or less undrawn material remained at either end of the spherulite separated by highly drawn interiors. Keller was concerned to demonstrate that spherulites do not deform affinely with respect to the sample nor is there a simple relationship between the deformations of the spherulite and those of its constituent lamellar units.

Although the precise nature of spherulites and their boundaries needs specification of many factors, lamellar and molecular, in these experiments, carried out on thin films, yielding between spherulites was related to reduced film thickness a factor which is pertinent to commercial film production. Study of polypropylene films stretched at high temperature to simulate bubble technology has shown [29] that film thickness is systematically reduced between spherulites because in these sites of final crystallization the supply of material runs out and can no longer replenish the volume shrinkage on crystallization. Moreover as, in these experiments, spherulites were consistently nucleated on die lines present in the unstretched precursor film, the mechanical properties of the final product were strongly dependent on sample history.

Peterlin's work, however, appeared to show that the complications of the early stages of draw were lost in a discontinuous transformation to morphologies characteristic of fibres [30]. The principal evidence for this came from SAXS data, supported by lamellar thickness measurements from line widths and nitric acid etching, indicating that, beyond a certain extension, the long period was invariant and depended only upon the draw temperature. This drastic change, at ~ 2.7 x draw, was claimed to occur in the neck, by a process of micronecking, producing microfibrils, a concept which, however, was always questioned. Supporting morphological studies [31] were invariably of longitudinal views, i.e. with the draw direction in the plane, on samples which had often been etched with fuming nitric acid mostly showing lamellae in transverse orientation. With the benefit of hindsight and more recent microscopic techniques, one can state with confidence that this real space model was drawn too sharply despite the essential correctness of many of the details and that such microfibrils are not generally present in drawn polymers. This has been confirmed by transverse views which have not only revealed novel lateral organization in drawn polyolefines [32] in which the highest melting regions are interconnected walls [33] rather than the postulated microfibrils but have also demonstrated unambiguously that memory of the initial morphology normal to the draw direction persists in polyethylene to at least 45x draw ratio [34]. The early work should not be criticized for reaching its plausible but erroneous conclusion with respect to microfibrils. It is rather a consequence of scattering data not themselves leading to a unique microstructure but having to be compared with the consequences of alternative real-space models. This tends to lead to the adoption of the simplest, usually lamellar, model consistent with the very limited data whereas in practice real-space morphologies have almost invariably been more complex and spatially inhomogeneous than the models suggested.

The data used include intensities, positions, orientation and line widths of SAXS together with intensities and orientational maxima, equivalent to pole figures, for WAXS. Models consistent with these observations are generated, then implicitly taken to apply to the whole sample, although as Meinel, Morosoff and Peterlin point out [35] it is not necessarily the case that SAXS and WAXS data refer to the same crystal populations as badly ordered lamellae will contribute only to the second. The radial distances of SAXS maxima are usually converted into long periods using Bragg's law, Peterlin having found that a more correct procedure for limited sequences of repeating units did not change values substantially [35]. The orientation of such maxima is taken to be that of the normal to lamellae whose thickness lies within the long period, sometimes assessed as the fraction of it equal to the crystallinity of the sample.

A decrease of equatorial long period has been represented as due to lateral compression or to the onset of fine chain slip. Disappearance of SAXS intensity at the equator and the appearance either of a four-point pattern or of meridional maxima, as for inclined or perpendicular lamellae respectively, clearly demonstrates differential deformation according to lamellar orientation. On the other hand, increase of long period is attributed, when it occurs, to extensibility of interlamellar, so-called 'amorphous' regions or, when it does not, to too high a density of tie-molecules. Dimensional changes on yield, which give unequal longitudinal strains on soft interlamellar regions and stiff lamellae arranged in series, will impose corresponding differential lateral tensile stresses. It has been supposed that, in consequence, voids will 'eventually' form between lamellae and the same interpretation as the cause of enhanced SAXS from 'cavitation' at yield. The deterioration of SAXS and WAXS, not least at higher orders has been seen as due to breaking of the original lamellae into smaller blocks which then fuse using the adiabatically generated heat of disruption. Notwithstanding a reasonable consensus of interpretation from different authors [30,5,6], there is evidently much detail which would benefit, in principle, from the support of microscopic studies to assess, for example, whether deformation does occur on the fine scale and as uniformly as has been presumed.

2.7   Microscopic Observations

As is well known systematic electron microscopy of bulk melt-crystallized polymers only became possible some twenty years ago with the introduction of chlorosulphonation of polyethylene [36] and permanganic etching [37,38] which is effective for all polyolefines. Before then the most informative data were usually from examination of thin films or fracture surfaces. For bulk polymer, or thick films, development and application of the latter technique have provided novel information complementary to the very many scattering studies. Nevertheless, it was possible before this time to observe lamellar deformation *in situ.* Using pressure-crystallized, high molecular weight polyethylene, whose lamellae were ~0.5µm thick, Attenburrow and Bassett [39] were able to observe the same groups of lamellae as they deformed plastically over 100% extension, using polarizing optical microscopy of specimens previously immersed in xylene for 24 h. and to show unequivocally that deformation involved separation of lamellae along the tensile axis, rotation of lamellar planes towards the draw direction and increased aspect ratio associated with intralamellar shear. Stress-whitening, attributed to crazing and/or voiding occurred but neither feature was specifically identified in the morphology. Thermal analysis, crystal thickness measurements (by nitration/GPC) and electron microscopy all showed [40]. that drawing produced a second population of lower melting lamellae which surrounded those which survived albeit highly sheared. These thinner lamellae could be melted out and recrystallized at a temperature intermediate between the two melting points as an oriented array (Figure 2.4) indicating the presence of nuclei in the matrix highly aligned along the draw direction [41].

Figure 2.4   Fine slip in anabaric polyethylene lamellae which have survived fivefold drawing. The striations on the edge of a lamella lie along the chain axis which has become steeply inclined to the respective normals. Partial melting on subsequent annealing at 130 °C has allowed the matrix to recrystallize in herringbone fashion on linear nuclei lying along the (near horizontal) draw direction. Replica of etched surface; bar = 500 nm.

A further point of importance is the survival at least to 5x draw, at 80 °C, of the original thick lamellae into the plastically-deformed product [40, 41]. These were formed initially at 5 kbar pressure by crystallization into the hexagonal polyethylene phase and can not be reproduced by recrystallization at atmospheric pressure. It follows that they had not been melted during draw by adiabatic heating as has been regularly proposed to be the case in cold drawing [42]. The alternative process, hot drawing, is one in which deformation is not confined to pre-existing structures but also involves entities formed during the process. There will be a change with increasing deformation temperature when adiabatic heating may bring additional processes and populations into play but, as many authors have shown, this is neither the cause of necking [4] nor of any inherent deformation mechanism.

2.8   Electron Microscopy

All these early studies were made before systematic study of melt-crystallized lamellar textures became possible. Morphological knowledge was then based on analogy with the well-ordered lamellae grown from solution, on fracture surface studies and examination of the remnants of specimens degraded with fuming nitric acid. This situation changed with the introduction of chlorosulphonation of polyethylene and permanganic etching of polyolefines. It then became clear that melt-crystallized specimens of high crystallinity were profusely lamellar and that textures developed on an open skeleton of first-forming, so-called dominant, lamellae filled in with later-growing subsidiary layers [43]. Lamellae did not, as had previously been widely assumed, grow together in parallel array. They were, therefore, subject to the same physical factors causing molecular chainfolding as in solution growth. This finding also implies looser interlamellar connection than is often supposed [5,6,44] – with the inference that it will lead to voiding - especially for dominant lamellae.

In addition, it was evident that lamellae were not all equivalent but that there were at least two populations differing in spatial location, namely dominant and subsidiary lamellae even in a linear polymer [45]. At the highest crystallization temperatures there is fractionation which tends to place the longest and most linear molecular sequences in the dominant lamellae although this differentiation declines with temperature effectively disappearing for the linear polymer. These two populations of dominant and subsidiary lamellae have since been found to respond to deformation differently, both as described above when the lamellar thickness is substantially different between the two populations [46] and when there is little or no difference [47]. With decreasing crystallinity in polyethylene, subsidiary lamellae decrease in number until only dominant lamellae remain. More complex situations, with more populations, have been observed in, for example, certain ethylene copolymers which had at least four populations differing not only in type of location but also in lamellar thickness [48].

Another important finding was that polyethylene lamellar profiles were found to differ with crystallization temperature and molecular length. Three were observed for quiescent growth: ridged with {201} facets, planar {201} sheets for higher temperatures and/or longer molecules and S- or C-profiles incorporating {201} regions for the dominant lamellae in banded spherulites [49]. These three were also found for crystallization under shear flow for low strain rates at the appropriate temperatures but only {001} fold surfaces were formed at high strain rates [50]. This combination of new knowledge and the new techniques which had given rise to it offered the prospect of better resolution and increased insight into the structure of deformed polyolefines. Examination of the development of orientation in pressure-crystallized Hifax, which has no evident higher textural order than lamellae showed abundant fine chain slip in lamellae (Figure 2.4) which caused their normals and the chain axis both to move towards the draw direction with increasing strain but from opposite sides in confirmation of previous studies using fracture surfaces [39,41]. Not infrequently the phenomenon shown in Figure 2.5 was present i.e. coarse shear by the lamellar thickness with the former top and bottom surfaces now coplanar [51].

This is a pretty indication of the effect of chainfolding which is what holds the two sheared portions together. There was little or no difference in general behaviour between different specimens crystallized at pressures between 5 and 2.1 kbar. For the former all the sample solidifies as thick lamellae of the hexagonal phase which convert into the orthorhombic structure on return to ambient conditions. Lower pressures give a decreasing proportion of thick lamellae within a matrix of thin lamellae crystallized directly as the orthorhombic phase [48]. The nature of deformation is thus not sensitive to the precise character of surrounding lamellae.

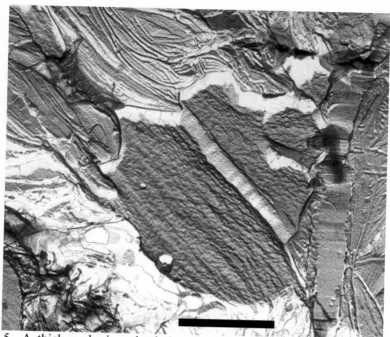

Figure 2. 5   A thick anabaric polyethylene lamella which has sheared by the lamellar thickness with the sheared portions kept together by the original chainfolding. Replica of etched surface; bar = 1 μm.

2.9   The Deformation of Banded Spherulites

The study of banded spherulites is particularly favourable for revealing how individual melt-crystallized lamellae have deformed [47]. This is because it is possible to infer the local crystallographic orientation by inspection: the **b** axis is always radial and the **c** axis emerges more or less normally from fold surfaces of undeformed spherulites. If one selects the largest spherulites for examination they will appear in diametral section, with **b** in the plane of the image. Such specimens may be examined, with increasing draw ratio, with respect both to spherulitic banding and the habit and integrity of lamellae. Inspection of appropriate areas of a spherulite will reveal the response for different crystallographic orientations: **b** is parallel to the tensile axis for radii along the draw direction but for perpendicular radii, **a** and **c** alternate with the bands in lying along the draw direction. The **c** axis will lie in planes of maximum shear stress for those radii making an angle of 45° to the tensile axis.   This variation of crystallography means that, initially, there will be correspondingly different responses to the applied strain on the scales of the banding and of the spherulite size [28]. It is a recent finding [34], contrary to previous opinion, that the former persists to high draw ratios, at least to 45 times extension.

In terms of banding, both linear and linear-low-density polyethylenes, the latter with 21 butyl branches per 1,000 C atoms, behave identically: along the draw direction bands extend affinely to ~60% strain then fall below this condition whereas in the orthogonal direction they contract affinely to 140% strain by which time spherulites have begun to behave inhomogeneously in their extreme regions along the tensile axis [47,52]. This takes two forms, one being the appearance of acuminate, i.e. pointed tips suggestive of the intersection of two shear planes (Figure 2.6). Additionally, less-deformed regions may be present which, at low magnification appear as detached islands within the texture [28] though at higher magnification lamellar continuity is evident across the apparent interface.

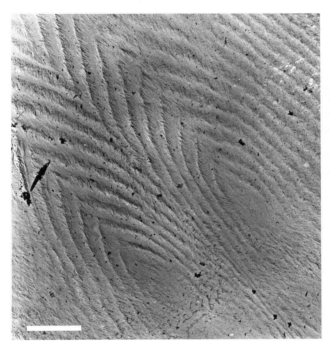

Figure 2.6 An intermediate stage in the tensile deformation of a polyethylene banded spherulite in which the once circular appearance of the bands has become sharply pointed with adjacent portions of a band lying parallel to intersecting shear planes. Replica of etched surface; bar = 5 μm..

In the early stages, to the yield point in linear polyethylene or, in the branched polymer, to the first yield point of two, at ~30% strain [53] and due to twinning [52], deformation of both bands and lamellae is elastic. Elliptical bands will become circular again over periods of days while lamellar profiles which, as described below, alter on stretching return to become indistinguishable from the original. The responsibility of the stretched molecular network for this behaviour is emphasized by the changed response of stretched samples to etching. For them, contrast between lamellar and interlamellar regions is much reduced in relation to both original and recovered specimens as would be expected of taut covalent sequences. If elastically-stretched polyethylene specimens are held at fixed length for long periods then they crack normal to the stretching direction [54]. This would be expected to happen once the lifetime of the entanglements in the original network has been exceeded.

The most-extended regions of the deformed spherulites are the most severely deformed. Lamellar profiles which for dominant lamellae initially were S- or C-shaped have become planar which is not the case around the waist. Electron diffraction confirms that whereas single crystal orientation still pertains in the waist at 2.2x draw ratio, there is fibre orientation in the region of the tips [47]. This difference produces a lower melting point for the latter which has been revealed by annealing and recrystallization. If this is done at high pressure the greater thickness of the recrystallized lamellae, which are aligned perpendicular the tensile axis, implying that they have grown from linear nuclei along this direction, makes them very conspicuous. At a particular annealing temperature, it is found that those areas with dominants near parallel to the tensile axis retain these intact while intervening subsidiary lamellae have melted [47]. This small difference in melting point between the two types of lamellae, which initially would have been the same thickness, indicates that only the subsidiary lamellae have deformed, probably because their inclination to the dominants makes them experience a greater shear stress. It also shows that mechanical properties vary systematically on three dimensional levels: those of the spherulite [28], the band period [34] and the sub-micron scale of the inter-dominant separation [46,47].

2.10   Lamellar Deformation

A convenient way in which to observe the differing responses of polyethylene lamellae within spherulites according to their orientation to the tensile axis is by studying stretching of blends of the linear polymer in a branched polyethylene matrix [55]. Figure 2.7a shows a typical isolated object of the linear polymer in a 5% blend crystallized at 123 °C which has the form of a well-developed sheaf or immature spherulite. When 2 cm gauge lengths of such blends are drawn, at temperatures to 100 °C at extension rates of 5 and 1,000 mm.min^{-1}, the spherulites deform within cylindrical envelopes (Figure 2.7b). Often they develop a four-pronged appearance in which the prongs are highly extended, with chain slip, in a cone around the tensile axis. At the waist there is compression resulting in zigzag profiles (Figure 2.7c) which are in phase along lines parallel to the tensile axis. These do not seem to have been reported before but must result from lamellar rotation because c shear would not reduce the lateral dimensions. Lamellae are seen to be broken up (in scanning electron microscopy, SEM, of etched surfaces in which bright contrast derives from topographical ridges on the surface) into regions of sub-micron dimensions (Figure 2.7d). This occurs not only in lamellae which have been extended longitudinally but also in those which have been compressed laterally. Such a straightforward demonstration of how lamellar response varies with orientation to the tensile axis reinforces the need to consider texture at dimensions higher than the lamellar level in analysing deformation of crystalline polymers. The experiments reported are still in progress at the time of writing but it may be that the variation of lamellar response with angle to the tensile axis is sufficient to account for the differing responses of dominant to subsidiary lamellae [46,47] and for the newly found alternating response of bands in spherulites [34] discussed in more detail below.

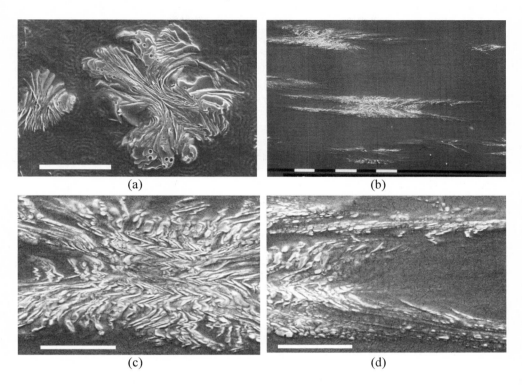

(a)   (b)

(c)   (d)

Fig 2.7 Immature spherulites in a 5% blend of linear polyethylene in branched polymer: a) undeformed; b) 4.5 x drawn at 75 °C; c) detail revealing zigzags of rotated lamellae in phase parallel to the draw direction; d) detail of the extremity of an object as in 7b showing how individual lamellae have become disrupted giving an interrupted appearance. SEM images; bars = 10 μm.

2.11   Memory Retention in Cold Drawing

A recent example which has shown how the original spherulitic size and banded texture affect the nature of the highly drawn product concerns the drawing, at 75 °C and initial strain rate of 5 min^{-1}, of polyethylenes of different catalytic origin [34]. Under these conditions it is the original microstructure which is being deformed, i.e. cold drawing is still operative. In longitudinal section, the observations agree with previous work in that little clear memory of the original texture remains at 5x draw and none is evident at 10 x extension. On stretching still further, a series morphology forms along the tensile axis which is fully developed, by ~35 x draw according to the precise material, consisting of alternating zones (Figure 2.8), conveniently referred to as Pisa structure because of its resemblance to the façade of the Leaning Tower. Interspaced between what are termed *entire* regions, without evident internal structure and 4-20 μm apart, are highly fibrillated zones. These latter are a consequence of etching and were initially crazes or similar regions of reduced density because sectioning at low temperatures shows equivalent contrast but a smooth surface [34]. Similar textures have been discussed previously for 'overdrawn' polypropylene [56]. Crazing and/or fibrillation results here from inhomogeneous draw at constant diameter in which the more highly extended regions suffer lateral tensile stresses to which they have yielded. The entire regions, whose dimensions and character seem invariant on extension are suggested to be the locations of the entanglements which are required to sustain the applied stress.

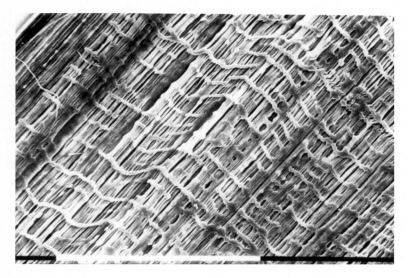

Figure 2.8 The so-called Pisa structure revealed by permanganic etching in longitudinal internal surfaces of linear polyethylene 45 x drawn at 75 °C. SEM image; bar = 100 μm.

It is, however, in transverse sections that the legacy of the original banding is seen. Unlike their longitudinal counterparts for which perceptible banding has all but disappeared for 5x draw, they show banding (Figure 2.9) which persists to the highest draw ratios. The SEM contrast in Figure 2.9 derives from the topology of the etched surfaces which in turn relates to the relative ease of attack reflecting the differences in deformation according to the two initial lamellar orientations, parallel and perpendicular to the tensile axis. The latter have etched more deeply than the former which have survived the better. These effects occur at much lower draw ratios than those for which the Pisa structure develops. That is an additional effect which is overwritten, in a texturally related way and depending on the particular grade of polyethylene used, on the spherulitic banding. Accordingly, at the higher draw ratios, transverse sections are through the Pisa structure which both before, but especially after annealing at temperatures into the melting range, is seen to be organized in bands of appropriate spacing. The implication is that in some, but not necessarily all, locations where lamellae have been parallel to the draw direction, lateral yield via crazing has been easier to initiate. That this should persist to the highest draw ratios examined (45x) negates the long-held view [30] that all memory of the original morphology is lost on formation of the fibrous microstructure. In consequence the morphology of the starting material will influence both microstructure and properties of the drawn product.

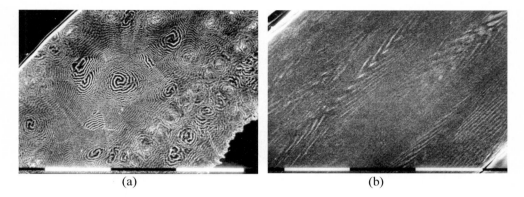

(a)   (b)

Figure 2.9   Contrasting microstructure after permanganic etching in SEM images of a) transverse and b) longitudinal sections of Sclair 2907 linear polyethylene with initial banded spherulitic microstructure drawn at 75 °C to 5.2 and 4.8 x extension respectively. While longitudinal texture is hardly apparent, transverse sections retain their banded appearance to at least 45 x extension. Bars = 100 μm.

2.12   Ordering Within Fibres

In contrast to the above fibres in which existing structures are deformed into the final product, commercial high modulus polyethylene fibres have been produced based on various procedures in which structures are produced and deformed concurrently [57,58,59], followed by heating into the melting region. The latter is attested by the presence of transverse lamellae, as in row structures, formed by local melting and recrystallization on linear nuclei microns long parallel to the fibre axis [32]. The question of the extent of longitudinal continuity is central to fibre properties and has been modelled in several ways, notably the Takayanagi [60] and Ward's crystal bridge [61] representations. A more recent treatment of the former kind is based upon the detection of two crystalline components of crystalline material [62]. .Direct imaging of morphology using thin films shows linear features which in {110} dark field imaging are continuous for ~100 nm [63] in agreement with X-ray line width measurements of the 002 reflection [64]. Permanganic etching, on the other hand, reveals both longitudinal continuity of high melting regions extending over several microns and novel lateral ordering [32,33]. There is no inherent contradiction between these longitudinal measures. So far as covalent continuity is concerned the dark field and line width data are a lower bound in that kinking – which is liable to happen due to longitudinal compression on cooling, as it does in liquid crystalline polymers - would not destroy such continuity but would limit the length measured in the two diffraction techniques to straight portions of chains.

The lateral ordering, present in melt-spun, gel-spun and melt-kneaded polyethylene fibres, is due to regions of deficit density i.e. high free volume which are more accessible to the etchant [65]. This free volume has the further consequence that it allows internal melting within fibres. Indeed, it is possible to concentrate the free volume into a central cylindrical hole, within higher melting surrounds, by partly melting and recrystallizing fibres [66]. If this process is followed through the melting range, the highest melting regions are revealed, in transverse section, to form a network of walls on the micron scale, varying characteristically with different manufacturing origins [33]. Longitudinally the appearance is of linear spines continuous over many microns which are the sections through the transverse network (Figure2.10). This morphology is consistent with having grown out from the highest melting walls of the network into the enclosed regions of melt then becoming subject to the difficulty of replenishing melt within what would be comparatively rigid boundaries once it had solidified and contracted. The ensuing lateral tensions would tend to produce the observed high free volume within the walls. The strong candidate to have nucleated crystallization is highly aligned portions of the entangled molecular network.

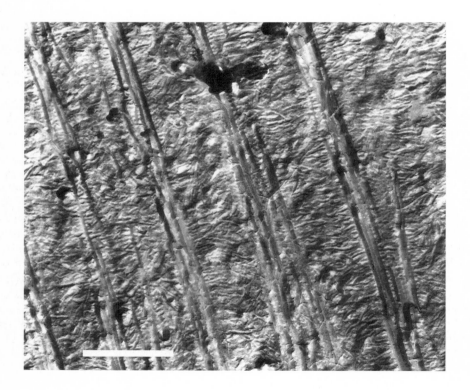

Figure 2.10   The internal microstructure, revealed by permanganic etching, of a partly melted Tekmilon high modulus polyethylene fibre. Residual high-melting walls, appearing as spines in section separate lower melting regions. Replica of etched surface; bar = 1 μm.

2.13   Disentanglement

Strategies for increasing ultimate draw ratio and high modulus with it generally rely on reducing entanglement density. This was, for example, the explicit motive behind the introduction of gel-spinning [58]. Alternatively, it has recently been proposed that ultradrawing is only possible when it is possible to translate chains through crystallites as in α–relaxations [67]. . In any event, entanglements must not be reduced too far. Conceptually the presence of entanglements provides the covalent continuity necessary for high strength. As has already been remarked, the high temperature drawing of PTFE needs to be carried out within the lifetime of its transient entanglements. Somewhat similarly, pressure annealing of solid polyethylene, which increases the lamellar thickness, improves mechanical properties so long as the material stays as the orthorhombic phase. Once conditions allow the hexagonal phase to form with its associated increase of lamellar thickness the entanglement network becomes too attenuated and is unable to sustain the stresses of subsequent processing [68,69].

2.14   Overview

With this finding we return to our starting point: that a network of covalent molecules extending through samples is an essential background within which particular deformation mechanisms operate; without a continuous network ductile samples fail. Interlamellar regions (whose properties alter with the degree of molecular alignment and the applied stress) support this network and transmit stress to lamellae. Within this matrix lamellae may be able to change their relative positions, increasingly so for lower crystallinity materials. Intralamellar deformation mechanisms crystals include dislocation glide, especially chain slip, mechanical twinning and martensitic transformation, all of which are involved in the one or more yield points in linear and branched polyethylenes. What has not been well established is where these mechanisms operate in the specimens. It remains common practice to combine SAXS data, which relate to lamellae, with the molecular orientations identified by WAXS into uniform textural models of lamellae and their stacks [5,6] without reference to higher levels of order in the original specimens. While it was thought that such details were irrelevant to the nature of the product [30] it may have seemed more reasonable to ignore the complications of inhomogeneous deformation of spherulites [28] but that can no longer be justified. Electron microscopic examination, especially after permanganic etching, has revealed two general intraspherulitic factors which need to be brought into a unified understanding of deformation in quiescently-crystallized polymers. One is the presence of different lamellar populations at the sub-micron   level,   and   their   differing   mechanical   responses,   generated   by   the dominant/subsidiary architecture which pervades quiescently crystallized materials. The second is the legacy of different initial lamellar orientations, as demonstrated by spherulitic banding, which is carried through to the final very highly drawn product. It may be that differing lamellar orientations to the tensile axis is sufficient to account for these findings but this remains to be confirmed. In fibres themselves the presence of characteristic lateral ordering has similarly still to be taken into account in modelling mechanical behaviour.

While there is general agreement that the characterization of such complex materials as synthetic polymers requires input from as many complementary methods as are available, in reality this has usually still to be put into practice with results often coming, and conclusions drawn, from too few techniques. The principal challenge is to relate the long-standing diffraction data, obtained in reciprocal space, to the new real-space knowledge of spatially-different responses revealed using electron microscopy i.e. in dominant and subsidiary populations as well as for spherulitic banding. A proper understanding of the subject requires that they be combined into an integrated whole.

Acknowledgements

Acknowledgement is due those one-time Reading colleagues, namely Drs T.C. Amornsakchai, A.M. Freedman, A.M. Hodge, S-Y Lee, M.M. Miraftab, A. Sneck and J. Teckoe, whose unpublished photographs are, respectively, Figures 2.8, 2.9; 2.5; 2.4; 2.7; 2.1; 2.2, 2.3; and 2.10.

References

1.   Stachurski, Z.H., Prog. Polym. Sci. 1997, 22, 407-474.
2.   Cottrell, A.H., *The Mechanical Properties of Matter,* 1964, Wiley.
3.   Bowden, P.B.; Young, R.J., J Mat. Sci. 1974, 9 2034-2051.
4.   Hadley, D.W.; Ward, I.M., *An Introduction to the Mechanical Properties of Solid Polymers* 1993, Wiley.
5.   Lin, L.; Argon, A.S., J Mat. Sci. 1994, 29, 294-323.
6.   Butler, M.F.; Donald, A.M.; Ryan, A.J., Polymer 1998, 39, 39-52.
7.   Bevis, M; Hull, D., J Mat. Sci. 1970, 5, 983
8.   Olf, H.G.; Peterlin, A., J. Polym. Sci. Phys. Edn. 1974, 12, 2209-2251.
9.   British patent 1355373, 1971.
10. Termonia, Y; Smith, P., Macromolecules 1988, 21 2184-2189.
11. Bucknall, C.B., *Toughened Plastics,* 1977, Applied Science.
12. Carder, D.R.; Bassett, D.C., Phil. Mag. 1973, 28, 535-545.
13. Hodge, A.M.; Bassett, D.C., unpublished work cited in Bassett, D.C. *Principles of Polymer Morphology*, 1981, Cambridge University Press, Figures 9.1 & 9.2.
14. Keller, A.; Rider, J.G., J. Mat. Sci. 1966 1, 389-398.
15. Bunn, C.W.; Alcock, T.C., Trans. Faraday Soc. 1945, 41, 317-325.
16. Brown, A., J. Appl. Phys. 1949, 20, 552-558.
17. Horsley, R.A.; Nancarrow, H.A., Brit. J. Appl. Phys. 1951, 2, 345-351.
18. Frank, F.C.; Keller, A.; O'Connor, A., Phil. Mag. 1958, 3, 64-74.
19. Keller, A., Phil Mag. 1957, 2, 1171-1175.
20. Keith, H.D.; Passaglia, E., J. Res. Natl. Bur. Stds., 1964, 68A 513-518.
21. Peterlin, A., J. Polym. Sci. A-2, 1969, 7, 1151-1163.
22. Haas, K.; Geil, P.H., J. Polym. Sci. A-2, 1966, 289-298.
23. Allan, P.; Bevis, M., Proc. Roy. Soc. A, 1974, 341, 75-90.
24. *Faraday Disc. R. Soc. Chem.* 1979, 68.
25. Bowden, P.B.; Young, R.J., J. Mat. Sci. 1974, 9, 2034-2051.
26. Bowden, P.B.; Young, R.J.; Ritchie, J.M.; Rider, J.G., J. Mat. Sci. 1973, 8, 23-36.

27. Galeski, A.; Bartczak, Z.; Argon, A.S.; Cohen, R.E., Macromolecules, 1992, 25, 5705-5718.
28. Hay, I.L.; Keller, A., Kolloid Z. 1965, 201, 43-74.
29. Olley, R.H.; Bassett, D.C., J. Macromol. Sci. Phys. 1994, B33, 209-227.
30. Peterlin, A., J. Mat. Sci. 1971, 6, 490-508.
31. Peterlin, A.; Sakaoku, A., Kolloid Z, 1966, 212, 51-
32. Abo el Maaty, M.I.; Olley, R.H.; Bassett, D.C., J Mat. Sci. 1999, 34, 1975-1989.
33. Teckoe, J, Bassett, D.C.; Olley, R.H, Polymer J. 1999, 31, 765-771.
34. Amornsakchai, T.; Bassett, D.C.; Olley, R.H,; Al-Hussein, M.O.M.; Unwin, A. P.; Ward, I.M., submitted to Polymer.
35. Meinel, G,; Morosoff, N.; Peterlin, A., J Polym. Sci. A-2, 1970, 8, 1723-1746.
36. Kanig, G., Kolloid Z., 1973, 251, 782-
37. Olley, R.H.; Hodge, A.M.; Bassett, D.C., J. Polym Sci. Phys. Edn. 1979, 17, 627-643.
38. Olley, R.H.; Bassett, D.C., Polymer, 1982, 23, 1707-1710.
39. Attenburrow, G.E.; Bassett, D.C., Polymer 1979, 20, 1312-1315.
40. Attenburrow, G.E.; Bassett, D.C., J. Mat Sci. 1977, 12, 192-200.
41. Hodge, A.M.; Bassett, D.C., unpublished work cited in Bassett, D.C. *Principles of Polymer Morphology*, 1981, Cambridge University Press, Figures 9.14 & 9.15.
42. Phillips, P.J.; Philpot, R.J., Polymer Comm. 1986, 27, 307-309.
43. Bassett, D.C.; Hodge, A.M.; Olley, R.H., Proc. Roy. Soc. A, 1981, 377, 39-81.
44. Pope, D.P.; Keller, A., J. Polym. Sci. Phys. Edn. 1975, 13, 533-566.
45. Hodge, A.M.; Bassett, D.C.,. Proc. Roy. Soc. A, 1978, 359, 121-132.
46. Freedman, A.M.; Bassett, D.C.; Vaughan, A.S.; Olley, R.H., Polymer 1986, 27, 1163-1169.
47. Bassett, D.C.; Freedman, A.M., Progr. Colloid Polym. Sci. 1993, 92, 23-31.
48. Bassett, D.C., *Principles of Polymer Morphology*, 1981, Cambridge University Press, Figure 5.5.
49. Bassett, D.C.; Hodge, A.M., Proc. Roy. Soc. A, 1981, 377, 25-37.
50. Hosier, I.L.; Bassett, D.C.; Moneva, I.T., Polymer, 1995, 36, 4197-4202.
51. Hodge, A.M., PhD thesis, University of Reading, 1978.
52. Janimak, J.J.; Bassett, D.C., unpublished work.
53. Brooks, N.W.; Duckett, R.A.; Ward, I.M., Polymer, 1992, 33, 1872-1880.
54. Petermann, J.; Schultz, J.M., J. Mat. Sci., 1978, 13, 50-54.
55. Lee, S-Y.; Bassett, D.C.; Olley, R.H., unpublished work.
56. Abo el Maaty, M.I.; Bassett, D.C.; Olley, R.H.; Dobb, J.G.; Tomka, J.G.; Wang, I-C., Polymer, 1996, 37, 213-218.
57. Capaccio, G; Ward, I.M., Polymer, 1974, 15, 233-
58. Smith, P.; Lemstra, P.J., Colloid Polym. Sci., 1980, 258, 891-
59. Prevorsek, D.C., in *The Handbook of Fiber Science and Technology, Vol. 3, High Technology Fibers* ed. Lewin M and Preston, J. 1996, Marcel Dekker, pp 1-170.
60. Takayanagi, M; Imada, I; Kajiyama, T., J. Polym. Sci. C, 1966, 15, 263-
61. Gibson, A.G.; Davies, G.R.; Ward, I.M., Polymer, 1978, 19, 683-693.
62. Wong, W.F.; Young, R.J., J. Mat. Sci., 1994, 29, 520-526.
63. Yang, D.C.; Thomas, E.L., J. Mat. Sci., 1984, 19, 2098-2110.
64. Clements, J.; Jakeways, R.; Ward, I.M., Polymer, 1978, 19, 639-644.
65. Olley, R.H.; Bassett, D.C.; Hine, P.J.; Ward, I.M., J. Mat. Sci., 1993, 28, 1107-1112.

66. Kabeel, M.A.; Bassett, D.C.; Olley, R.H.; Hine, P.J.; Ward, I.M., J. Mat. Sci., 1995, 30, 601-606.
67. Hu, W-G.; Schmidt-Ross, K., Acta Polym. 1999, 59, 271-285.
68. Shahin, M.M.; Olley, R.H.; Bassett, D.C.; Maxwell, A.S.; Unwin, A.P.; Ward, I.M., J. Mat. Sci., 1996, 31, 5541-5549.
69. Maxwell, A.S.; Unwin, A.P.; Ward, I.M., Abo el Maaty, M.I.; Shahin, M.M.; Olley, R.H.; Bassett, D.C., J. Mat. Sci., 1997, 32, 567-574.

3   Characterization of Orientation

K. C. Cole and A. Ajji
Industrial Materials Institute, National Research Council Canada
75 De Mortagne Blvd., Boucherville, Québec, Canada J4B 6Y4

3.1   Molecular Orientation and Its Definition

Orientation of polymers enhances many of their properties, in particular mechanical, impact, barrier, and optical. The products of orientation processes can generally be classified into three categories: fibres, films, and parts (sheets, bottles, rods, etc.). The fibre spinning process is the most simple and induces uniaxial orientation in the polymer. Films, on the other hand, constitute a major application and most often involve biaxial orientation, which is an added advantage allowing enhancement of properties in two directions, avoiding any weakness in the transverse direction. The most widely used biaxial orientation processes for films are tubular film blowing and cast film biaxial orientation (or tentering). Other processes such as blow molding, compression and injection molding, and thermoforming also produce mostly biaxial orientation. Finally, the solid state deformation process may involve either uniaxial or biaxial orientation in the parts and shapes formed. In pipes, for example, the orientation is usually biaxial whereas in wire production it is uniaxial and in roll-drawing planar.

Thus the precise knowledge of the polymer orientation produced by the different processes mentioned above is critical for establishing the process conditions and the final properties of the oriented polymer. There is a long history of investigations of orientation in polymers (1, 2). The simplest representation of the orientation is that corresponding to uniaxial symmetry, which simply defines the orientation in terms of the angle θ that the polymer chains make with respect to the draw direction, according to the well known Hermans orientation function:

$$f \; = \; \frac{3\langle \cos^2 \theta \rangle - 1}{2} \tag{1}$$

where the brackets indicate an average over all the angles. However, when the orientation is other than uniaxial, or when crystalline phases are involved, the definition of orientation is more complicated (3-8). Generally speaking, it is necessary to describe the orientation of molecular units on an atomic scale with respect to the macroscopic sample in which these units are found. The first step is to define a coordinate system for the sample. Fortunately, most industrial samples possess a certain degree of symmetry which allows us to define a coordinate axis system in terms of three orthogonal directions designated as M (machine), T (transverse), and N (normal), as shown in Figure 3.1. Within this sample there are structural units on a molecular scale (for example, the molecular chain, a segment of the molecular chain, or a unit cell of a crystalline structure). A second coordinate axis system (a, b, c) can be assigned to the structural unit. For example, in the case of the crystalline phase of polyethylene, which possesses an orthogonal unit cell, the obvious choice is the three axes of the unit cell. For amorphous phases, the axes are chosen with respect to a monomeric unit of the polymer chain, generally with one axis along the chain axis of the polymer. Figure 3.2 shows as an illustration poly(ethylene terephthalate), which in its most regular conformation is close to planar. In the choice of axes shown in the figure, the c axis is along the polymer chain, the b axis is perpendicular to the benzene rings, and the a axis is in the plane of the benzene rings.

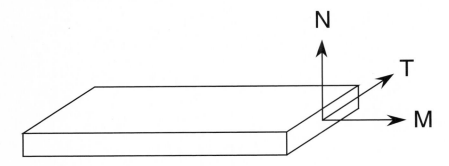

Figure 3.1 Definition of sample coordinate axes.

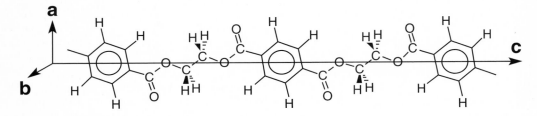

Figure 3.2 Possible definition of coordinate axes for poly(ethylene terephthalate).

For a single structural unit, the orientation with respect to the sample geometry can be specified in terms of the three Euler angles θ, ϕ, and ψ made between the two sets of coordinate axes, as shown in Figure 3.3. The c axis of the molecule is tilted away from the M axis by the angle θ, the direction in which it is tilted is defined by ϕ, and there can also be a rotation of the a and b axes around the c axis by the angle ψ. For an ensemble of molecules, the structural units are found in a variety of orientations (i.e. values of θ, ϕ, and ψ). The overall orientation can be visualized in terms of a probability distribution. The probability of finding a structural unit oriented within the generalized solid angle $\sin\theta\, d\theta\, d\phi\, d\psi$ is given by the distribution function $N(\theta,\phi,\psi)$. This function may be expressed as the sum of a series of generalized spherical harmonics:

$$N(\theta,\phi,\psi) \;=\; \sum_{l=0}^{\infty}\sum_{m=-l}^{+l}\sum_{n=-l}^{+l} P_{lmn} Z_{lmn}(\cos\theta)\, e^{-im\phi} e^{-in\psi} \tag{2}$$

where the Z_{lmn} are a generalization of the Legendre functions. To fully define the orientation distribution, which can be rather complex, all the coefficients P_{lmn} must be determined.

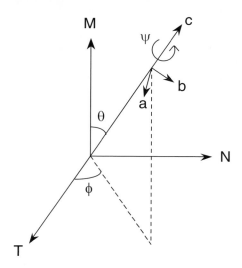

Figure 3.3. Definition of orientation in terms of the Euler angles.

If both the sample and the structural units have at least orthorhombic symmetry (point group D_2), then the P_{lmn} are non-zero only when m and n are even. This is the case for many real situations. Techniques like infrared spectroscopy and birefringence measurements allow the determination of only those P_{lmn} with $l = 2$. These are known as the second-order moments of the orientation distribution function. Higher-order moments ($l = 4$, 6, 8, etc.) can be obtained by other techniques like Raman spectroscopy, nuclear magnetic resonance, and X-ray scattering, but the procedure is considerably more complex. Hence many studies of orientation are limited to the second-order moments. In many cases this is sufficient, but in a highly oriented polymer the second-order moments can approach saturation values (0.9 to 1), and higher moments are required to fully characterize the polymer. The four second-order moments are related to the following averages over the Euler angles:

$$P_{200} = \tfrac{1}{2}\langle 3\cos^2\theta - 1\rangle \tag{3a}$$

$$P_{220} = \tfrac{1}{4}\langle (1-\cos^2\theta)\cos 2\phi\rangle \tag{3b}$$

$$P_{202} = \tfrac{1}{4}\langle (1-\cos^2\theta)\cos 2\psi\rangle \tag{3c}$$

$$P_{222} = \tfrac{1}{4}\langle (1+\cos^2\theta)\cos 2\phi\cos 2\psi - 2\cos\theta\sin 2\phi\sin 2\psi\rangle \tag{3d}$$

Expressions for the fourth-order moments are given in References 5, 6, and 9.

As mentioned, the four second moments give a good general picture of the orientation in most cases, and this is the most elegant way of quantitatively defining the orientation, but the physical significance of the P_{2mn} coefficients is not easy to visualize. Another way of describing the orientation is in terms of the angles θ_{iJ} between each of the axes of the molecule coordinate system ($i =$ a, b, or c) and each of the axes of the sample coordinate system ($J =$ M, T, or N). There are nine such angles in all and the average values of their squared cosines are related to the P_{2mn} coefficients as follows (6):

$$\left\langle \cos^2 \theta_{cM} \right\rangle = \tfrac{1}{3} + \tfrac{2}{3} P_{200} \tag{4a}$$

$$\left\langle \cos^2 \theta_{cT} \right\rangle = \tfrac{1}{3} - \tfrac{1}{3} P_{200} + 2 P_{220} \tag{4b}$$

$$\left\langle \cos^2 \theta_{cN} \right\rangle = \tfrac{1}{3} - \tfrac{1}{3} P_{200} - 2 P_{220} \tag{4c}$$

$$\left\langle \cos^2 \theta_{aM} \right\rangle = \tfrac{1}{3} - \tfrac{1}{3} P_{200} + 2 P_{202} \tag{4d}$$

$$\left\langle \cos^2 \theta_{aT} \right\rangle = \tfrac{1}{3} + \tfrac{1}{6} P_{200} - P_{220} - P_{202} + P_{222} \tag{4e}$$

$$\left\langle \cos^2 \theta_{a} \right\rangle = \tfrac{1}{3} + \tfrac{1}{6} P_{200} + P_{220} - P_{202} - P_{222} \tag{4f}$$

$$\left\langle \cos^2 \theta_{b} \right\rangle = \tfrac{1}{3} - \tfrac{1}{3} P_{200} - 2 P_{202} \tag{4g}$$

$$\left\langle \cos^2 \theta_{bT} \right\rangle = \tfrac{1}{3} + \tfrac{1}{6} P_{200} - P_{220} + P_{202} - P_{222} \tag{4h}$$

$$\left\langle \cos^2 \theta_{bN} \right\rangle = \tfrac{1}{3} + \tfrac{1}{6} P_{200} + P_{220} + P_{202} + P_{222} \tag{4i}$$

These angles are not independent. For a given axis in one coordinate system, the sum of the squared cosines with respect to the three axes of the other system is equal to unity. (For example,
$$\cos^2 \theta_{cM} + \cos^2 \theta_{cT} + \cos^2 \theta_{cN} = 1$$
and $\cos^2 \theta_{aM} + \cos^2 \theta_{bM} + \cos^2 \theta_{cM} = 1$.) Hence there are only four independent parameters, corresponding to the four P_{2mn} coefficients, and all nine angles can be related to these.

The value of $<\cos^2 \theta_{i,j}>$ ranges from 0 if the axes i and J are perpendicular, to 1/3 for random orientation, to 1 when the axes i and J are parallel. For each angle $\theta_{i,j}$ it is also possible to calculate a Hermans-type orientation function f according to Equation 1, where the value of f ranges from $-1/2$ if the axes i and J are perpendicular, to 0 for random orientation, to 1 when the axes i and J are parallel. Thus the orientation can also be expressed in terms of nine orientation functions, which are related to the P_{2mn} coefficients as follows:

$$f_{cM} = P_{200} \tag{5a}$$

$$f_{cT} = -\tfrac{1}{2} P_{200} + 3 P_{220} \tag{5b}$$

$$f_{cN} = -\tfrac{1}{2} P_{200} - 3 P_{220} \tag{5c}$$

$$f_{aM} = -\tfrac{1}{2} P_{200} + 3 P_{202} \tag{5d}$$

$$f_{a} = \tfrac{1}{4} P_{200} - \tfrac{3}{2} P_{220} - \tfrac{3}{2} P_{202} + \tfrac{3}{2} P_{222} \tag{5e}$$

$$f_{aN} = \tfrac{1}{4} P_{200} + \tfrac{3}{2} P_{220} - \tfrac{3}{2} P_{202} - \tfrac{3}{2} P_{222} \tag{5f}$$

$$f_{b} = -\tfrac{1}{2} P_{200} - 3 P_{202} \tag{5g}$$

$$f_{bT} = \tfrac{1}{4} P_{200} - \tfrac{3}{2} P_{220} + \tfrac{3}{2} P_{202} - \tfrac{3}{2} P_{222} \tag{5h}$$

$$f_{bN} = \tfrac{1}{4} P_{200} + \tfrac{3}{2} P_{220} + \tfrac{3}{2} P_{202} + \tfrac{3}{2} P_{222} \tag{5i}$$

Inversely, the P_{2mn} coefficients can be expressed in terms of the orientation functions in different ways, for instance:

$$P_{200} \quad = \quad f_{cM} \quad = \quad -(f_{aM}+f_{bM}) \quad = \quad f_{aT}+f_{aN}+f_{bT}+f_{bN} \tag{6a}$$

$$P_{202} \quad = \quad \tfrac{1}{6}(f_{aM}-f_{bM}) \quad = \quad \tfrac{1}{6}(-f_{aT}-f_{aN}+f_{bT}+f_{bN}) \tag{6b}$$

$$P_{220} \quad = \quad \tfrac{1}{6}(f_T-f_N) \quad = \quad \tfrac{1}{6}(-f_{aT}+f_{aN}-f_{bT}+f_{bN}) \tag{6c}$$

$$P_{222} \quad = \quad \tfrac{1}{6}(f_{aT}-f_{aN}-f_{bT}+f_{bN}) \tag{6d}$$

White and Spruiell (10) have proposed yet another method of describing the orientation, which is used mainly for the case of biaxial orientation. They define the following set of biaxial orientation factors (using the same notation already defined):

$$f_{cM}^{B} \quad = \quad 2\langle\cos^2\theta_{cM}\rangle+\langle\cos^2\theta_{cT}\rangle-1 \quad = \quad P_{200}+2P_{220} \tag{7a}$$

$$f_{cT}^{B} \quad = \quad 2\langle\cos^2\theta_{cT}\rangle+\langle\cos^2\theta_{cM}\rangle-1 \quad = \quad 4P_{220} \tag{7b}$$

$$f_{aM}^{B} \quad = \quad 2\langle\cos^2\theta_{aM}\rangle+\langle\cos^2\theta_{aT}\rangle-1 \quad = \quad -\tfrac{1}{2}P_{200}-P_{220}+3P_{202}+P_{222} \tag{7c}$$

$$f_{aT}^{B} \quad = \quad 2\langle\cos^2\theta_{aT}\rangle+\langle\cos^2\theta_{aM}\rangle-1 \quad = \quad -2P_{220}+2P_{222} \tag{7d}$$

$$f_{bM}^{B} \quad = \quad 2\langle\cos^2\theta_{bM}\rangle+\langle\cos^2\theta_{bT}\rangle-1 \quad = \quad -\tfrac{1}{2}P_{200}-P_{220}-3P_{202}-P_{222} \tag{7e}$$

$$f_{bT}^{B} \quad = \quad 2\langle\cos^2\theta_{bT}\rangle+\langle\cos^2\theta_{bM}\rangle-1 \quad = \quad -2P_{220}-2P_{222} \tag{7f}$$

As before, these are not all independent, and four parameters are sufficient to define the orientation. It is important to note the difference between the orientation functions defined in Equations. 5 and 7. The two sets are related by the following equations (where the indices specifying the a, b, or c axis are dropped for simplicity):

$$f_M^{B} \quad = \quad \tfrac{2}{3}(2f_M+f_T) \tag{8a}$$

$$f_T^{B} \quad = \quad \tfrac{2}{3}(2f_T+f_M) \tag{8b}$$

$$f_M \quad = \quad f_M^{B}-\tfrac{1}{2}f_T^{B} \tag{9a}$$

$$f_T \quad = \quad f_T^{B}-\tfrac{1}{2}f_M^{B} \tag{9b}$$

$$f_N \quad = \quad -\tfrac{1}{2}(f_M^{B}+f_T^{B}) \tag{9c}$$

The approach just described is necessary when the orientation of crystals or specific molecular units is to be described. In many cases, however, it can be assumed that there is no preferential orientation of chains around the chain axis, i.e. that the polymer structural units may be considered as having cylindrical symmetry. In other words, there is no need to specifically consider the a and b axes of the molecule and we need to consider only the orientation of the polymer chain (c) axis. In such a case, the Euler angle ψ is randomly distributed and the two coefficients P_{202} and P_{222} are equal to zero. This approach can be applied to amorphous polymers or when an average orientation over both amorphous and crystalline phases is being determined. For *general biaxial* orientation, the M, T, and N directions are all different. When only the chain axis c is considered, it is possible to define the biaxial orientation in terms of Hermans-type orientation functions based on $\left\langle \cos^2 \theta_{cM} \right\rangle$, $\left\langle \cos^2 \theta_{cT} \right\rangle$, and $\left\langle \cos^2 \theta_{cN} \right\rangle$:

$$f_{cM} = \frac{3\left\langle \cos^2 \theta_{cM} \right\rangle - 1}{2} = P_{200} \tag{10a}$$

$$f_{cT} = \frac{3\left\langle \cos^2 \theta_{cT} \right\rangle - 1}{2} = -\tfrac{1}{2}P_{200} + 3P_{220} \tag{10b}$$

$$f_{cN} = \frac{3\left\langle \cos^2 \theta_{cN} \right\rangle - 1}{2} = -\tfrac{1}{2}P_{200} - 3P_{220} \tag{10c}$$

Because the sum of the three squared cosines is equal to unity, the three orientation functions are not independent and their sum $f_{cM} + f_{cT} + f_{cN}$ is equal to zero. Hence two quantities are sufficient to define the biaxial orientation. These could be f_{cM} and f_{cT}, or P_{200} and P_{220}, or f_{cM}^B and f_{cT}^B as defined in Equations. 7a and 7b. When f_{cM} is equal to f_{cT}, we have *equibiaxial* orientation. For typical biaxial orientation, as in a film stretched in two perpendicular directions, f_{cM} and f_{cT} are both positive, although not necessarily equal, while f_{cN} is negative. In the case of *uniaxial* orientation, the polymer chains are oriented toward the machine direction and there is no difference between the T and N directions. As a result, P_{220} is equal to zero and the single quantity f_{cM} of Equation 10a, equivalent to the well known Hermans orientation function, is all that is required to describe the uniaxial orientation. The quantity f_{cM} is positive, while f_{cT} and f_{cN} are both equal to $-\tfrac{1}{2}f_{cM}$. Table 3.1 gives values of the various parameters for ideal cases, with the assumption of cylindrical symmetry for the polymer unit.

Many techniques can be used for the characterization of the orientation states of polymers. One can mention birefringence, infrared spectroscopy, wide angle X-ray diffraction, small angle X-ray scattering, small angle light scattering, small angle neutron scattering, Raman spectroscopy, fluorescence spectroscopy, nuclear magnetic resonance, and ultrasonic measurements. Each of these techniques has its advantages and disadvantages, which are described briefly in Table 3.2. The remainder of the chapter will be devoted to a description of the different techniques. The object is not to provide an extensive review, bur rather to give an overview of the techniques along with selected examples and references containing more details. More attention will be devoted to the more widely used techniques, namely birefringence, infrared spectroscopy, and X-ray methods.

Table 3.1 Values of orientation parameters corresponding to some ideal cases

Orientation Parameter	Type of Orientation			
	None	Perfect uniaxial, machine direction	Perfect uniaxial, transverse direction	Perfect equibiaxial
$\left\langle \cos^2 \theta_M \right\rangle$	$\frac{1}{3}$	1	0	$\frac{1}{2}$
$\left\langle \cos^2 \theta_T \right\rangle$	$\frac{1}{3}$	0	1	$\frac{1}{2}$
$\left\langle \cos^2 \theta_N \right\rangle$	$\frac{1}{3}$	0	0	0
P_{200}	0	1	$-\frac{1}{2}$	$\frac{1}{4}$
P_{220}	0	0	$\frac{1}{4}$	$\frac{1}{8}$
f_M	0	1	$-\frac{1}{2}$	$\frac{1}{4}$
f_T	0	$-\frac{1}{2}$	1	$\frac{1}{4}$
f_N	0	$-\frac{1}{2}$	$-\frac{1}{2}$	$-\frac{1}{2}$
f_M^B	0	1	0	$\frac{1}{2}$
f_T^B	0	0	1	$\frac{1}{2}$

Table 3.2. Summary of different techniques for characterization of orientation.

Technique	Phase(s) characterized	Orientation distribution function moments	Comments
Birefringence	Average over crystalline and amorphous	2	Simple and rapid; refractometry can be used for any sample, other light techniques require transparent samples; can be used on-line.
Infrared spectroscopy	Crystalline or amorphous	2	Overlap of IR bands can cause problems in some cases; the angle of the transition moment of the band used must be known; can be used on-line.
Raman spectroscopy	Crystalline or amorphous	2 and 4	Some depolarization problems may be encountered; tedious calculation procedure.
Fluorescence	Amorphous	2 and 4	Requires the use of a dye; difficult correlation between fluorescence axes and polymer chain axis; may be used on-line.
NMR	Crystalline or amorphous	2, 4, 6 and 8	Polymer structure must be known; requires a high magnetic anisotropy; may resolve intra and inter molecular interactions.
WAXS	Crystalline	All	Difficult to determine the background intensity; some interference from amorphous phase may occur; very good accuracy.
SAXS	Crystalline	All	Allows determination of the long period; same comments as for WAXS.
SANS	Amorphous	2	Allows determination of the gyration radius; requires deuterated samples; limited equipment availability.
Ultrasonic	Average	2	Requires a structural model; may need thick samples; applicable on-line.
SALS	Crystalline	None	Qualitative technique.

3.2   Birefringence

Birefringence is the anisotropy of refractive index resulting from a variation in the polarizability along the different directions within the sample. Like infrared spectroscopy, it is an optical technique that depends on the interaction of light with the material. In contrast with infrared spectroscopy, measurements are usually performed with visible light. Common wavelengths are the sodium D line (589 nm), the mercury green line (545 nm), and the helium-neon laser red line (633 nm).

The fundamental property determining the interaction of electromagnetic radiation with a polymer is the complex index of refraction:

$$\hat{n} \;\; = \;\; n - i\,k \tag{11}$$

where the real part n is the refractive index and the imaginary part k is the absorption index. These are known as the optical constants of the material. Both vary with the frequency (and hence wavelength) of the radiation. Another property of great importance in treating the interaction of light with a polymer is the molecular polarizability (11, 12). With certain assumptions involving the local electric field, the Lorentz-Lorenz or Clausius-Mosotti equation relates the polarizability to the optical constants as follows:

$$\hat{\phi} \;\; = \;\; \tfrac{4}{3}\pi N\hat{\alpha} \;\; = \;\; \frac{\hat{n}^2 - 1}{\hat{n}^2 + 2} \tag{12}$$

In this equation $\hat{\alpha}$ is the complex molecular polarizability, N is the number of molecular species per unit volume, and $\hat{\phi}$ is a polarizability function proportional to $\hat{\alpha}$. The real and imaginary parts of $\hat{\phi}$ are given by:

Real:
$$\phi' \;\; = \;\; 1 - \frac{3(n^2 - k^2 + 2)}{(n^2 - k^2 + 2)^2 + 4n^2k^2} \tag{13a}$$

Imaginary:
$$\phi'' \;\; = \;\; \frac{6nk}{(n^2 - k^2 + 2)^2 + 4n^2k^2} \tag{13b}$$

For birefringence, the pertinent quantity is the real part of the polarizability function. Since the measurements are done in the visible range, where k can usually be neglected, Equation 13a reduces to:

$$\phi \;\; = \;\; \frac{n^2 - 1}{n^2 + 2} \tag{14}$$

Birefringence refers to the difference in ϕ between any two of the three directions M, T, and N. Thus it is necessary to determine the refractive index with respect to all three directions, i.e. n_{M}, n_{T}, and n_{N}. The orientation is related to the birefringence through the following equations (6):

$$\frac{2\phi_M - \phi_T - \phi_N}{\phi_M + \phi_T + \phi_N} = \frac{2\Delta\alpha}{3\alpha_0}P_{200} + \frac{2\delta\alpha}{\alpha_0}P_{202} \tag{15a}$$

$$\frac{\phi_T - \phi_N}{\phi_M + \phi_T + \phi_N} = \frac{4\Delta\alpha}{3\alpha_0}P_{220} + \frac{2\delta\alpha}{3\alpha_0}P_{222} \tag{15b}$$

where:

$$\Delta\alpha = \alpha_c - \tfrac{1}{2}(\alpha_a + \alpha_b) \tag{16a}$$

$$\delta\alpha = \alpha_a - \alpha_b \tag{16b}$$

with α_a, α_b, and α_c being the components of the polarizability tensor in the directions a, b, and c of the polymer structural unit. (It is assumed that these coincide with the principal axes of the polarizability tensor; if not, the analysis is more complicated.) For a detailed analysis, it is necessary to estimate these values by calculations based on a molecular model. Even then, since there are only two known quantities, it is impossible to determine all four P_{2mn} values unless results from other techniques are also used.

Letting $\phi_0 = \tfrac{1}{3}(\phi_M + \phi_T + \phi_N)$, Equation 15 can be used to calculate the following expressions for the birefringence values:

$$\phi_M - \phi_N = \frac{\phi_0\Delta\alpha}{\alpha_0}(P_{200} + 2P_{220}) + \frac{\phi_0\delta\alpha}{\alpha_0}(3P_{202} + P_{222}) \tag{17a}$$

$$\phi_M - \phi_T = \frac{\phi_0\Delta\alpha}{\alpha_0}(P_{200} - 2P_{220}) + \frac{\phi_0\delta\alpha}{\alpha_0}(3P_{202} - P_{222}) \tag{17b}$$

$$\phi_T - \phi_N = \frac{\phi_0\Delta\alpha}{\alpha_0}(4P_{220}) + \frac{\phi_0\delta\alpha}{\alpha_0}(2P_{222}) \tag{17c}$$

For the greatest accuracy, the measured values of the index of refraction n should be converted to ϕ according to Equation 14. However, the quantity ϕ can be approximated quite well by a linear function of n, so the difference $\phi_I - \phi_J$ is proportional to $n_I - n_J$. Consequently, birefringence is often expressed in terms of Δn rather than $\Delta\phi$. Equation 17 can be used to relate the measured birefringence values Δn to the orientation as follows:

$$\Delta n_{MN} = n_M - n_N = \Delta^0(P_{200} + 2P_{220}) + \delta^0(3P_{202} + P_{222}) \tag{18a}$$

$$\Delta n_{MT} = n_M - n_T = \Delta^0(P_{200} - 2P_{220}) + \delta^0(3P_{202} - P_{222}) \tag{18b}$$

$$\Delta n_{TN} = n_T - n_N = \Delta^0(4P_{220}) + \delta^0(2P_{222}) \tag{18c}$$

where $\Delta^0 = n_c - \tfrac{1}{2}(n_a + n_b)$ and $\delta^0 = n_a - n_b$, with n_a, n_b, and n_c corresponding to the indices of refraction in the a, b, and c directions of a perfectly oriented sample. In terms of the Hermans or Spruiell-White orientation factors, the relationships are as follows:

$$\Delta n_{MN} = \Delta^0 \cdot \tfrac{2}{3}(f_{cM} - f_{cN}) + \delta^0 \cdot \tfrac{1}{3}(f_{aM} - f_{aN} - f_{bM} + f_{bN}) \tag{19a}$$

$$\Delta n_{MT} = \Delta^0 \cdot \tfrac{2}{3}(f_{cM} - f_{cT}) + \delta^0 \cdot \tfrac{1}{3}(f_{aM} - f_{aT} - f_{bM} + f_{bT}) \tag{19b}$$

$$\Delta n_T = \Delta^0 \cdot \tfrac{2}{3}(f_{cT} - f_c) + \delta^0 \cdot \tfrac{1}{3}(f_{aT} - f_a - f_{bT} + f_b) \tag{19c}$$

$$\Delta n_{MN} = \Delta^0 \cdot f_{cM}^B + \delta^0 \cdot \tfrac{1}{2}(f_{aM}^B - f_{bM}^B) \tag{20a}$$

$$\Delta n_{MT} = \Delta^0 \cdot (f_{cM}^B - f_{cT}^B) + \delta^0 \cdot \tfrac{1}{2}(f_{aM}^B - f_{aT}^B - f_{bM}^B + f_{bT}^B) \tag{20b}$$

$$\Delta n_{TN} = \Delta^0 \cdot f_{cT}^B + \delta^0 \cdot \tfrac{1}{2}(f_{aT}^B - f_{bT}^B) \tag{20c}$$

The relationship expressed in Equation 20 is consistent with that given in another form by Choi et al. (13). It should be stressed that, because the birefringence measurement is a global one, in the case of a semicrystalline polymer the measured values correspond to a weighted average over the contributions from the amorphous and crystalline phases.

The situation is considerably simplified if cylindrical symmetry is assumed for the polymer unit, i.e. only the chain (or c) axis orientation is considered. This is tantamount to assuming that differences in orientation between the a and b axes are negligible. In the above equations, the terms involving δ^0 vanish and the equations provide the means for calculating various orientation parameters from the measured birefringence values:

$$P_{200} = \frac{n_M - \tfrac{1}{2}(n_T + n_N)}{\Delta^0} = \frac{\tfrac{1}{2}(\Delta n_{MT} + \Delta n_{MN})}{\Delta^0} \tag{21a}$$

$$P_{220} = \frac{\Delta n_{TN}}{4\Delta^0} \tag{21b}$$

$$f_{cM} = \frac{n_M - \tfrac{1}{2}(n_T + n_N)}{\Delta^0} = \frac{\tfrac{1}{2}(\Delta n_{MT} + \Delta n_{MN})}{\Delta^0} \tag{22a}$$

$$f_{cT} = \frac{n_T - \tfrac{1}{2}(n_M + n_N)}{\Delta^0} = \frac{\tfrac{1}{2}(\Delta n_{TN} + \Delta n_{MT})}{\Delta^0} \tag{22b}$$

$$f_{cN} = \frac{n_N - \tfrac{1}{2}(n_M + n_T)}{\Delta^0} = \frac{\tfrac{1}{2}(-\Delta n_{MN} - \Delta n_{TN})}{\Delta^0} \tag{22c}$$

$$f_{cM}^B = \frac{\Delta n_{MN}}{\Delta^0} \tag{23a}$$

$$f_{cT}^B = \frac{\Delta n_{TN}}{\Delta^0} \tag{23b}$$

The quantity Δ^0 is the well known maximum or intrinsic birefringence of the polymer, i.e. the value corresponding to perfect orientation of the chain axes. Equation 23 underlines the very simple relationship that exists under these circumstances between the birefringence and the biaxial orientation factors proposed by White and Spruiell (defined in Equation 7). To characterize the biaxial orientation, two independent birefringences must be determined and two parameters are required to define the orientation (for example, f_{cM}^B and f_{cT}^B, or f_{cM} and f_{cT}). However, in a sample that is uniaxially oriented in the machine direction M, the T and N directions are equivalent, in which case $n_T = n_N$, $f_{cT} = f_{cN} = -\tfrac{1}{2}f_{cM}$, and $f_{cT}^B = 0$. Only one parameter ($f_{cM} = f_{cM}^B$) is required to define the orientation.

Thus, the orientation in the material can be described by the refractive index ellipsoid (indicatrix), which is defined by the refractive indices n_M, n_T, and n_N in the three axial directions (machine, transverse, and normal, respectively), which in turn originate from the different polarizabilities in the three directions as a result of molecular chain alignment. It should be mentioned that birefringence is a global measurement and does not discriminate between the different phases present (amorphous and crystalline).

Many techniques have been used for birefringence measurements. The first approach to be used was refractometry, which consists in making measurements of the three different refractive indices using a polarizer and by rotating the sample (14). The birefringence is calculated as the difference between the different refractive indices. Orientation functions can also be calculated, with some assumptions, directly from the refractive indices (7). In refractometry a contact liquid (generally α-bromonaphthalene) is used to avoid reflection at the interface between the prism and the sample. This technique, however, has several limitations: in many cases the use of a contact liquid is not appropriate or possible; it is a reflection technique; it is tedious; and it cannot be applied for on-line monitoring.

Another approach measures the birefringence directly using optical methods with polarized light (from the retardation in polarized light on going though an oriented sample). Different optical set-ups may be used. The most complete (described below) is the one that uses a multiwavelength source and detector (array detector), which allow the determination of the absolute birefringence without a limit on its order, as obtained with monochromatic methods. These techniques are based on the principle of light retardation when polarized in a certain direction, due to the difference in light propagation speeds in the direction of polarization and the direction perpendicular to it. They include monochromatic polarized light retardation (15, 16), multiwavelength polarized light retardation (17-19) and some qualitative techniques (fringes and light scattering). Monochromatic polarized light techniques are suitable for low orientations (low retardation), which are encountered in some cases, but are useless for moderately or highly oriented films, sheets or shapes. This technique was used for biaxial orientation measurements (20, 21) in polymers with low degrees of orientation, mostly in the molten state.

To overcome the limitations described above, particularly for moderate to highly biaxially oriented or thick samples, the use of the multiwavelength polarized light technique was extended to measure the absolute biaxial orientation in biaxially oriented films, sheets, and shapes (22, 23). This was achieved by directing at least two beams at different angles in order to measure the biaxial birefringence for transparent or translucent materials. One light source, one array detector, and data acquisition and analysis software that takes into account the effect of wavelength and the change of material optical properties with wavelength are required. The wavelength range used is from 350 nm to 900 nm. For multilayer materials, provided they are significantly optically different, the difference in birefringence dependence on wavelength (optical dispersion) is exploited to discriminate among the contributions from different layers.

The technique is illustrated in Figure 3.4. A multiwavelength light source is directed using a fibre to a fibre switcher which directs light alternately to one of two optical fibres at a fixed and known time interval (adjustable, from 20 ms). At the end of the optical fibre, a collimating lens and rotating polarizer are fixed and the light beam is directed at two different incident angles to the sample to be measured. After the sample, and directed to the light path, a set of polarizers and focusing beam lenses collect the light into a bifurcated optical fibre which takes the light to a photodiode array detector. The detector is connected through an acquisition card to a computer for data acquisition and analysis. Data is acquired in the form of intensity or transmittance as a function of wavelength at different time intervals. With both polarizers oriented in the vertical position and the sample having a plane XY perpendicular to the light path, for example the machine-transverse plane, the governing equation for the light intensity can be written as:

$$ I \;\propto\; \cos^2\left[\frac{\pi \Delta n_0 d}{\lambda} f(\lambda)\right] \tag{24} $$

where I is the light intensity, Δn_0 is a birefringence constant, d is the thickness of the sample, λ is the wavelength (we will call the term $\Delta n_0 \cdot d$ the retardation and denote it by Γ_0), and $f(\lambda)$ is the variation of birefringence with wavelength, which depends on the material (optical dispersion). The birefringence in the XY plane for a wavelength λ_0 (in most cases $\lambda_0 = 589.6$ nm, corresponding to the sodium light) can thus be determined as:

$$ \Delta n_{XY}(\lambda_0) \;=\; \Delta n_0 f(\lambda_0) \tag{25} $$

For the $f(\lambda)$ function, different dependencies have been proposed, but the simplest and most convenient and widely used form is:

$$ f(\lambda) \;=\; \alpha + \frac{\beta}{\lambda^2} + \frac{\delta}{\lambda^4} + \cdots \tag{26} $$

In most cases the first two terms are sufficient for calculations. The α, β, and δ constants depend on the material. For a multilayer material, containing two or more significantly optically different materials (that is, the dependencies of their refractive indices as a function of wavelength are significantly different), there will be $f_1(\lambda)$, $f_2(\lambda)$, etc., each associated with a retardation in the corresponding material. The cosine argument becomes:

$$ \frac{\pi d}{\lambda}\left[\Delta n_{01} f_1(\lambda) + \Delta n_{02} f_2(\lambda) + \cdots\right] \tag{27} $$

Since the functions f_1, f_2, ... are known, the regressions can be made on the different Δn_{0i}. The same equations apply in the case of oblique incidence. The retardation obtained in this case will depend on the angle of incidence. The different birefringences can then be obtained from measurements at two angles. Let us assume Γ_1 to be the retardation obtained for an angle θ_1 and Γ_2 to be that for θ_2. The angles θ_1 and θ_2 (between 0 and 90°) are the angles between the normal to the MT plane and the incident light beam, by rotating the MT plane around the T direction. Calculations lead to the following equations for the birefringences:

$$ \Delta n_{MN} \;=\; \frac{n}{d\,(\sin^2\theta_2 - \sin^2\theta_1)}\left[\Gamma_1(n^2 - \sin^2\theta_1)^{\frac{1}{2}} - \Gamma_2(n^2 - \sin^2\theta_2)^{\frac{1}{2}}\right] \tag{28} $$

$$\Delta n_{MT} = \frac{\Gamma_1}{nd}(n^2 - \sin^2\theta_1)^{\frac{1}{2}} - \frac{\sin^2\theta_1}{n^2}\Delta n_{MN} \tag{29}$$

$$\Delta n_{TN} = \Delta n_{MN} - \Delta n_{MT} \tag{30}$$

where n designates the average refractive index. Since in the intensity equation the sine or cosine functions are squared, it is not possible to know the sign of the birefringence by a single measurement. However, through the comparison of the retardation values at two different angles, this sign can be determined. In fact, it can be shown that for angles between $0°$ and $90°$, following the measurement procedure mentioned above and in most cases, if $\theta_2 > \theta_1$ the retardation Γ_1 should be greater than Γ_2. If Γ_2 is greater than Γ_1, it means that Γ_2 is negative and Γ_1 may be positive or negative and a measurement at another angle close to the first one is necessary to assess its sign. For other normals or rotations around another direction, that is any angle, more general calculations can be made.

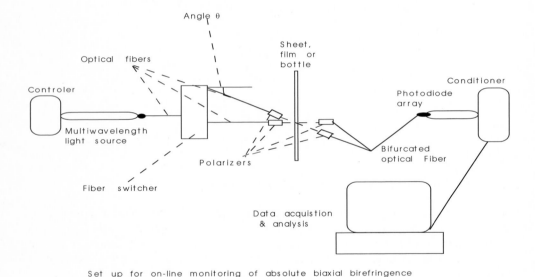

Set up for on-line monitoring of absolute biaxial birefringence

Figure 3.4. Sketch of the setup used for birefringence measurements.

The technique can be used to determine the degree of orientation in a wide range of products as well as for on-line monitoring of orientation in polymers. Two examples follow below.

3.2.1 Biaxially oriented polyethylene terephthalate (PET) film

Figure 3.5 shows the results obtained for a PET film (thickness of 90 μm) produced by the biaxial orientation process. The figure shows both the experimental and regression results at three angles (the third angle was used to confirm the values). The best parameters found by regression yielded the following values for the different birefringences at a wavelength of 589 nm using the equations above: $\Delta n_{MT} = +0.0322$, $\Delta n_{MN} = +0.1203$ and $\Delta n_{TN} = +0.0881$. The average refractive index taken for the calculations was 1.640. It can be concluded that this film is highly oriented in both the machine and the transverse directions with a lower orientation in the transverse one.

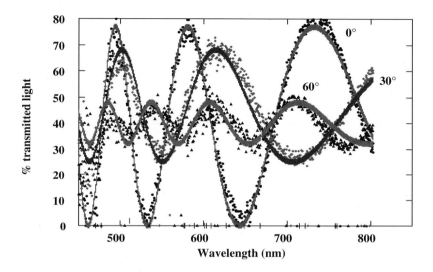

Figure 3.5. Experimental and calculated results for biaxially oriented PET film (90 μm thick).

3.2.2 PET orientation and relaxation monitoring

Figure 3.6a shows the results obtained for uniaxially oriented PET during its orientation at 80°C with a draw rate of 2 cm/min. The data were acquired every 5 seconds and are presented in terms of birefringence as a function of draw ratio. A transition point is clearly seen around a draw ratio of 2.5, which indicates the onset of stress-induced crystallization with a marked increase of birefringence. The following Figure 3.6b shows the relaxation of orientation of a film drawn to a draw ratio of 2 (as in the previous figure) and kept at the same temperature between the grips of the machine while data were acquired every 10 s. This sample is completely amorphous and it can be seen that the decrease of birefringence is linear on the basis of a logarithmic time scale. This indicates that relaxation follows an exponential form as would be predicted from molecular models. The average longest relaxation time can thus be determined by this measurement.

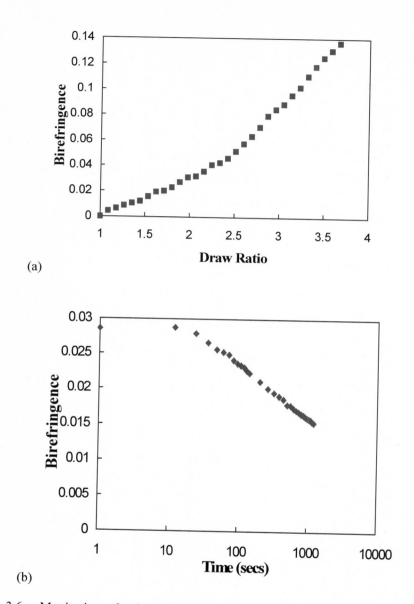

(a)

(b)

Figure 3.6.   Monitoring of orientation in PET by birefringence: (a) development of orientation during drawing at 80°C and 2 cm/min;   (b) relaxation of orientation under tension after drawing to a draw ratio of 2 at 80°C.

3.3 Vibrational Spectroscopy

3.3.1 General

Vibrational spectroscopy includes both infrared (IR) and Raman spectroscopy (24, 25). Like birefringence, these are optical techniques that rely on the interaction of light with the material being examined. Raman spectroscopy makes use of monochromatic laser light with a wavelength in the visible (e.g. argon ion, 514 nm) or near-infrared (e.g. Nd-YAG, 1.064 µm) range. Infrared spectroscopy uses multiwavelength radiation, usually in the mid-infrared range (2.5 to 25 µm) but sometimes in the near-infrared (0.8 to 2.5 µm) or far-infrared (25 to 200 µm) range. Both IR and Raman spectroscopies give information on the vibrational modes of the chemical bonds and groups that are present in the sample being studied. As a result, they possess the important ability to be able to characterize the orientation of specific bonds, molecular conformations, and phases (e.g. amorphous and crystalline).

3.3.2 Transmission infrared spectroscopy

Polymers absorb infrared radiation at specific wavelengths that correspond to the frequencies of vibration of the chemical bonds and groups that are present in the molecule. Thus, the optical constant of main interest in Equation 11 is not the refractive index n but rather the absorption index k. When it is possible, IR measurements are best made in the transmission mode. A transmission IR spectrum typically expresses the transmittance T of a sample as a function of wavenumber $\overline{\nu}$, where $\overline{\nu}$ (in cm^{-1}) is the reciprocal of the wavelength and $T(\overline{\nu})$ is the ratio of the transmitted intensity I_t with respect to the incident intensity I_0. If the effects of reflection at the sample surfaces are neglected, the following relation applies:

$$A = -\log_{10} T = \frac{4\pi}{\lambda_0 \ln 10} kb \qquad (31)$$

where A is the absorbance, λ_0 is the wavelength of the incident radiation, and b is the sample thickness. When transmission measurements are not possible, reflection methods can also be used to determine k. To characterize the orientation in a sample like that shown in Figure 3.1, it is necessary to measure k (or the related quantity A) in the three directions of interest, i.e. k_M, k_T, and k_N. This is done by using a linearly polarized infrared beam with the electric vector in the particular direction being measured. This is usually easy for the M and T directions, but more difficult for the N direction. In some cases, depending on the sample, k_N can be determined by making measurements with the incident beam at an angle other than normal. In other cases, thin slices must be microtomed in different directions and individually analyzed.

Since infrared dichroism is based on absorption of the radiation, it is appropriate to use the imaginary or "loss" component of the molecular polarizability function, as defined in Eq. 13b. Strictly speaking, it is this quantity that should be used in calculations of dichroism, because it is the most closely related to the molecular properties (26), and hence it is used in the equations given below. However, unless the bands being analyzed are particularly intense, ϕ can be considered to be proportional to k to a good approximation, so k (or A) can usually be substituted for ϕ in these equations without causing undue error.

Each vibrational mode of the polymer molecule possesses a transition moment $\vec{M}$, corresponding to the change in dipole moment $\vec{\mu}$ resulting from the vibration; $\vec{M}$ is a vector quantity that makes an angle α with respect to the polymer chain axis. The intensity of infrared absorption is proportional to the squared dot product of this vector with that of the electric field of the polarized infrared radiation. The values of ϕ are related to the P_{2lm} coefficients of the orientation distribution function by (6):

$$\frac{2\phi_M - \phi_T - \phi_N}{\phi_M + \phi_T + \phi_N} = 2p_{200}(\alpha)P_{200} + 4[p_{200}(\alpha)-1]P_{202} \quad (32a)$$

$$\frac{\phi_T - \phi_N}{\phi_M + \phi_T + \phi_N} = 4p_{200}(\alpha)P_{220} + \tfrac{4}{3}[p_{200}(\alpha)-1]P_{222} \quad (32b)$$

where: $\quad P_{200}(\alpha) = \tfrac{1}{2}(3\cos^2\alpha - 1) \quad (33)$

Assuming α is known, Equation 32 involves four unknowns and only two known quantities. However, if the polymer unit is assumed to have cylindrical symmetry (as discussed above in connection with birefringence), P_{202} and P_{222} are equal to zero, so P_{200} and P_{220} can be determined from the following equations:

$$P_{200} = \frac{1}{2p_{200}(\alpha)} \cdot \frac{2\phi_M - \phi_T - \phi_N}{\phi_M + \phi_T + \phi_N} \quad (34a)$$

$$P_{220} = \frac{1}{4p_{200}(\alpha)} \cdot \frac{\phi_T - \phi_N}{\phi_M + \phi_T + \phi_N} \quad (34b)$$

The orientation functions defined in Equations. 10 and 7 are given by:

$$f_{cM} = \frac{1}{p_{200}(\alpha)} \cdot \frac{\phi_M - \tfrac{1}{2}(\phi_T + \phi_N)}{\phi_M + \phi_T + \phi_N} \quad (35a)$$

$$f_{cT} = \frac{1}{p_{200}(\alpha)} \cdot \frac{\phi_T - \tfrac{1}{2}(\phi_M + \phi_N)}{\phi_M + \phi_T + \phi_N} \quad (35b)$$

$$f_{cN} = \frac{1}{p_{200}(\alpha)} \cdot \frac{\phi_N - \tfrac{1}{2}(\phi_M + \phi_T)}{\phi_M + \phi_T + \phi_N} \quad (35c)$$

$$f_{cM}^B = \frac{1}{p_{200}(\alpha)} \cdot \frac{\phi_M - \phi_N}{\phi_M + \phi_T + \phi_N} \quad (36a)$$

$$f_{cT}^B = \frac{1}{p_{200}(\alpha)} \cdot \frac{\phi_T - \phi_N}{\phi_M + \phi_T + \phi_N} \quad (36b)$$

Equations. 34, 35, and 36 represent three different but equally valid ways of quantifying the biaxial orientation. Regardless of the method, two independent parameters are required. However, if the orientation is uniaxial, then $\phi_T = \phi_N$ and Equation 35a reduces to the well known equation:

$$f_{cM} = \frac{1}{p_{200}(\alpha)} \cdot \frac{\phi_M - \phi_T}{\phi_M + 2\phi_T} = \frac{2}{3\cos^2\alpha - 1} \cdot \frac{D-1}{D+2} \qquad (37)$$

where $D = \phi_M/\phi_T$ is the dichroic ratio.

Even if the polymer unit does not have cylindrical symmetry, if a vibration can be found for which $\alpha = 0$, then $p_{200}(\alpha) = 1$, the P_{202} and P_{222} terms in Equation 32 vanish and P_{200} and P_{220} can be determined. Then if a vibration with $\alpha = 90°$ is found, P_{202} and P_{222} can also be calculated (7).

In the case of biaxial orientation, the absorption must be measured with the electric field of the radiation polarized in the three orthogonal directions M, T, and N. For thin films, this can be accomplished by means of the "tilted film" technique (27-31). First the film is inserted into the spectrometer in the usual way, perpendicular to the IR beam, and spectra S_M and S_T are measured with the polarization in the M and T directions respectively. Then the film is tilted by a known angle (for example, 45°) as shown in Figure 3.7 and spectra are again measured with the polarization in two directions (in the plane of the page and perpendicular to it). Because of refraction, the beam passes through the film at an angle β given by:

$$\sin\beta = \frac{\sin 45°}{n} \qquad (38)$$

where n is the refractive index of the polymer. The two spectra obtained in this way correspond to:

$$S_{MN} = \frac{S_M \cos^2\beta + S_N \sin^2\beta}{\cos\beta} \qquad (39)$$

$$S_T' = \frac{S_T}{\cos\beta} \qquad (40)$$

The S_T' spectrum is redundant because it is the same as the S_T spectrum except for the factor $\cos\beta$. The S_{MN} spectrum is used to calculate the S_N spectrum according to:

$$S_N = \frac{\cos\beta}{\sin^2\beta}(S_{MN} - S_M \cos\beta) \qquad (41)$$

It is then possible to calculate the structural factor spectrum corresponding to an isotropic (unoriented) sample of the same material:

$$S_0 = \tfrac{1}{3}(S_M + S_T + S_N) \qquad (42)$$

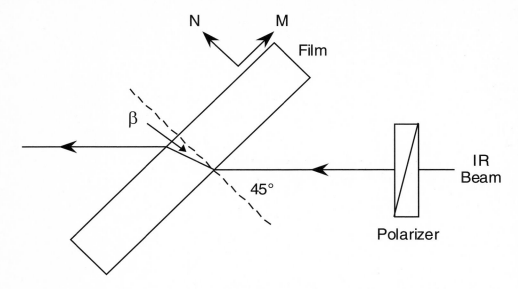

Figure 3.7 Illustration of the "tilted film" technique

The orientation of a given molecular axis i (= a, b, or c) with respect to a given sample direction J (= M, T, or N) can be determined from the dichroism of a specific peak in the IR spectra by the following equation (equivalent to Equation 35):

$$f_{iJ} = \left(\frac{\phi_J}{\phi_0} - 1\right) \cdot \frac{1}{3\cos^2\alpha_i - 1} \tag{43}$$

where ϕ_J is the intensity of the peak in the spectrum S_J corresponding to polarization in the J direction, $\phi_0 = \frac{1}{3}(\phi_M + \phi_T + \phi_N)$ is the intensity in the structural factor spectrum, and α_i is the angle that the transition moment of the vibration giving rise to the peak makes with the axis whose orientation is being defined (often the chain axis, but the approach can be applied to other axes as well).

These concepts can be illustrated with some examples based on polyethylene films. Figure 3.8 shows the M, T, N, and SF (structural factor) spectra obtained by use of the "tilted film" technique for a low density polyethylene (LDPE) film. Figure 3.9 shows the 750-690 cm^{-1} region in more detail. The film (about 50 μm thick) was prepared by uniaxially drawing an unoriented film at room temperature to a draw ratio between 3 and 4. The spectra illustrate two problems associated with transmission spectra. First, even for thin films the strongest bands in the spectrum are often saturated ($A > 2$ or $T < 1\%$). Second, reflection at the polymer-air interfaces sometimes results in regularly spaced interference fringes that obscure the weaker bands. The effect of the fringes is amplified in the N spectrum because it is calculated by difference. In spite of these problems, considerable information can usually be obtained from the medium-intensity bands. Comparison of the M and T spectra shows them to be quite different, an indication of strong dichroism and orientation.

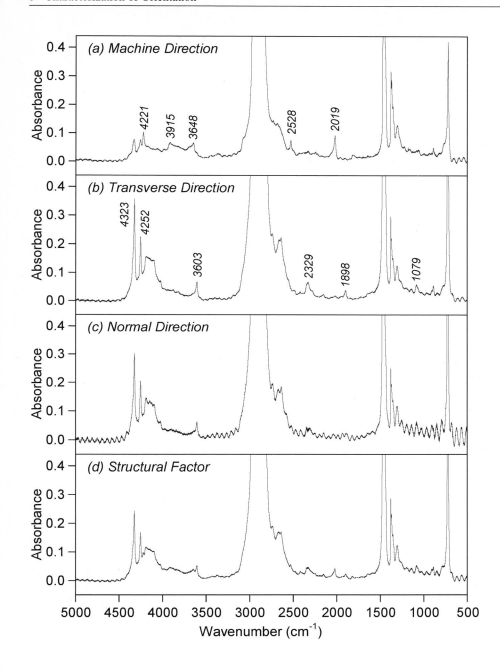

Figure 3.8. IR spectra of uniaxially stretched LDPE film.

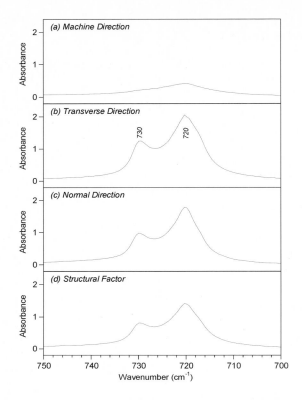

Figure 3.9 IR spectra (CH$_2$ rocking band) of uniaxially stretched LDPE film.

Certain peaks are present in the M spectrum but weak or absent in the T spectrum; they correspond to vibrations whose transition moment is approximately parallel to the chain axis ($\alpha \sim 0°$). Others are present in the T spectrum but weak or absent in the M spectrum; they correspond to vibrations whose transition moment is approximately perpendicular to the chain axis ($\alpha \sim 90°$). All the peaks are present in the structural factor spectrum, which matches that of an unoriented film. The fact that the N spectrum is virtually the same as the T spectrum confirms that the orientation is uniaxial. In such a case it is not necessary to use the tilted film technique and orientation functions can be calculated from the M and T spectra according to Equation 37. When this was done for the measurable peaks in Figure 3.8, the results in Table 3.3 were obtained. It can be seen that the orientation varies widely, depending on which conformations contribute to the peak in question. Similar results have been reported in the literature (32). The 1894 cm^{-1} peak arises only from the crystalline phase, which is found to be highly oriented, with the c axis of the unit cell (corresponding to the chain axis) aligned along the draw direction. The peaks at 2016 and 730-720 cm^{-1} also arise from the crystalline phase but include as well a contribution from the amorphous phase, in particular from chain segments involving extended *trans* units. The latter are somewhat less well oriented than the crystalline phase, so the orientation functions for these peaks are slightly lower, although still quite significant. The remaining peaks in the table arise only from the amorphous phase and show a still lower degree of orientation. The exact value depends on the importance of the contribution from the *gauche* conformers, which in general are poorly oriented.

Table 3.3   **Orientation functions calculated from the spectra of Figure 3.8 for an LDPE film**

Peak	Origin	f_{cM}
1894	CH$_2$ rocking combination band, crystalline phase only, $\alpha = 90°$	0.85
2016	CH$_2$ combination (twist + rock), crystalline + amorphous (extended *trans*), $\alpha = 0°$	0.82
730-720	CH$_2$ rocking, crystalline + amorphous (> 4 *trans* units), $\alpha = 90°$	0.67
1078	C–C stretch, amorphous only, *gauche* and *trans* (extended tie chains), $\alpha = 90°$	0.32
1368, 1352	CH$_2$ wagging, amorphous (GTTG and GTTG' conformations), $\alpha = 0°$	0.14
1303	CH$_2$ wagging, amorphous (GTG and GTG' conformations, $\alpha = 0°$	0.04

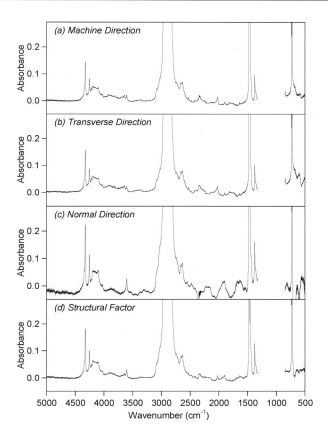

Figure 3.10 IR spectra (with Fluorolube region excluded) of biaxially stretched HDPE film.

Figures 3.10 and 3.11 show another set of PE spectra, in this case for a high density polyethylene (HDPE) film that was biaxially oriented by a tentering-type process. To eliminate interference fringes, the films were sandwiched between two potassium bromide windows along with a thin layer of perfluorinated paraffin oil ("Fluorolube") to ensure good optical contact. Although most of the Fluorolube peaks could be removed from the spectra by subtraction, the 1325-850 cm^{-1} region remains unusable because of the strong bands occurring there, so it is erased in Figure 3.10. In contrast with Figure 3.8, in Figure 3.10 the M and T spectra are similar while the N spectrum is different. This confirms the biaxial nature of the orientation. The CH$_2$ rocking band at 730-720 cm^{-1} is particularly interesting (Figure 3.11).

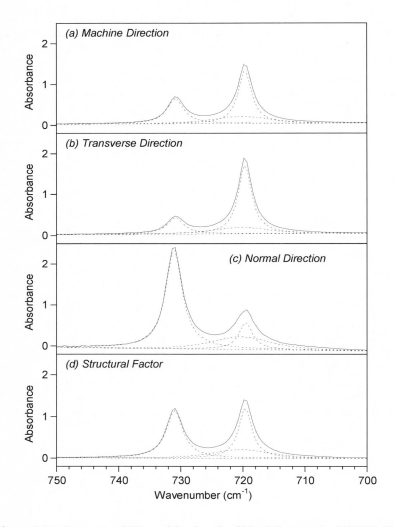

Figure 3.11 IR spectra (CH$_2$ rocking band) of biaxially stretched HDPE film.

Because there are two polymer chains in the orthorhombic unit cell of crystalline PE, crystal field splitting (in-phase and out-of-phase vibration of the two chains) gives rise to two sharp peaks at 730 and 720 cm^{-1}.   In both cases the transition moment of the vibration is perpendicular to the polymer chain or c-axis of the unit cell, but for the 730 cm^{-1} peak it lies paralllel to the a-axis and for the 720 cm^{-1} peak it lies parallel to the b-axis. This makes it possible in theory to use these peaks to study the crystal orientation. However, they have been considered of limited value for this purpose because the 720 cm^{-1} peak is subject to interference from a broader underlying peak arising from sequences of more than four *trans* units in the amorphous phase (33). Furthermore, under certain circumstances a monoclinic structure is present and gives rise to a peak near 716 cm^{-1}. For the spectra of Figure 3.11, the monoclinic peak appears to be negligible and the rocking band can be decomposed by curve fitting into the two peaks of the orthorhombic crystalline phase plus the one peak arising from the amorphous phase. From the areas of the individual peaks it is then possible to calculate orientation functions for both the crystalline and amorphous phases, through the use of Equation 43. These are given in Table 3.4. For both phases the orientation function of the chain axis is positive in both the M and T directions, as expected for biaxial orientation.   However, for the crystalline phase the a and b axes show quite different orientation, with the a axis showing a strong preference for alignment in the thickness direction. The difference between the a and b axes is evident from Figure 3.11, where the intensity ratio of the 730 and 720 cm^{-1} peaks is different in the M, T, and N spectra. Such a difference is not observed for the uniaxially drawn LDPE sample (Figure 3.9). In that case, drawing at room temperature resulted in deformation of the orthorhombic phase to produce a significant amount of monoclinic structure, as can be seen from the shoulder at 716 cm^{-1} in the spectra. This would make the curve fitting procedure more difficult to apply.

The two examples just described clearly illustrate the ability of IR spectroscopy to provide highly specific information on the uniaxial or biaxial orientation of the different phases and molecular conformations present in a polymer.   Usually, the overall orientation as determined by IR correlates well with that obtained by birefringence, and when the crystalline orientation can be specifically determined it correlates with that obtained by X-ray diffraction. The relative rapidity of IR measurements also makes it possible to perform rheo-optical measurements, i.e. real-time spectroscopic measurements of the orientation of specific molecular groups while the sample is mechanically deformed (34, 35). Most of the large body of work that has been done in IR spectroscopy has made use of the mid-infrared region (4000-400 cm^{-1}). It can be seen from Figures 3.8 and 3.10 that the near-infrared region (12000-4000 cm^{-1}) is also sensitive to orientation, but the overtone and combination bands that occur here overlap significantly and their assignment is not always clear. Nevertheless, the near-infrared region has attracted growing attention recently for a wide range of applications, including the characterization of orientation. It has two particular advantages: the bands that occur there are intrinsically less highly absorbing, so it can be applied to thicker samples (of the order of 1-5 mm), and fibre optic probes are readily available to bring the IR beam to the sample when the reverse is not possible (35). The far-infrared region has also been exploited for orientation measurement, but to a much lesser extent (36).

Table 3.4   **Orientation functions calculated from the spectra of Figure 3.11 for an HDPE film**

Phase	Orientation Function	Value
Crystalline	f_{aM}	−0.23
"	f_{aT}	−0.33
"	f_{aN}	+0.56
"	f_{bM}	+0.02
"	f_{bT}	+0.15
"	f_{bN}	−0.17
"	f_{cM}	+0.21
"	f_{cT}	+0.18
"	f_{cN}	−0.39
Amorphous	f_{cM}	+0.20
"	f_{cT}	+0.15
"	f_{cN}	−0.35

Although near-infrared and far-infrared measurements extend somewhat the range of thickness over which films can be successfully analyzed, in general the transmission IR technique is limited to rather thin samples. It can also be applied to fibres, either individually (with the use of an IR microscope) or collectively (by carefully arranging the fibres side by side). Because fibres have uniaxial symmetry, only two spectra are required, with polarization parallel and perpendicular to the fibre axis. However, fibres constitute a sample of nonuniform thickness, and this can have an effect on the shapes of the peaks in the spectrum, particularly the more intense ones, so caution must be exercised in analyzing the results. For thicker samples like sheets, bottles, rods, or injection-moulded parts it is difficult to make bulk measurements by IR. One approach that has been used to overcome this problem is to microtome thinner specimens from the thick sample and measure them in transmission (37). This makes it possible to characterize any spatial variation of orientation within the sample, for example a gradient across the thickness dimension. Biaxial orientation can be studied by microtoming and measuring slices in different planes (MT, MN, TN). One problem with this technique is that the microtoming can affect the orientation of the surface of the specimen (38). The error can be minimized by using the thickest slices that can be analyzed (36). An alternative to microtoming thick samples is to limit the orientation measurement to the surface of the sample by using infrared reflection rather than transmission. This can be done in different ways as described in the following sections.

3.3.3   Attenuated total reflection (ATR) infrared spectroscopy

Attenuated total reflection IR, also known by the general term internal reflection, has been in use for a long time (39-41). In this technique, the infrared beam passes through a prism made from a material with a high index of refraction. Commonly used materials are thallium bromoiodide, or KRS-5 (n = 2.4) and germanium (n = 4). The prism is designed so that the beam is totally internally reflected one or more times at the interface between the prism and the surrounding material. Total reflection occurs if the angle of incidence is greater than a certain critical angle that depends on the relative refractive indices. When the surrounding material is nonabsorbing, like air, the reflection is total. If an absorbing sample like a polymer is placed in close optical contact with the prism surface, the evanescent wave that extends slightly beyond the interface is partly absorbed by the polymer, and the reflection is "attenuated". By taking the ratio of the beam reflected under these conditions with respect to that reflected when air is present, a spectrum is obtained that usually resembles the transmission spectrum of a thin sample (i.e. it is dependent mainly on the absorption index k). However, care must be exercised because under certain circumstances the refractive index n can make a significant contribution and distort the shape of the absorption peaks. This effect is more pronounced when the refractive index of the polymer approaches that of the prism, or when the angle of incidence is close to the critical angle. The ATR technique probes the surface of the sample to a depth of a few micrometres. The exact value of the effective penetration depth d_p depends on the materials and the angle of incidence and is also proportional to the wavelength of the radiation. For a typical polymer and an angle of $45°$, d_p varies over the wavenumber range 4000-500 cm^{-1} from 0.7 to 5.6 μm for a KRS-5 prism and from 0.17 to 1.4 μm for a germanium prism. By making measurements on the same sample with different prism materials or different angles of incidence, it is possible to probe different depths and detect any variations within the first few micrometres from the sample surface. Also, when it is possible to make transmission and ATR measurements on the same sample, the surface can be compared with the bulk.

Biaxial orientation can be characterized by ATR (42-44). As for transmission, it is necessary to acquire spectra corresponding to polarization in the three directions M, T, and N. This is done by measuring four spectra, one of which is redundant but can serve as a check. The spectra are obtained by using four different possible combinations of sample positioning and beam polarization with respect to the plane of incidence of the reflection; to be precise, both the sample's M direction and the beam's electric vector can be either parallel or perpendicular to this plane. Further details along with the equations needed to calculate the M, T, and N spectra are given in the literature (42-44). The ATR method suffers from one important problem, which is that the overall intensity of the spectrum (or the "effective sample thickness") depends on the degree of contact between the sample and the prism. For the highest accuracy, perfect optical contact is required, but this is impossible to achieve in practice so there is always some error (45). For biaxially oriented samples, problems arise when the sample is removed from the prism, turned, and replaced in order to change the alignment of the M direction; the degree of contact is unlikely to be the same for the two positions. Different approaches have been developed to minimize the problem. For example, some polymers possess one or more bands that are known to be insensitive to orientation, and they can be used as a reference to normalize all four spectra to the same effective thickness; the 1410 cm^{-1} band of poly(ethylene terephthalate), or PET, is such a band (46). It was used by Walls and Coburn (47, 48) in their ATR study of drawn PET films, in which they showed that, on drawing, the *gauche* glycol conformers present in amorphous PET do not orient but are converted into *trans* conformers with a significant degree of orientation. Furthermore, by making measurements with both KRS-5 and germanium prisms they found that the *trans* conformers are somewhat more numerous and more highly oriented at the surface. Sung and coworkers (42) developed a way to avoid repositioning of the sample with respect to the prism by using a prism that can be rotated together with the sample, so that the degree of contact does not change. In a recent improvement on this technique, the use of a circular-shaped prism allows one to make measurements over a range of angles of rotation, thus giving more detailed information on the orientation. For example, Sung *et al.* (49, 50) studied the angular distribution of orientation in both uniaxially and biaxially drawn PET films and observed a peanut-shaped distribution in the uniaxially drawn samples. Palm (51) showed how the technique can be used to obtain more accurate values of k_M, k_T, and k_N in the case of uniaxially drawn poly(vinylidene fluoride), or PVDF. Although this technique reduces error by maintaining the same degree of contact for all the spectra, errors still exist if the contact is not perfect because of surface roughness (45). This can be especially important for uniaxially drawn samples, where the surface texture can be quite different in the directions parallel and perpendicular to the draw direction (43). Yet another approach to minimizing error related to prism-sample contact consists of using the ratio of two different peaks in the same spectrum as an indicator of orientation. The peaks should have significantly different values of the angle α, preferably $0°$ and $90°$. This approach was successfully applied to polypropylene (PP) by Mirabella (52), who showed that the surface orientation in biaxially oriented PP films was not significantly different from that of the bulk (as measured by transmission). The same was found to be true for uniaxially drawn PP films (37). In another study of drawn PP, Chen and Fina (53) used a range of angles of incidence in order to vary the effective depth of penetration and perform a depth-profiling study over the first 2 μm from the surface. They found a high degree of orientation, particularly for the crystalline phase, but negligible variation with respect to depth. However, the fraction of crystalline phase decreased somewhat with depth. ATR can also be applied to fibres. For instance, one study was accomplished by winding PET yarn around metal plates to form a parallel uniform layer of fibres that could be pressed against the ATR prism (54).

3.3.4   External reflection infrared spectroscopy

It is also possible to characterize polymer samples through the use of external reflection. In this technique, instead of passing through a prism the IR beam arrives directly onto the sample surface from the air, and the reflected beam is detected and ratioed against that reflected from a highly reflecting reference. Depending on the morphology of the sample, the reflection may be specular, diffuse, or a combination of both, and the sampling accessory and reference material are chosen accordingly. With a thick polymer sample, a large fraction of the incident energy is absorbed by or transmitted through the sample; only the small fraction reflected at the front surface is detected, so the noise level in the spectrum is relatively high. Historically this meant that external reflection was used much less than transmission or ATR, but as a result of the tremendous improvements in instrumental sensitivity over the last twenty years it is now a quite viable technique. Recently it has been shown to be useful for characterizing orientation, particularly in cases where transmission and ATR are difficult to apply, like thick nondeformable samples (55, 56).

The ideal sample for external reflection measurement of orientation is infinitely thick and has a very smooth flat surface. The reflection is then purely specular and front-surface, and is governed by the laws of Fresnel. Unfortunately, whereas in transmission and ATR the contributions of the refractive index n can often be neglected, this is never the case in front-surface reflection; the reflectance depends strongly on both n and the absorption index k. Fortunately, however, because these two quantities are related to each other by the Kramers-Kronig relation (57), it is possible to extract the n and k spectra from the reflectance spectrum. Software is commercially available for this purpose. In most cases, the k spectrum can be used to characterize the orientation, but once the n and k spectra are obtained, they can also be used to calculate the imaginary polarizability function ϕ by means of Equation 13b. As mentioned previously, the ϕ spectrum is very similar in appearance to the k spectrum, although not identical.

Like ATR, external reflection is not without problems. Although prism-sample contact is not a factor, surface imperfections are. By causing diffusion of the reflected beam they reduce the overall intensity of the spectrum. The easiest way to correct for this is to use an insensitive peak as a reference, as described for ATR. However, there are limits to the degree of roughness that can be handled, because at some point scrambling of the polarization will cause significant error. Another important source of possible error is reflection from the back surface of the sample. If this occurs, part of the reflected radiation will have passed through the sample and will introduce a transmission-like component into the spectrum, and possibly interference fringes as well. This makes it very difficult, if not impossible, to extract accurate n and k spectra. Back-surface reflection is more of a problem with thinner samples. It can be reduced by abrading the back surface to diffuse the reflected radiation. One way of practically eliminating both back-surface reflection and surface roughness effects is to mount the sample in epoxy resin and polish the surface to be analyzed (58). Since the resin and the sample have similar refractive indices, there is very little reflection at the back surface. The polishing procedure can alter the surface material, but if taken to a fine finish (0.05 µm) the disturbed material is mostly removed and any that remains is negligible in comparison with the sampling depth.

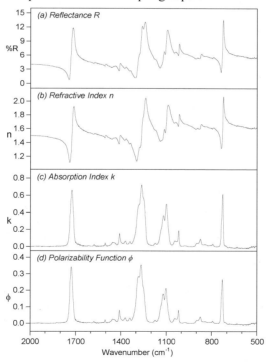

Figure 3.12   The measured reflectance spectrum of unoriented amorphous PET and the spectra derived from it.

Figure 3.12 shows the external reflectance spectrum of a sample of unoriented amorphous PET sheet prepared in this manner, along with the calculated n, k, and ϕ spectra. The strong influence of the refractive index on the reflectance spectrum is obvious. The k and ϕ spectra are very similar to each other but differ slightly in the details of the strongest bands. Figure 3.13 shows the ϕ spectra obtained in a similar manner, with parallel and perpendicular polarization, from a sample of initially amorphous PET sheet that was drawn uniaxially at 80°C to a draw ratio of 3.8. The strong dichroism observed for many of the peaks in the spectrum has been widely used to follow orientation in PET (56, 60, 61). The figure also includes the structural factor spectrum calculated by means of Equation 42 on the assumption that $\phi_N = \phi_T$ for uniaxial drawing.

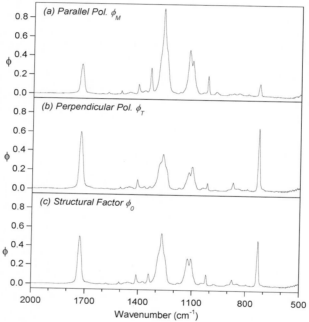

Figure 3.13.   Polarizability function spectra for a sample of amorphous PET drawn to λ = 3.8.

The differences between this spectrum and the ϕ spectrum of the undrawn sample (Figure 3.12) are mainly due to the conversion of *gauche* glycol conformers into oriented *trans* conformers, resulting in strain-induced crystallization. The ϕ spectra of Figures. 3.12 and 3.13 show exactly the same features as the transmission spectra of similar samples (59), with one important advantage: they clearly show the most intense bands in the spectrum, which are generally saturated in transmission spectra of PET unless the thickness is less than about 5 μm. Even in ATR spectra of PET the stronger bands are subject to distortion as a result of the refractive index rising to values approaching the refractive index of the prism (46). Thus under certain circumstances external reflection can provide more complete information than transmission and ATR. For instance, it has been used to show up a significant difference at the molecular level between the strain-induced crystalline phase produced by drawing of PET at 80°C and the crystalline structure produced by thermal annealing (62).

For external reflection, methods have not yet been developed to characterize biaxial orientation in a manner analogous to the tilted film technique in transmission or the rotatable prism technique in ATR. However, this can be achieved by making measurements on polished sections in the MT, TN, and MN planes (63). For insufficiently wide samples it is possible to use an infrared microscope (63, 64). In the case of PET a method has been developed (63) that uses the ratio of two peaks in the reflectance spectrum (similar to that developed by Mirabella for ATR of PP). This approach was combined with milling and polishing to follow the gradient in biaxial orientation across the thickness of a blow-moulded bottle.

Generally speaking the external reflection technique works best for polymers that possess intrinsically strong absorption bands in the 2000-500 cm^{-1} region, like polyesters, polyamides, liquid crystal polymers, and polyetheretherketone (55, 58, 65, 66). It can also work to some extent for materials with medium-intensity bands, like polyethylene and polystyrene (58, 67), but is of little use for less strongly absorbing polymers like poly(vinyl chloride).

3.3.5 Photoacoustic infrared spectroscopy

Like external reflection, photoacoustic (PAS) infrared spectroscopy suffers from a higher noise level than transmission and ATR, so it does not have as long a history of use. It is also a surface technique, but whereas reflection methods typically probe a depth on the order of the wavelength of the radiation, PAS goes an order of magnitude deeper. The principle is quite different from that of reflection (68). The sample is placed in a closed chamber and irradiated through a window with infrared radiation modulated at a frequency in the acoustic range. In a Fourier transform infrared spectrometer, the modulation frequency is proportional to the wavenumber of the radiation. The sample absorbs energy at certain frequencies and this energy is converted into heat that is then transferred to the gas surrounding the sample. This creates pressure fluctuations in the gas with modulation frequencies corresponding to the absorbed energy. Such pressure fluctuations constitute sound, which is detected by a microphone and analyzed in order to obtain the infrared spectrum of the sample. In contrast to external reflection, the strongest bands are often saturated but the weaker ones are generally better detected. Surface morphology is less critical than in reflection, and the presence of strongly absorbing filler like carbon black is not a problem.

The possibility of using a polarized infrared beam to determine orientation by PAS was demonstrated with PET in 1982 (69). However there have been surprisingly few reports of its application. Jasse and Noël used it to map the orientation over the surface of an injection-molded plaque of liquid crystalline polyesteramide (70), the same material that they studied by external reflection (65). More recently it was used to follow orientation of the *trans* fraction in PET fibers (71).

3.3.6   Raman spectroscopy

Like infrared spectroscopy, Raman spectroscopy probes the vibrational states of polymer molecules, but the mechanism is different (5, 24, 25). Infrared spectroscopy involves the direct absorption of radiation whose frequency corresponds to the difference in energy between the ground vibrational state and a higher quantum level. In Raman spectroscopy, the sample is irradiated with a rather intense beam of monochromatic (usually laser) radiation with a frequency in the visible or near-infrared range. Upon "collision" with the molecules, a small fraction of the photons are scattered elastically, i.e. without any change in energy (Rayleigh scattering). An even smaller fraction (less than one in a million) are scattered inelastically, i.e. with a change in energy that corresponds to the difference between two vibrational energy states. The inelastically scattered radiation is analyzed and plotted as a function of frequency to give a spectrum that resembles an IR absorption spectrum. Infrared spectroscopy depends on the fact that the molecular vibrations lead to variations of the dipole moments that are permanently present in the molecule, and it is most sensitive to vibrations of polar bonds or groups (for example C–H and C=O bonds). Raman spectroscopy depends on dipole moments that are induced by the electric field of the radiation, and the determining factor is variation in polarizability, as defined by the differential polarizability tensor.   As a result Raman spectroscopy is also sensitive to vibrations of nonpolar bonds or groups (e.g. C–C and C=C bonds) and thus can provide complementary information to IR spectroscopy.

The determination of orientation by Raman spectroscopy is more complicated than by IR (24, 25).   The spectrum can be measured with various combinations of direction and polarization. The most commonly used geometries are 90° (in which the laser beam is directed onto the sample surface at a grazing angle and the scattered radiation is analyzed in a direction normal to the surface) and 180° (in which both the incident beam and the direction of analysis are normal to the surface).   Whatever the geometry, spectra are recorded with different combinations of incident beam polarization and analyzer polarization. For a given Raman scattering peak, the intensities in the different spectra are related to different components of the polarizability tensor. These in turn depend on the molecular polarizability of the scattering group responsible for the peak in question as well as on the molecular orientation.   The rather complicated equations describing these relationships are given in the literature (72).   One particular advantage of Raman spectroscopy over IR is that it is possible to obtain the fourth-order moments of the orientation distribution in addition to the second-order moments. A disadvantage is that many commercial polymers show strong fluorescence (sometimes from additives or impurities) that masks the Raman effect. However, this problem is often greatly reduced when the laser wavelength is longer, as in many modern Fourier transform Raman systems that employ a near-infrared laser as the radiation source.

Among the first quantitative studies of orientation in polymers by Raman spectroscopy is the work on PET by Bower and co-workers (73, 74). For samples uniaxially drawn at 80°C they studied five different Raman bands with irradiation from the 488 nm line of an argon ion laser, and were able to determine both P_{200} and P_{400} as a function of draw ratio. The results were in good agreement with the behaviour predicted by a model in which the orientation takes place primarily through the stretching of a rubberlike network with about six equivalent random links between adjacent crosslink points.   In later work Raman spectroscopy was combined with IR, X-ray, and birefringence measurements to develop a detailed picture of biaxial orientation in PET, including preferential alignment of benzene rings of the crystalline phase in the plane of the film (75, 76). A recent study of a liquid crystalline polyester has also demonstrated the power of Raman spectroscopy in combination with IR and wide-angle X-ray diffraction (77). Other examples include poly(methyl methacrylate) (78), poly(vinyl chloride) (79), polyethylene (80, 81), and recent FT Raman studies of polyethylene (82) and polypropylene (83). As for FTIR, the rapidity of the FT Raman technique also facilitates the study of polymers in real time during deformation; examples include PET (84) and PVDF (35). One interesting application of Raman spectroscopy depends on the fact that certain peaks show a noticeable shift in position when the groups responsible for them are subjected to stress. Thus, for example, the 1616 cm^{-1} band can be used as a monitor of stress distribution at the molecular level in oriented PET (85). In other examples of interesting applications, a strongly Raman-active dye molecule was dissolved in polyethylene and used to probe orientation in the amorphous phase (86), and Raman microscopy was used to study orientation and crystallinity in uniaxially drawn PET (87).

3.4   Other Spectroscopic Techniques

3.4.1   Fluorescence

Polarized fluorescence intensity measurements yield information about the orientation of the transition moment of a fluorescent dye in both absorption and emission (88, 89). Such measurements have been used to determine molecular orientation in polymer melts, and in solid polymer films and fibers (89-93). To characterize polymer systems, a dye is incorporated into the polymer matrix by mixing at low levels of concentration; less than 100ppm by weight is usually sufficient. In some cases, the polymer molecule itself contains fluorescent moieties that eliminate the need for adding a fluorescent dye (90). Also, the fluorescent dye can be covalently bonded or "tagged" to the polymer molecule (89). The fluorescent probe can undergo some motion in the excited state, which complicates the calculations. The orientation measurement employs excitation light polarized with a fixed spatial relationship with respect to the principal orientation axes of the film or sheet. The simplest case is when the absorption and emission dipole moments coincide. In order to get the biaxial orientation factors, five intensity measurements are needed and no sample tilt is required. In these measurements, polarizer, analyser, and incidence angle are changed in order to get the five intensity values needed. A detailed theoretical description and measurement procedure can be found in Ref. 90. Since in most cases the dye is dispersed within the amorphous phase, it is thus the orientation of this phase that is measured. This technique allows the determination of the second and fourth moments of the orientation distribution function (90).

For uniaxial orientation, two intensity values are needed, and they can be obtained with one incident angle by changing the polarizer and analyser positions, as illustrated in Figure 3.14. In this figure z is defined as the direction of resin flow, $I_{vv} = I_{zz}$ and $I_{vh} = I_{zx}$ are respectively the vertically and horizontally polarized fluorescent light that is produced by vertically polarized excitation light. Molecular models yield a relationship between r (the ratio defined in Figure 3.14) and fluorescent dye orientation (89-91, 93). Recent advances led to the development of a fluorescent sensor that could be used on line for measurement of anisotropy (94, 95). It uses a compact set-up incorporating all the optics for the measurements. However, it is not clear how it is possible to determine biaxial orientation factors using this sensor.

Great care must be taken however when performing fluorescence measurements and the corresponding calculations to determine the orientation moments. The first factor that can influence the measurements is birefringence. In fact, for an oriented sample, the refractive indices in the three principal directions are different. This will have an effect on the incidence angle inside the material; it can be minimized by using specific conditions (90). The reflection and transmission coefficients at the air-sample interface can vary with the direction of polarization with respect to the principal axes, and this can generate up to 5% error. Another effect of birefringence is to introduce a phase difference between the two polarizations of light propagating in the same direction along two different principal axes. This can be avoided by proper choice of the experimental method (90).

Crystallinity can also be a source of some uncertainty. In fact, it causes depolarization of both the excitation light and the emitted fluorescence. This effect appears when light scattering phenomena occur, which can be easily verified using a small angle light scattering device.

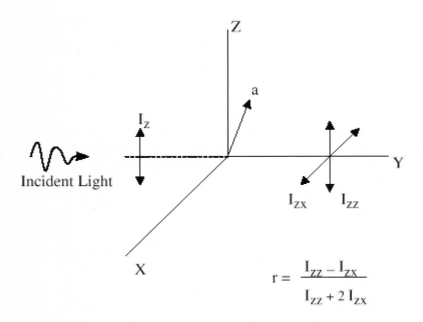

Figure 3.14. Diagram illustrating the measurement of orientation by fluorescence.

3.4.2   Nuclear magnetic resonance (NMR) spectroscopy

The technique of nuclear magnetic resonance is applied to atomic nuclei that possess an odd number of either protons or neutrons. This gives rise to a nonzero value of nuclear spin and consequently a magnetic moment. The most widely studied nuclei are the proton, ^{1}H, and the carbon isotope ^{13}C, whose natural abundance is only 1.1%. Both of these have a spin of ½. When placed in a magnetic field, such nuclei exist in two quantum energy states, which may be thought of as aligned with the field (lower energy) and aligned against the field (higher energy). The energy difference between the two states is proportional to the magnetic field and typically corresponds to that of electromagnetic radiation in the radiofrequency (rf) range of 10-100 MHz. Thus, when a sample is exposed simultaneously to rf energy and a magnetic field, under certain conditions "resonance" will occur, energy will be absorbed, and nuclei will pass from the lower to the higher energy state. The required condition is:

$$v_0 = \gamma H_0 \qquad\qquad (44)$$

where v_0 is the radiofrequency, H_0 is the magnetic field strength, and γ is the magnetogyric ratio of the nucleus. Since the latter is a constant for a given type of nucleus, one might expect resonance to occur at only a single frequency for a given applied magnetic field. In actual fact, the magnetic field experienced by a nucleus differs from the applied field because of various interactions. For example, the electrons in the chemical bonds around the nucleus produce a shielding effect that leads to a "chemical shift", with the result that the resonance condition is slightly different for nuclei in different chemical environments. In addition, the resonance is affected by two types of coupling between neighbouring nuclei, namely scalar coupling through chemical bonds and direct through-space dipole-dipole interactions. The latter are particularly important in solid samples. Both the chemical shift and the dipolar coupling depend on the orientation of the molecule with respect to the applied magnetic field. Thus, NMR can be used to probe the orientation of particular atoms in the molecule, and in this respect is even more specific than infrared and Raman spectroscopy. The theoretical aspects of the technique are too complicated to give here, but are given in various references (96-99).

In simple NMR measurements on solid samples, the dipole-dipole interactions result in very broad lineshapes. Starting in the late 1960s, Ward et al. applied broad line NMR to oriented PE, PMMA, PVC, and PET (5, 100). The second moment of the NMR lineshape depends on the second- and fourth-order moments P_{200} and P_{400} of the orientation distribution function defined in Equation 2. The latter two quantities can be determined provided the molecular geometry is defined. This is usually relatively easy to do for the crystalline phase but more difficult for the amorphous phase with its lower conformational regularity. In general, the results obtained by Ward et al. showed good correlation with the expected behaviour and with results obtained by other techniques. From the fourth moment of the NMR lineshape it was also possible to determine P_{600} and P_{800}, although with less accuracy than for P_{200} and P_{400}.

In the thirty years following these experiments, the NMR technique has undergone a number of major advances. First and foremost is the replacement of continuous wave NMR by Fourier transform NMR, which uses a sequence of rf pulses to excite the sample and follows the free induction decay of the magnetization. By manipulation of both the pulse sequences and the detection conditions, it is possible to perform complex experiments that provide a wealth of data, including information on the mobility of molecular groups. It is no longer necessary to deal with the broad lines previously associated with solid samples; high-resolution solid-state spectra can be obtained through the use of dipolar decoupling (DD) and magic angle spinning (MAS). The intensity of ^{13}C spectra can be enhanced through cross polarization (CP), which transfers magnetization from the abundant ^1H nuclei to the less abundant ^{13}C nuclei. Based on the use of the techniques just mentioned, highly sophisticated approaches, including "multidimensional NMR", have been developed to study polymer orientation and molecular dynamics. Details are given in recent books (96-99); we limit ourselves here to a few examples that illustrate their power.

PET is one of the polymers that has attracted considerable interest. Henrichs (101) used two-and three-dimensional ^{13}C NMR to study biaxially oriented films of PET and was able to measure all four second-order orientation moments (P_{200}, P_{202}, P_{220}, and P_{222}). He was also able to distinguish two different components: one with slowly relaxing protons (characteristic of a crystalline structure) in which the aromatic ring planes lie close to, but not in, the plane of the film, and another with fast relaxing protons (characteristic of a more mobile amorphous structure) in which the ring planes are less oriented. The technique used by Henrichs was extended by Spiess and coworkers and named DECODER ("Direction Exchange with Correlation for Orientation-Distribution Evaluation and Reconstruction"). It was applied to fibres and films (both uniaxial and biaxial) of PET (102). Detailed information on the shape of the orientation distribution functions could be obtained, including moments as high as $P_{14,0,0}$. For fibres, the distribution was not Gaussian but at least bimodal, consisting of a narrow component (about 60%) corresponding to highly ordered chains and a broader component (about 40%) corresponding to a lower degree of order. Tzou et al. (103) used two-dimensional rotor-synchronized CP-MAS ^{13}C NMR ("ROSMAS") to study melt-extruded PET fibres. This technique gives a less detailed picture of the shape of the distribution function than does DECODER, but provides higher spectral resolution, i.e. better separation of peaks from different morphological components in the sample. Thus they were able to separately characterize the orientation, in terms of P_{200} and P_{400}, for two components in PET fibres: a more ordered component with a narrow carbonyl peak, corresponding to the crystalline phase, and a less ordered component with a broader carbonyl peak. In a given sample, the two components gave similar values for P_{200} but different values for P_{400}. However the trends were difficult to explain, because of the complexity of the PET morphology. Finally, a recent paper by Clayden et al. (104) has demonstrated a good degree of consistency between results from 2D-MAS ^{13}C NMR, FTIR microscopy, and birefringence, for uniaxially drawn PET film. The DECODER technique has also been successfully applied to a liquid crystal polyester, for which detailed orientation distribution functions were determined for both a melt-extruded monofilament and a highly ordered melt-spun fibre (105).

Other polymers have also been characterized. For example, Tzou et al. (106) have applied the ROSMAS technique to ultrahigh molecular weight polyethylene fibres and analyzed the results in terms of three morphological components: a predominant highly ordered crystalline component, a minor less-ordered crystalline component, and a poorly ordered amorphous component. They have also studied films and fibres of Nylon 6 (107) and determined P_{200} and P_{400} as a function of draw ratio; both carbonyl and aliphatic carbons gave similar results. As a last example we mention a recent study (108) involving the use of deuterium (^{2}H) NMR to determine the average orientation of chain segments in an elastomer, poly(butadiene). In this case the free induction decay separates the contributions to the orientation arising from the network constraint (crosslinks) and from chain segment interactions, and it was possible to compare the experimental results with theoretical predictions. Excellent agreement was observed except at the highest deformation studied (λ = 1.83); this was attributed to non-affine deformation of the effective crosslink points at high elongations.

3.5. X-Ray Diffraction

3.5.1 General

X-ray diffraction (XRD) can yield all the moments of the orientation distribution of the crystalline phase. With the same notation for the drawing directions as shown in Figure 3.3 and taking O as the origin of the frame, the general description of the orientation of the crystallographic planes can be performed by measuring the diffracted intensity using the pole figure accessory. If axis c in Figure 3.3 is taken to represent the normal to the (hkl) crystallographic plane, its orientation can be defined in terms of the two independent angles θ and ϕ; θ is the angle between the normal of the crystallographic plane (hkl) and the direction OM, and ϕ is the angle between the projection of the (hkl) plane normal on the ONT plane and the OT direction. If $I(\theta, \phi)$ is the intensity representing the relative amount of plane normals oriented in the θ,ϕ direction, the different $<\cos^2 \theta_{hkl,J}>$ can be determined as:

$$\left\langle \cos^2 \theta_{hkl,\mathrm{M}} \right\rangle = \frac{\int_0^{2\pi}\int_0^{\pi/2} I(\theta,\phi)\cos^2\theta \sin\theta \, d\theta \, d\phi}{\int_0^{2\pi}\int_0^{\pi/2} I(\theta,\phi)\sin\theta \, d\theta \, d\phi}$$

$$\left\langle \cos^2 \theta_{hkl,\mathrm{T}} \right\rangle = \frac{\int_0^{2\pi}\int_0^{\pi/2} I(\theta,\phi)\sin^3\theta \cos^2\phi \, d\theta \, d\phi}{\int_0^{2\pi}\int_0^{\pi/2} I(\theta,\phi)\sin\theta \, d\theta \, d\phi} \tag{45}$$

$$\left\langle \cos^2 \theta_{hkl,\mathrm{N}} \right\rangle = \frac{\int_0^{2\pi}\int_0^{\pi/2} I(\theta,\phi)\sin^3\theta \sin^2\phi \, d\theta \, d\phi}{\int_0^{2\pi}\int_0^{\pi/2} I(\theta,\phi)\sin\theta \, d\theta \, d\phi}$$

where $\theta_{hkl,J}$ is the angle between the (hkl) normal and the J direction (J = M, T, or N). By using the equations given above, the second moments of the orientation function distribution can be determined.

The pole figure approach is the best technique to characterize and present the orientation of crystallographic planes. In fact, the orientation modes of a system can be described by using the orientation of the normals with respect to given crystallographic planes, ρ_{hkl}. Let us consider an isolated crystal of a polymer surrounded by a sphere. The normal vector ρ_{hkl} of a crystallographic plane can be prolonged by a line until it touches the surface of the sphere. The intersection point is called pole P_{hkl}. This can be performed for normals to the different crystallographic planes. This construction represents the spherical projection of the crystal and gives a precise representation of its spatial orientation. However, the use of spherical diagrams is difficult and instead, planar stereographic projections are used. This implies the projection of the latitudinal and longitudinal circles on a plane. The final projection of all the circles forms a stereographic network. Figure 3.15 illustrates this approach.

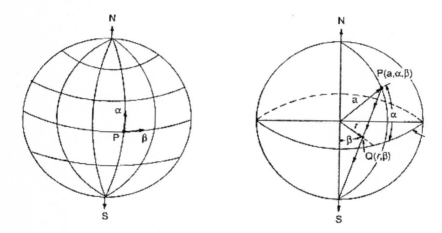

Figure3.15   (a).Position of a pole specified by the angles on a sphere. (b) Projection of a pole on an equatorial plane.

The poles of a particular family of a polymer crystallographic plane form an orientation distribution. Since this distribution is large and diffuse, usually the orientation of only one crystallographic axis is represented on a pole figure. The relative concentration of the poles is indicated on the pole figure using contour maps. Hence, if a preferential orientation is present in the sample, the curves tend to concentrate in certain regions of the projection. In the case of an isotropic distribution, a uniform distribution is observed. Generally the pole figure is defined using the sample's three reference axes, which are usually MD, TD and ND. Examples of pole figures for extreme cases are presented in Figure 3.16. It is obvious that these pole figures allow a rapid distinction between the different orientational states.

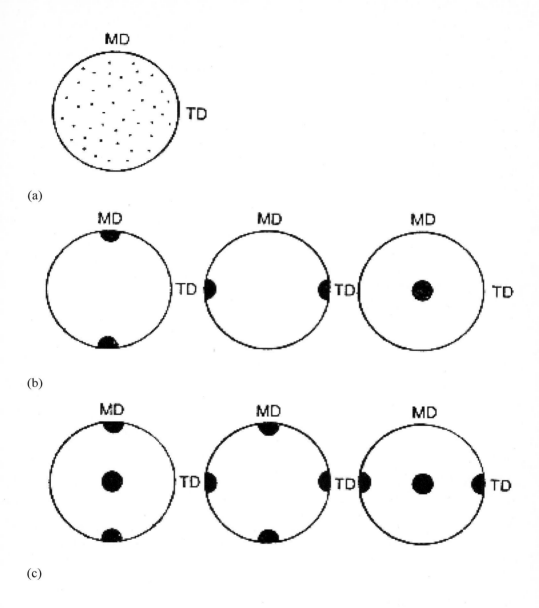

Figure 3.16 (a) The case of isotropic orientation, (b) Three cases of perfect uniaxial orientation, (c) Cases of bimodal orientation.

Pole figures are constructed in a quantitative way, thus allowing an analytical description of orientation. The experimental evaluation of the $<\cos^2\theta_{hkl,J}>$ can be done from the normal vectors ρ_{hkl} distribution of the pole figure, or more directly, from the intensity distribution $I(\theta,\phi)$ according to Equation 45 above.

In the following, the evolution of pole figures for the second moment of the crystalline orientation functions will be illustrated for a polyethylene film drawn to different draw ratios. The crystalline planes of interest are (200), (020), and (002).

Figure 3.17a shows the pole figures for the undrawn film. It is apparent that the a axis is preferentially aligned in the MD/TD plane, with a slight tendency towards the MD direction. The b axis is strongly oriented normal to the film plane. Diffraction from the (002) plane is weak, but it is still possible to infer that the c axis is randomly oriented about the perimeter of the MD/TD plane.

Thus, the initial film possesses a strong b axis orientation normal to the film plane and the a and c axes lie within this plane. At a low draw ratio of 1.04 (Figure 3.17b), even though the deformation was uniaxial, the resulting change in the pole figures is not a uniaxial pattern. The initial planar a axis orientation has separated into four maxima in the MD/TD plane, the b axis orientation seems to have become more reinforced and defined along the normal direction, and the c axis indicates a tendency towards orientation in the MD direction. These indicate that two preferred orientation states have developed; one population of crystals has the a axis aligned predominantly along MD and the second seems to have the c axis along MD, the b axis along ND and the a axis along TD.

As the draw ratio is increased to 1.12 (Figure 3.17c), the prior a axis orientation along MD is completely removed and the axis becomes randomly oriented in the ND/TD plane. For the b axis, a rotation of its orientation of about 35° confined to the second and fourth quadrants of its pole figure has taken place about the ND axis. The c axis begins to broaden about the MD direction. At a draw ratio of 1.25 (Figure 3.17d), little change is observed for the a axis which remains strongly oriented in the ND/TD plane. The b axis now shows two series of bands of high orientation, symmetric about each MD pole. The c axis continues to broaden around the MD. It is evident for the two last draw ratios that only the b and c axes are experiencing re-orientation.

For a draw ratio of 1.40 (Figure 3.17e), the a axis orientation is unchanged, while the b and c axes continue to re-orient. Orientation of the b axis along the transverse direction is becoming more pronounced and the band of high orientation of the c axis is beginning to converge upon the MD pole. For the much higher draw ratio of 4.23 (Figure 3.17f), the pole figures are representative of a uniaxial system. The a and b axes are randomly oriented in the ND/TD plane and the c axis is aligned in the MD direction. More details can be found in Refs. 109 and 110. It is obvious that the information gained from the pole figures is very helpful in understanding the deformation mechanisms and evolution, in addition to the determination of the moments of the orientation function.

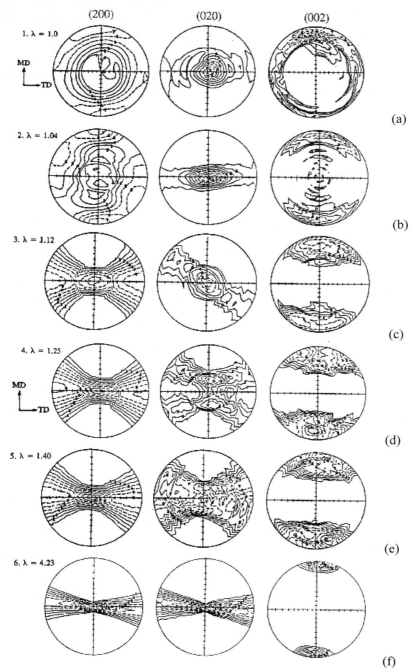

Figure 3.17 Pole figures for the second moment of the crystalline orientation functions for a polyethylene film drawn to different draw ratios

Higher moments of the crystalline orientation function, $P_{2n}(\cos\theta)_c$, involve different combinations of even powers of $<\cos\theta>$, i.e. $<\cos^{2n}\theta>$ as indicated in the first part of this chapter. These different averages can be determined from the intensity $I(\theta,\phi)$ as follows:

$$\left\langle \cos^{2n}\theta_{hkl,\mathrm{M}} \right\rangle = \frac{\displaystyle\int_0^{2\pi}\int_0^{\pi/2} I(\theta,\phi)\cos^{2n}\theta\sin\theta\,d\theta\,d\phi}{\displaystyle\int_0^{2\pi}\int_0^{\pi/2} I(\theta,\phi)\sin\theta\,d\theta\,d\phi}$$

$$\left\langle \cos^{2n}\theta_{hkl,\mathrm{T}} \right\rangle = \frac{\displaystyle\int_0^{2\pi}\int_0^{\pi/2} I(\theta,\phi)\sin^{2n}\theta\cos^{2n}\phi\sin\theta\,d\theta\,d\phi}{\displaystyle\int_0^{2\pi}\int_0^{\pi/2} I(\theta,\phi)\sin\theta\,d\theta\,d\phi} \tag{46}$$

$$\left\langle \cos^{2n}\theta_{hkl,\mathrm{N}} \right\rangle = \frac{\displaystyle\int_0^{2\pi}\int_0^{\pi/2} I(\theta,\phi)\sin^{2n}\theta\sin^{2n}\phi\sin\theta\,d\theta\,d\phi}{\displaystyle\int_0^{2\pi}\int_0^{\pi/2} I(\theta,\phi)\sin\theta\,d\theta\,d\phi}$$

where 2n is the order of the orientation moment.

3.5.2   Amorphous orientation from X-Ray diffraction

Recently, more general use of X-ray diffraction for the determination of the amorphous orientation of polymers has started to appear in a systematic manner. For historical reasons, the noncrystalline phase is usually referred to as amorphous, although it is not truly amorphous. Unlike in a gas, the position and/or the orientation of segments of chains in the amorphous phase can be correlated over short distances (~1 nm). By convention, scattering from crystallites smaller than about three unit cells (~2.5 nm) is considered amorphous (111). In this section we discuss the methods for describing amorphous orientation, and the measurement of this orientation by means of XRD.

An example of the two-dimensional (2-D) XRD pattern from a semicrystalline polymer (~35% crystalline, polyethylene terephthalate PET) is shown in Figure 3.18a. The information about the amorphous phase is contained in the weak amorphous halo underneath the intense crystalline peaks. This amorphous halo can be easily seen in the diffraction pattern of the oriented amorphous PET (Figure 3.18b) where there are no crystalline peaks. The scattering from an unoriented amorphous phase is isotropic around the azimuth, whereas that from an oriented amorphous phase is intense along the equator, and weaker along the meridian, giving rise to an amorphous crescent. The amorphous orientation can be calculated from the azimuthal width of this crescent, i.e. from the intensity distribution of this amorphous halo along the azimuth, using methods described in the literature (112-121), basically using equations similar to those given above, without a reference to a crystalline axis. In polymers such as polyethylene and polyamides, one amorphous peak is adequate to describe the amorphous scattering in the angular range of $10°–35°$ 2θ (with Cu Kα) (112). In PET, however, there are at least two peaks (at $17°$ and $22°$ 2θ) within this amorphous halo; of these two, the $22°$ peak gives more reproducible results (113). Figure 3.19 shows the variation in the intensity within this $22°$ peak as a function of the azimuthal angle ϕ. The peak at $\phi = 0°$ in Figure 3.19 corresponds to the equatorial peak in the amorphous scattering in Figures 3.18a and 3.18b. The intensity of the peak decreases away from the equator, and eventually reaches a constant value, i.e. the baseline (at $\phi > 60°$ in Figure 3.19). As the orientation is increased, the intensity or the height of the baseline decreases, and the width of the peak above the baseline decreases. At high degrees of orientation, as commonly encountered with crystalline peaks, the baseline intensity reduces to zero, and all the scattered intensity will be concentrated within the peak at $\phi = 0°$.

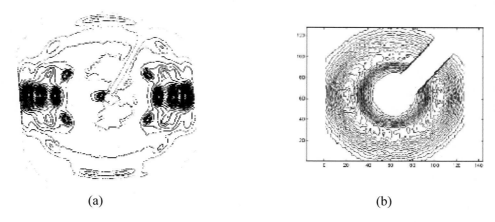

(a)   (b)

Figure 3.18   (a)   Two dimensional X-ray diffraction pattern of oriented semi-crystalline PET;   the draw direction is vertical (meridional direction).   (b)   Two dimensional X-ray diffraction pattern of oriented amorphous PET.

Two parameters are required to characterize an intensity distribution like the one shown in Figure 3.19: the area and the width of the peak (at $\phi = 0°$) above the base line. The peak-area normalized to the total amorphous scattering is a measure of the fraction of the oriented or the anisotropic component, or the *percent* amorphous orientation. The peak-width is a measure of the *degree* of amorphous orientation (of the anisotropic component) (116). The peak in Figure 3.19 is sometimes attributed to a distinct third phase, an intermediate phase, whose structure is between that of an ideal, completely disordered amorphous phase and the three-dimensionally ordered crystalline phase (114). However, it is difficult to justify introducing a new phase when the properties (e.g. the glass transition) of this phase cannot be distinguished from the rest of the amorphous phase, and its spatial distribution cannot be clearly described. It is more likely that the entire intensity arises from a single, continuous, amorphous phase in which islands of oriented arrangement of densely packed straight segments of polymer chains exist in a matrix of unoriented, loosely packed, disordered chain segments (112, 115). The two parameters, the percent (or the fraction) and the degree of orientation describe two characteristics of the amorphous phase. The intensity or the height of the baseline is a measure of the isotropic component, and the area of the peak above the base line characterizes the anisotropic component.

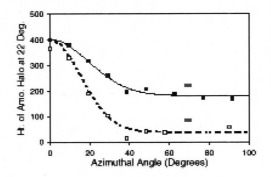

Figure 3.19   Intensity along the azimuthal angle of the second amorphous halo in two PET samples: undrawn fiber with low degree of orientation and drawn one with a higher degree of orientation.

3.5.3   Synchrotron X-Ray diffraction

Finally, in recent years, more and more research and applications involving high intensity X-ray sources have appeared in the literature (122-126). In fact, increased intensity over lab sources makes possible real-time studies under conditions similar to those encountered during processing. Applications resulting from the highly collimated radiation include mapping spatial variations of structure in films or injection molded plaques.

Synchrotron radiation could be ~10^8 times more intense than conventional lab sources, depending on the application. Radiation from insertion devices (ID) is intense, highly collimated (10-30 µrad), monochromatic, and tunable over a broad wavelength range (3 keV < E < 50 keV). This higher intensity makes possible real-time studies of polymers under non-equilibrium conditions, which may mimic those to which materials are subjected during processing. The high collimation makes scattering experiments from very small volumes a reality, and this may be useful in the mapping of spatial variations of crystallinity or orientation over a specimen on the micrometre or sub-micrometre length scale. For example, such studies may be useful for characterizing orientation as a function of thickness in injection moulded parts or blown films, cases for which large stress and temperature gradients result in marked structural variations at the surface. The option of changing the wavelength provides distinct advantages in that the sample absorption can be varied over a significant range, making possible, for example, experiments in thick containing vessels or on highly absorbing samples. One advantage of ID-based radiation over lab sources for oriented materials is the capability of performing ultra-small-angle X-ray scattering experiments on a pinhole machine, as data collection can then be performed with a 2-D detector. The capability of obtaining 2-D desmeared data is granted by the combination of high intensity and high collimation.

Londono et al. (122) illustrated the use of this synchrotron SAXS technique in conjunction with a mechanical deformation device. A Kevlar® fibre was mounted on an Instron mechanical test frame on the beam line. The CCD detector was placed 1.2 m away from the sample. The beam was 0.3 mm in height, and the wavelength was 1.07 Å. A typical fan of radiation was observed about the equatorial streak. This streak arises from spatial correlations of fibrils or needle-like voids aligned along the fiber axis, and the fan is due to misorientation of these elongated objects. Cuts through the fan of scattered intensity were performed by azimuthal scans at different values of s ($s = 2\sin\theta/\lambda$, where 2θ is the scattering angle), and these scans (intensity vs. azimuthal angle) consisted of peaks which were fitted with Pearson VII profiles. The method of analysis was suggested by Ruland (127). Ideally, the results should be straight lines, where the misorientation is given by the square root of the slope, and the height of fibrils by the square root of the reciprocal of the intercept (122, 127). This expectation was fulfilled for the sample at maximum strain. The other two sets of data suggest that there may be a correlation between size and orientation for unstrained samples. Data were actually collected continuously with a time resolution of about 4 seconds, and about 0.5 s per exposure. The analysis shown corresponded to three individual frames, of a total of 130 frames collected in the full experiment. This experiment was possible because of the intensity of the ID-beam line.

3.6   Ultrasonic and Other Techniques

Ultrasonic techniques are also good candidates for the determination of the anisotropy of a polymer sample. It is however difficult to relate ultrasonic measurements to the other techniques since they do not give the orientation function as such but rather the elastic constants. These can be helpful in direct determination of mechanical anisotropy and on-line process monitoring. In this section, we will give only a brief outline from the recent literature on applications of ultrasonic techniques in measurements of anisotropy. More details can be found in Refs. 128-131.

Enderby et al. (128) and Hine et al. (129) developed an automated ultrasonic immersion system based on ultrasonic transducers operating at a frequency of 2.25 MHz, investigating both tensile and shear velocities in fibre-reinforced composite samples. The automation of the system allowed fast and efficient large area scanning to be performed. Both large-scale readings and spatial variations in the tensile velocities were investigated and also, all of the three-dimensional elastic moduli of the fibre-reinforced polymer composites could be deduced by monitoring the variation of ultrasonic velocity with angle of incidence on orthogonal planes. The results indicated that the system yielded most of the three-dimensional elastic moduli to absolute accuracies of better than ±1%, and in the fast scanning mode detected spatial variations of the time-of-flight of ultrasound with a fractional timing error of ±0.001%.

Gorbatsevich (130), on the other hand, described the physical principles and instruments for the acoustic polarization method for determining parameters of elastic anisotropy in solid media samples. He analyzed the propagation of shear (transverse) waves through anisotropic media. Transducers that can transmit purely shear linear-polarized waves, an acoustopolariscope, an ultrasonic device, and a special coupling medium are needed.

Measurements were made by rotating the sample first with the polarization vectors of the source and receiver aligned parallel, and rotating again with the polarization vectors orthogonal. In the first stage of measurements one can determine the spatial orientation of the anisotropic symmetry axes and examine the presence of linear anisotropic absorption and some other effects. The second stage deals with measurements of velocities of longitudinal (compression) and shear waves in the direction of the symmetry axes and in other directions. In the last stage the order of calculation of the whole set of constants of elasticity and the determination of the type of cubic, hexagonal, tetragonal, and orthorhombic elastic symmetry in the samples are given. This method may be useful for determining properties and testing samples of concrete, plastics, woods, ceramics, composites, and other materials.

Levesque et al. (131) used ultrasonics for the determination of the mechanical properties of oriented semicrystalline polymers through time-of-flight measurements of elastic waves propagating in various directions within the material. While being nondestructive, such a method allowed to obtain more mechanical moduli with a better accuracy than the conventional tensile tests, especially regarding the shear properties and the Poisson's coefficients. Until now, the approach used to interpret the data was approximate and not rigorous. Their work presents a self-consistent rigorous approach for interpreting time-of-flight data based on the group velocity, including allowance for lateral displacement of the transmitted beam. Results for roll-drawn PET with various draw ratios are illustrated in Figures 3.20 and 3.21. The samples were considered to have transversely isotropic symmetry. Comparisons made with conventional tensile tests indicated differences which were interpreted in terms of viscoelastic effects considering both the amorphous and crystalline phases.

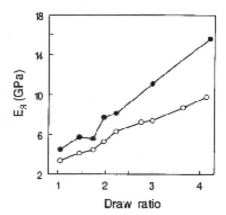

Figure 3.20.   Longitudinal modulus from ultrasonics (filled symbols) and tensile testing (open symbols) for PET as a function of draw ratio (131).

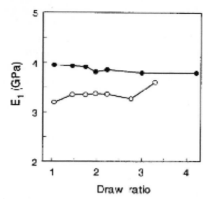

Figure 3.21   Transverse modulus from ultrasonics (filled symbols) and tensile testing
                   (open symbols) for PET as a function of draw ratio (131).

Choy et al. (132) used the ultrasonic method to measure the elastic moduli of a liquid
crystalline polyesteramide (LCP) and polycarbonate/LCP in-situ composites as functions of
draw ratio λ from 1 to 15. The elastic moduli of the composites showed a dependency on
draw ratio similar to those of the LCP. The strong increase in the longitudinal modulus with
increasing draw ratio arises from the higher aspect ratio of the LCP domains in the
composites and the improved molecular orientation within the domains. The reinforcement
effect on the other moduli was much weaker, with the transverse modulus and C44 of the
composites only 5 to 30% higher than those of polycarbonate.

Finally, other indirect techniques can give some information on orientation, but as for the
ultrasonic method, no determination of the orientation function is possible. Such techniques
include in particular thermal conductivity (133-135) and far infrared radiation (136, 137).

References

1.   J. L. White and M. Cakmak, *Advances in Polymer Technology*, **8**, 27-61 (1988).
2.   J. L. White and M. Cakmak, "Orientation", in *Encyclopedia of Polymer Science and
     Engineering, Second Edition*, Vol. 10, pp. 595-618, John Wiley & Sons, New York
     (1987).
3.   R. J. Samuels, *Structured Polymer Properties*, John Wiley, Toronto (1974).
4.   R. Zbinden, *Infrared Spectroscopy of High Polymers*, Academic Press, New York
     (1964).
5.   I. M. Ward, *Adv. Polym. Sci.*, **66**, 81-115 (1985).
6.   D. A. Jarvis, I. J. Hutchison, D. I. Bower, and I. M. Ward, *Polymer*, **21**, 41-54 (1980).
7.   I. Karacan, A. K. Taraiya, D. I. Bower, and I. M. Ward, *Polymer*, **34**, 2691-2701
     (1993).
8.   I. Karacan, D. I. Bower, and I. M. Ward, *Polymer*, **35**, 3411-3422 (1994).
9.   S. Nomura, "Oriented Polymers", in *Comprehensive Polymer Science*, Vol. 2, pp. 459-
     485, Pergamon Press, Oxford (1989) [G. Allen and J. C. Bevington (Series Eds.), C.
     Booth and C. Price (Vol. Eds.)]

10. J. L. White and J. E. Spruiell, *Polym. Eng. Sci.*, **21**, 859-868 (1981).
11. A. Cunningham, G. R. Davies, and I. M. Ward, *Polymer*, **15**, 743-748 (1974).
12. A. Cunningham, I. M. Ward, H. A. Willis, and V. Zichy, *Polymer*, **15**, 749-756 (1974).
13. K.-J. Choi, J. E. Spruiell, and J. L. White, *J. Polym. Sci.: Polym. Phys. Ed.*, **20**, 20-47 (1982).
14. R. J. Samuels, *J. Appl. Polym. Sci.*, **26**, 1383-1412 (1981).
15. R. D. L. Marsh, J. C. Duncan, and S. Brister, *J. Thermal Analysis*, **45**, 891 (1995).
16. H. U. Hoppler, A. Dinkel, and I. Tomka, *Polymer*, **36**, 3809 (1995).
17. V. Abetz and G. G. Fuller, *Rheol. Acta*, **29**, 11-15 (1990).
18. K. Hongladarom and W. R. Burghardt, *Macromolecules*, **26**, 785 (1993).
19. F. Beekmans and A. P. de Boer, *Macromolecules*, **29**, 8726 (1996).
20. T. Takahashi and G. G. Fuller, *Rheol. Acta*, **35**, 97 (1996).
21. K. Hongladarom and W. R. Burghardt, *Macromolecules*, **27**, 483 (1994).
22. A. Ajji and J. Guèvremont, U.S. Patent No. 5,864,403 (1999).
23. A. Ajji, J. Guèvremont, R. G. Matthews, and M. M. Dumoulin, *Proc. SPE ANTEC '98*, 1588-1592 (1998).
24. B. Jasse and J. L. Koenig, *J. Macromol. Sci.–Rev. Macromol. Chem.*, **C17**, 61-135 (1979).
25. H. W. Siesler and K. Holland-Moritz, *Infrared and Raman Spectroscopy of Polymers*, Practical Spectroscopy Series Vol. 4, Marcel Dekker Inc., New York, 1980.
26. J. E. Bertie, S. L. Zhang, and C. D. Keefe, *J. Mol. Struct.*, **324**, 157 (1994).
27. P. G. Schmidt, *J. Polym. Sci. A*, **1**, 1271 (1963).
28. J. L. Koenig, S. W. Cornell, and D. E. Witenhafer, *J. Polym. Sci.: Part A-2*, **5**, 301-313 (1967).
29. J. P. Sibilia, *J. Polym. Sci.: Part A-2*, **9**, 27-42 (1971).
30. A. Garton, D. J. Carlsson, and D. M. Wiles, *Appl. Spectrosc.*, **35**, 432-435 (1981).
31. L. J. Fina and J. L. Koenig, *J. Polym. Sci.: Part B: Polym. Phys.*, **24**, 2509-2524 (1986).
32. A. R. Wedgewood and J. C. Seferis, *Pure & Appl. Chem.*, **55**, 873-892 (1983).
33. B. E. Read and R. S. Stein, *Macromolecules*, **1**, 116-126 (1968).
34. C. Marcott, A. E. Dowrey, and I. Noda, *Anal. Chem.*, **66**, 1065A-1075A (1994).
35. H. W. Siesler, Chapter 4 (pp. 138-166) in *Oriented Polymer Materials*, S. Fakirov (Ed.), Hüthig & Wepf Verlag, Zug (1996).
36. R. J. Samuels, *Polymer Preprints*, **29**(1), 581-583 (1988).
37. M. Houska and M. Brummell, *Polym. Eng. Sci.*, **27**, 917-924 (1987).
38. L. Lundberg, Y. Sjönell, B. Stenberg, B. Terselius, and J.-F. Jansson, *Polymer Testing*, **13**, 441-459 (1994).
39. N. J. Harrick, *Internal Reflection Spectroscopy*, Harrick Scientific Co., Ossining NY, 1967.
40. F. M. Mirabella, Jr., *Internal Reflection Spectroscopy: Review and Supplement*, Harrick Scientific Co., Ossining NY, 1985.
41. M. W. Urban, *Attenuated Total Reflectance Spectroscopy of Polymers: Theory and Practice*, Polymer Surfaces and Interfaces Series, American Chemical Society, Washington DC, 1996.
42. C. S. P. Sung and J. P. Hobbs, *Chem. Eng. Commun.*, **30**, 229-250 (1984).
43. F. M. Mirabella, Jr., *Appl. Spectrosc.*, **42**, 1258-1265 (1988).

44. V. Premnath, *Macromolecules*, **28**, 5139-5143 and 7584 (1995).

45. M. K. Gupta, D. J. Carlsson, and D. M. Wiles, *J. Polym. Sci.: Polym. Phys. Ed.*, **22**, 1011-1027 (1984).

46. N. J. Everall and A. Bibby, *Appl. Spectrosc.*, **51**, 1083-1091 (1997).

47. D. J. Walls, *Appl. Spectrosc.*, **45**, 1193-1198 (1991).

48. D. J. Walls and J. C. Coburn, *J. Polym. Sci.: Part B: Polym. Phys.*, **30**, 887-897 (1992).

49. P. Yuan and C. S. P. Sung, *Macromolecules*, **24**, 6095-6103 (1991).

50. K. H. Lee and C. S. P. Sung, *Macromolecules*, **26**, 3289-3294 (1993).

51. K. Palm, *Vibr. Spectrosc.*, **6**, 185-191 (1994).

52. F. M. Mirabella, Jr., *J. Polym. Sci.: Polym. Phys. Ed.*, **22**, 1283-1291 and 1293-1304 (1984).

53. G.-C. Chen and L. J. Fina, *J. Appl. Polym. Sci.*, **48**, 1229-1240 (1993).

54. A. K. Jain and V. B. Gupta, *J. Appl. Polym. Sci.*, **41**, 2931-2939 (1990).

55. A. Kaito and K. Nakayama, *Macromolecules*, **25**, 4882-4887 (1992).

56. K. C. Cole, J. Guèvremont, A. Ajji, and M. M. Dumoulin, *Appl. Spectrosc.*, **48**, 1513-1521 (1994).

57. M. Claybourn, P. Colombel, and J. Chalmers, *Appl. Spectrosc.*, **45**, 279 (1991).

58. H. Ben Daly, K. C. Cole, B. Sanschagrin, and K. T. Nguyen, *Polym. Eng. Sci.*, **39**, 1982-1992 (1999).

59. U. Hoffmann, F. Pfeifer, J. C. Merino, S. Okretic, J. M. Pastor, J. C. Rodriguez-Cabello, N. Völkl, M. Zahedi, and H. W. Siesler, *Appl. Spectrosc.*, **47**, 1531-1539 (1993) and **48**, 417 (1994).

60. J. Guèvremont, A. Ajji, K. C. Cole, and M. M. Dumoulin, *Polymer*, **36**, 3385-3392 (1995).

61. A. Ajji, J. Guèvremont, K. C. Cole, and M. M. Dumoulin, *Polymer*, **37**, 3707-3714 (1996).

62. K. C. Cole, A. Ajji, and E. Pellerin, *Proc. Intl. Symp. On Orientation of Polymers: Application to Films and Fibers (SPE RETEC)*, Boucherville, QC, Canada, Sept. 23-35, 1998, pp. 531-538.

63. K. C. Cole, H. Ben Daly, B. Sanschagrin, K. T. Nguyen, and A. Ajji, *Polymer*, **40**, 3505-3513 (1999).

64. N. J. Everall, J. M. Chalmers, A. Local, and S. Allen, *Vibr. Spectrosc.*, **10**, 253-259 (1996).

65. S. Bensaad, B. Jasse, and C. Noël, *Polymer*, **34**, 1602-1605 (1993).

66. J. A. J. Jansen, F. N. Paridaans, and I. E. J. Heynderickx, *Polymer*, **35**, 2970-2976 (1994).

67. H. Ben Daly, B. Sanschagrin, K. T. Nguyen, and K. C. Cole, *Polym. Eng. Sci.*, **39**, 1736-1751 (1999).

68. B. Jasse, *J. Macromol. Sci.—Chem.*, **A26**, 43-67 (1989).

69. K. Krishnan, S. Hill, J. P. Hobbs, and C. S. P. Sung, *Appl. Spectrosc.*, **36**, 257-259 (1982).

70. B. Jasse and C. Noël, *Proc. SPE ANTEC '91*, 955-957 (1991).

71. J. C. Rodriguez-Cabello, J. Santos, J. C. Merino, and J. M. Pastor, *J. Polym. Sci.: Part B: Polym. Phys.*, **34**, 1243-1255 (1996).

72. D. I. Bower, *J. Polym. Sci.: Polym. Phys. Ed.*, **10**, 2135-2153 (1972).

73. J. Purvis, D. I. Bower, and I. M. Ward, *Polymer*, **14**, 398-400 (1973).

74. J. Purvis and D. I. Bower, *J. Polym. Sci.: Polym. Phys. Ed.*, **14**, 1461-1484 (1976).
75. D. A. Jarvis, I. J. Hutchinson, D. I. Bower, and I. M. Ward, *Polymer*, **21**, 41-54 (1980).
76. P. Lapersonne, D. I. Bower, and I. M. Ward, *Polymer*, **33**, 1266-1276 (1992).
77. G. Voyiatzis, G. Petekidis, D. Vlassopoulos, E. I. Kamitsos, and A. Bruggeman, *Macromolecules*, **29**, 2244-2252 (1996).
78. J. Purvis and D. I. Bower, *Polymer*, **15**, 645 (1974).
79. D. I. Bower, J. King, and W. F. Maddams, *J. Macromol. Sci.—Phys.*, **B20**, 305-318 (1981).
80. M. Pigeon, R. E. Prud'homme, and M. Pézolet, *Macromolecules*, **24**, 5687-5694 (1991).
81. C.-P. Lafrance, P. Chabot, M. Pigeon, R. E. Prud'homme, and M. Pézolet, *Polymer*, **34**, 5029-5037 (1993).
82. P. A. Bentley and P. J. Hendra, *Spectrochim. Acta Part A*, **51**, 2125-2131 (1995).
83. M. Arruebarrena de Báez, P. J. Hendra, and M. Judkins, *Spectrochim. Acta Part A*, **51**, 2117-2124 (1995).
84. J. C. Rodríguez-Cabello, J. C. Merino, M. R. Fernández, and J. M. Pastor, *J. Raman Spectrosc.*, **27**, 23-29 (1996).
85. L. J. Fina, D. I. Bower, and I. M. Ward, *Polymer*, **29**, 2146-2151 (1988).
86. N. Everall, J. Chalmers, and P. Mills, *Appl. Spectrosc.*, **50**, 1229-1234 (1996).
87. N. J. Everall, *Appl. Spectrosc.*, **52**, 1498-1504 (1998).
88. J. R. Lakowicz, *Principles of Fluorescence Spectroscopy*, Plenum Press, New York, 1983.
89. A. J. Bur, R. E. Lowry, C. L. Thomas, S. C. Roth, and F. W. Wang, *Macromolecules*, **25**, 3503 (1992).
90. P. Lapersonne, J. F. Tassin, P. Sergot, L. Monnerie, and G. LeBourvellec, *Polymer*, **30**, 1558 (1989).
91. L. L. Chappoy, D. Spaseska, K. Rasmussen, and D. B.DuPre, *Macromolecules*, **20**, 680 (1979).
92. J. H. Nobbs, D. I. Bower and I. M. Ward, *Polymer*, **15**, 287 (1974).
93. B. Erman, J. P. Jarry, and L. Monnerie, *Polymer*, **28**, 727 (1987).
94. A. J. Bur and S. C. Roth, *Proc. SPE ANTEC '98*, 2090-2094 (1998).
95. A. J. Bur and S. C. Roth, *Proc. SPE ANTEC '99*, 2342-2346 (1999).
96. J. L. Koenig, *Spectroscopy of Polymers*, American Chemical Society (1992).
97. V. J. McBrierty and K. J. Packer, *Nuclear Magnetic Resonance in Solid Polymers*, Cambridge University Press (1993).
98. K. Schmidt-Rohr and H. W. Spiess, *Multidimensional Solid-State NMR and Polymers*, Academic Press (1994).
99. H. W. Spiess, Chapter 5 (pp. 234-268) in *Structure and Properties of Oriented Polymers*, 2nd ed., I. M. Ward (ed.), Chapman & Hall (1997).
100. I. M. Ward, *J. Polym. Sci.: Polym. Symp.*, **58**, 1-21 (1977).
101. P. M. Henrichs, *Macromolecules*, **20**, 2099-2112 (1987).
102. B. F.Chmelka, K. Schmidt-Rohr, and H. W. Spiess, *Macromolecules*, **26**, 2282-2296 (1993).
103. D. L. Tzou, P. Desai, A. S. Abhiraman, and T.-H. Huang, *J. Polym. Sci.: Part B: Polym. Phys.*, **33**, 63-69 (1995).
104. N. J. Clayden, J. G. Eaves, and L. Croot, *Polymer*, **38**, 159-163 (1997).

105. M.-Y. Liao and G. C. Rutledge, *Macromolecules*, **30**, 7546-7553 (1997).
106. D. L. Tzou, T.-H. Huang, P. Desai, and A. S. Abhiraman, *J. Polym. Sci.: Part B: Polym. Phys.*, **31**, 1005-1012 (1993).
107. D. L. Tzou, H. W. Spiess, and S. Curran, *J. Polym. Sci.: Part B: Polym. Phys.*, **32**, 1521-1529 (1994).
108. M. E. Ries, M. G. Brereton, P. G. Klein, I. M. Ward, P. Ekanayake, H. Menge, and H. Schneider, *Macromolecules*, **32**, 4961-4968 (1999).
109. R. Pazur, A. Ajji, and R. E. Prud'homme, *Polymer*, **34**, 4004-4014 (1993).
110. R. J. Pazur and R. E. Prud'homme, *Macromolecules*, **29**, 119-128 (1996).
111. W. Ruland, *Acta Cryst.*, **14**, 1180-1185 (1961).
112. N. S. Murthy, C. Bednarczyk, H. Minor, and S. Krimm, *Macromolecules*, **26**, 1712-1721 (1993).
113. N. S. Murthy, C. Bednarczyk, P. B. Rim and C. J. Nelson, *J. Appl. Polym. Sci.*, **64**, 1363-1371 (1997).
114. Y. Fu, W. R. Busing, Y. Jin, K. A. Affholter, and B. Wunderlich, *Makromol. Chem.*, **195**, 803-822 (1994).
115. W. Ruland, *Pure and Appl. Chem.*, **18**, 489-515 (1969).
116. N. S. Murthy, S. T. Correale, and R. A. F. Moore, *J. Appl. Polym. Sci., Appl. Polym. Symp.*, **47**, 185-197 (1991).
117. N. S. Murthy and K. Zero, *Polymer*, **38**, 2277-2280 (1997).
118. P. J. Harget and H. J. Oswald, *J. Polym. Sci., Polym. Phys.*, **17**, 531-534 (1979).
119. G. Wu, J.-D. Jiang, P. A. Tucker and J. A. Cuculo, *J. Polym. Sci. Polym. Phys.*, **34**, 2035-2047 (1996).
120. A. Ajji, J. Brisson, K. C. Cole, and M. M. Dumoulin, *Polymer*, **36**, 4023-4030 (1995).
121. N. S. Murthy and K. Zero, *Proc. SPE ANTEC '99*, 1672-1676 (1999).
122. J. D. Londono, R. V. Davidson, R. A. Leach, and R. Barton, Jr., *Proc. SPE ANTEC '99*, 2357-2360 (1999).
123. I. W. Hamley, J. P. A. Fairclough, N. J. Terrill, A. J. Ryan, P. M. Lipic, F. S. Bates, and E. Towns-Andrews, *Macromolecules*, **29**, 8835-8843 (1996).
124. A. J. Ryan, J. P. A. Fairclough, I. W. Hamley, S.-M. Mai, and C. Booth, *Macromolecules*, **30**, 1723-1727 (1997).
125. J. A. Pople, I. W. Hamley, J. P. A. Fairclough, A. J. Ryan, and C. Booth, *Macromolecules*, **31**, 2952-2956 (1998).
126. I. W. Hamley, J. A. Pople, J. P. A. Fairclough, A. J. Ryan, C. Booth, and Y.-W. Yang, *Macromolecules*, **31**, 3906-3911 (1998).
127. R. Perret and W. Ruland, *J. Appl. Cryst.*, **2**, 209 (1969).
128. M. D. Enderby, A. R. Clarke, M. Patel, P. Ogden, and A. A. Johnson, *Ultrasonics*, **36**, 245-249 (1998).
129. P. J. Hine, N. Davidson, R. A. Duckett, and I. M. Ward, *Compos. Sci. Technol.*, **53**, 125-131 (1995).
130. F. F. Gorbatsevich, *Ultrasonics*, **37**, 309-319 (1999).
131. D. Levesque, N. Legros, and A. Ajji, *Polym. Eng. Sci.*, **37**, 1833 (1997).
132. C. L. Choy, K. W. E. Lau, Y. W. Wong, and A. F. Yee, *Polym. Eng. Sci.*, **36**, 1256-1265 (1996).
133. D. B. Mergenthaler, M. Pietralla, S. Roy, and H. G. Kilian, *Macromolecules*, **25**, 3500 (1992).

134. D. Greig, in *Developments in Oriented Polymers—1*, I. M. Ward (Ed.), Applied Science Publishers, Essex, UK, 1982.
135. A. Ajji, N. Legros, and M. M. Dumoulin, *Proc. Composites '96 and Oriented Polymers Symposium*, Boucherville, Canada, Oct. 9-11, 1996, pp. 742-752.
136. B. Drouin, R. Gagnon, C. Bacon, and J. Pouyet, *Proc. Composites '96 and Oriented Polymers Symposium*, Boucherville, Canada, Oct. 9-11, 1996, pp. 99-106.
137. B. Drouin and R. Gagnon, *Polym. Eng. Sci.*, in press.

4. Solid State Processing of Fibers

W Bessey* and M Jaffe**
*SpaceNet Products Inc, USA
**Rutgers University and the New Jersey Institute of Technology, USA

4.1 Introduction

Fibers are intrinsically one-dimensional articles; i.e. for all practical purposes, they have infinite aspect ratios. The uniqueness of fibers lies in this anisotropy: anisotropy of geometry, anisotropy of properties and anisotropy of microstructure. Fiber processing is the art and science of producing and controlling the anisotropic geometry and molecular microstructure. It is the control of the microstructure that allows the polymer processor to design fiber properties and performance.

4.1.1 Background

The scale of fiber production is unique. Each year more than 30 million metric tons of polymer are converted to fiber with a market value of more than $60 billion. The total length of a year's fiber production is on the order of 20 light years or 30,000 kms for each person on earth. On the other hand the diameter of a typical textile staple fibers is in the 10 –20 micron range, smaller than human hair and comparable to the wavelength of infrared radiation.

The extremes of very large lengths and very small diameters have major processing implications. A single manufacturing unit can produce only a miniscule portion of the total product required. Consequently, fiber production involves a very large number of machines and fiber production is capital intensive. In addition, the extreme lengths of fiber produced requires the fiber producer to put great emphasis on minimizing the number of threadline breaks to preserve attractive production economics.

4.1.2 Common polymers used in man made fibers

Table 4.1 lists the physical properties, major applications and production volumes of the leading man made fibers. In addition, there is a multitude of small volume specialty and "high performance" fibers not included in the table. More complete listings can be found on the Internet [1].

Table 4.1: Common Man Made Fibers

Polymer	Glass transition temp. °C	Melting temp °C	Tensile Strength GPa	Tensile Modulus GPa	Density (g/cc)	Typical Uses	Volume 10^6 tonnes /yr
Polyester (PET)	70	265	1.1	15	1.4	Clothing Tire Cord	16
Polypropylene (PP)	<0	160	0.4	3	.98	Carpets Industrial	>5
Nylons (66 & 6)	50	260	1.0	6.0	1.14	Carpets Clothing	4
Acrylics (PAN)	104		0.5	18	1.18	Carpets Clothing	2.5
LCP Polyester	100	300	3.0	65	1.40	Industrial	<1
Polyethylene (PE -gel spun)	<0	140	3.1	105	.97	Ballistics	<1
Aramid	.>300	.>300	3.1	123	1.45	Ballistics Fire Protection	<1
Glass (inorganic)	>1000	>1000	3.4	73	2.6	Composite	
Steel (inorganic)	>1000	>1000	2.8	200	7.8	Tire Belts	

4.2   Overview of Fiber Processing

Fiber manufacturing consists of four generic processes.

- All polymeric fibers start as liquids, either melts or solutions. Fluid mechanics is beyond the scope of this book. However, the nature of the liquid, melt or solution, anisotropic or liquid crystalline, has a profound impact on all subsequent fiber processing and cannot be ignored.
- *Spinning* is the process of creating solid fibers from the liquid. If the liquid is a melt, the process is termed *melt spinning*. If the liquid is a solution and the solvent has a high vapor pressure, i.e. it flashes off after extrusion, the process is termed *dry spinning*. If the liquid is a solution and the solvent has a low vapor pressure, i.e. it must be removed in a "non-solvent" bath, the process is termed *wet spinning*. The basis for all structure and property development is created during spinning.
- Drawing is the process of stretching spun fibers to increase their molecular orientation and therefore their tenacity and modulus. Drawing reduces the fiber's entropy.
- Heat treating or annealing is the process of relaxing residual local stresses and locking-in or stabilizing the fiber's properties and microstructure. Heat-treating lowers the fibers internal energy.

The result of this series of processes is to establish the fiber geometry and microstructure, i.e. the molecular orientation, crystallinity, crystalline morphology, and inter chain connectivity. The fiber microstructure, in turn, defines the fiber's mechanical and thermal properties or "performance". The changes in fiber microstructure during each process are generally irreversible below the melting temperature and therefore cumulative. The commercial combinations of polymer liquids and spinning process and their impact on drawing and heat-treating are summarized in Table 4.2

Table 4.2 Overview of Fiber Processing

Process	Traditional Melt spinning	Liquid Crystal Melt spinning	Dry spinning	Wet spinning	Dry Jet/ Wet Spinning
Typical Materials	Flexible chain polymers such as Polyester and nylon	Aromatic polyester copolymers	Flexible chain polymers such as cellulose acetate	Flexible chain polymers such as polyacrylanitrile	Aromatic polyamindes or aramids
Starting Liquid	Isotropic polymer melt	Liquid crystal polymer melt	Isotropic polymer solution	Isotropic polymer solution	Liquid crystal polymer solution
Spinning *Extrusion:*	through very small holes	through very small holes	through very small holes	through very small holes	through very small holes
Stretch: Solid-liquid Transition	~100X via supercooling	~10X via crystalization	~ 10X via solvent evaporation	~ 100X via solvent extraction	~ 10X solvent extraction
Drawing	Stretching ~500%	Stretching ~1%	Negligible	Stretching ~500%	Stretching ~1%
Heat Treating	~ Tm-50°C	~ Tm-50°C	No	>Tg	~ 500°C

The importance of processing on fiber microstructure and properties cannot be overstated. Processing chemically different polymers in the same way generally produces fibers with similar physical properties. Conversely, processing the same polymer in different ways can yield dramatically different fiber properties. Table 4.3 shows the "reasonably" obtainable fiber tenacities for several combinations of polymers and processing routes. Comparing the various combinations illustrates the importance of process route and microstructure in determining macroscopic properties.

Table 4.3: The impact of process and structure on tensile Properties.

Polymer	Process	Morphology	Achievable Strength GPa
PET Polyester	Melt spun & drawn	Micro-fibrillar	1
Nylon 66	Melt spun & drawn	Micro-fibrillar	1
Polyethylene	Melt spun & highly drawn	Micro-fibrillar	1
Polypropylene	Melt spun & drawn	Micro-fibrillar	1
PET Polyester	Low speed melt spun & crystallized	Spherulitic	<0,2
PET Polyester	Very high stress melt spinning	Large crystal micro-fibrillar	0.4
Meta –aramid	Isotropic solution spun and drawn	Micro-fibrillar	<0,5
Para-aramid (LCP)	Liquid crystal Spun & annealed	Aligned	3+
Aromatic LCP polyester	Liquid crystal Spun & annealed	Aligned	3+
UHMWPE	Gel Spinning & drawing	Extended chain crystals	3+

Note: the strengths indicated for PP and PE represent demonstrated but currently non-commercial yarn variants.

- Melt spun and subsequently drawn polyester, polyamide, polypropylene, and polyethylene all develop similar microfibrillar structures and exhibit comparable maximum strengths.

- Conversely, if unoriented polyester spun yarn is allowed to crystallize prior to the introduction of molecular orientation, it becomes weak, brittle and difficult to draw. Polyester fiber spun under very high stresses develops both orientation and crystallinity during spinning and has properties that are intermediate to those obtained from the traditional melt spun product and the pre-crystallized version.

- Meta and para aramids are chemically similar, differing only in the geometry of moiety attachments. However the polymer chains of the meta variant is highly coiled and entangled and will not form a liquid crystal. Its tensile properties are a fraction of those obtained from the para aramids.

▪   LCP polyamides (para aramids) and LCP polyesters are both spun from liquid crystalline phase and have comparable strengths. Both are significantly stronger than their more flexible commodity kin that are spun from isotropic melts.

▪   Conventionally melt spun and drawn polyethylene fiber has the mechanical performance of a garbage bag while gel spun and superdrawn polyethylene is one of the strongest and stiffest materials known.

The individual steps of fiber processing are discussed in more detail below.

4.3   The Liquid State

Traditional flexible chain polymer melts and solutions are unstructured and viscous, comprised of highly entangled molecular chains. Molecular relaxation times are short compared to process times so inhomogeneities quickly damp out. More than 99 % of all fibers are produced from such isotropic liquids. Melt spinning is inherently cheaper and easier than solution processing, and is the dominant fiber spinning process. Polyester, polypropylene and nylon are all melt spun and constitute approximately 80% of all organic man-made fibers. Polymers that lack the thermal stability necessary for melt spinning are solution spun.

To maximize fiber tensile properties, one would ideally like all of the polymer chains to be straight and aligned along the fiber axis. The chain entanglements in isotropic polymer liquids are the primary physical aspect of the polymer limiting the deformability of the polymer chains. Less flexible, more rod-like molecules can reduce the level of entanglements and permit higher achievable molecular orientation. This approach leads to the families of nematogenic, liquid crystalline polymers, LCP's. LCP's may be melts (thermotropic) or solutions (lyotropic). The nematic state is defined as possessing orientational order, but neither positional or conformational order. Flory [2] and others [3] have shown that the nematic state is a consequence of geometrically driven packing requirements in both melts and solutions as the molecular chains become more rodlike. Note that the liquid crystal state exists only at certain temperatures and solution concentrations. While LCP's represent less than 1% of the fiber production they are on the leading edge of fiber technology and are the subject of much ongoing research.

An alternative route to high orientation is to carefully "comb" the entanglements from a flexible polymer during fiber processing. "Gel spinning" of high molecular weight linear polyethylene is an example of such a process [4]; see also Chapter 5.3.

4.4   The Spinning Process

All commercial polymer fibers are formed from a liquid starting state, either a melt or a solution. Spinning may thus appear to be more liquid than solid state processing. However, on closer examination it becomes obvious that the polymer rapidly solidifies in the spinline, There is no ambivalence that the polymer is a liquid at the spinneret. There is also no ambivalence that the polymer at the end of the spinning process is a solid. In between, the polymer must pass through a difficult to characterize nonequilibrium state or "mesophase". This liquid-solid transition is the single most important step in fiber processing.

4.4.1   Spinning technology

Figure 4.1 shows a simplified schematic of a melt spinning process. The polymer enters the spinning machine as a liquid. A specific quantity of polymer is metered into the spinning pack using a positive displacement gear pump. The gear pump is capable of producing the pressures of 100's of bars necessary to force the polymer through the downstream filter and spinneret orifices. The spin pack serves several functions. It holds the spinneret and serves to distribute the polymer uniformly across the spinneret plate. The actual fiber begins life as an individual fluid stream extruded through one of many small orifices in the spinneret plate. Filtration is also standard in spin packs. As a rule of thumb, the filtration level should be < 1/10 the fiber diameter or approximately 1-2 micron for commodity textile fibers. Below the spinneret the filament streams are rapidly cooled and solidified with the aid of controlled airflow. A lubricant or fiber finish is usually applied to the fibers to facilitate further processing. The fibers then pass onto the surface of rotating rollers or godets. The surface speed of the rollers determines the spinning speed. The fiber diameter or denier is determined by the mass flow rate, set by the metering pump, the number of filaments or spinneret holes, and the speed of the godet. After the godet the fibers may be wound up as spun yarns or passed on to an integrated drawing process. In many high speed spinning processes, the winder itself may serve as the godet.

Spinneret hole diameters are on the range of 100 to 600 microns. The lower limit on hole diameter is determined primarily by the fiber manufacturer's ability to clean and maintain the holes so that spinnerets can be reused. Common spun yarn fiber diameters are much smaller, circa 10 to 40 microns. Consequently fibers must stretch significantly, circa 25:1 to 400:1, in the spinning process. The spinneret hole diameter and the fiber diameter uniquely determine the stretch ratio or "drawdown". It is independent of the process speed.

The number of holes in a spinneret can range from one to tens of thousands. As indicated in Figure 4.1, multiple spun filaments are typically combined to form multifilament threadlines or yarns or tows. If the product is to be used as a yarn then the number of holes in the spinneret will be a small integral number times the number of filaments in the yarn. If the fiber is to be chopped and sold as a cotton or wool like staple, large numbers of filaments are combined into a tow and processed together. In staple spinning spinneret hole counts will be as high is operationally possible.

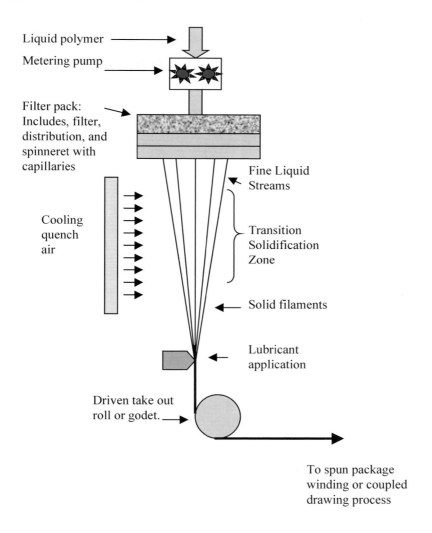

Figure 4.1   Schematic of melt spinning

Spinline stretching necessarily creates molecular orientation. Much of the molecular stretching occurs close to the spinneret while the filament is moving slowly and is clearly a liquid. In this region relaxation times are short compared to process times and the molecular orientation fades. Significant stretching also occurs further from the spinneret where the fiber is below the melting temperature and relaxation times are significantly longer. As relaxation times approach and then exceed characteristic process times (the reciprocal of the strain rate) more and more of the molecular orientation persists. It is commonly observed that as the molecular orientation occurring in the spinline increases, the crystallinity of the fiber also increases

Increasing spinning speed has a dramatic effect on process times such that the net molecular orientation increases dramatically as spinning speed increases. Spinning temperature, molecular weight, and cooling rates all have a second order effect on molecular orientation.

Figure 4.2 shows the impact of spinning speed on polyester spun yarn birefringence[1] and crystallinity. At the lowest speeds, the yarns exhibit little orientation or crystallinity. Increasing process speed increases orientation. At some level of orientation, crystallinity also begins to increase. At very high speeds both orientation and crystallinity saturate. Further increases in spinning stress lead to increasing levels of local inhomogeneity, most prominently the formation of radial variations in structure typically characterized as "sheath-core".

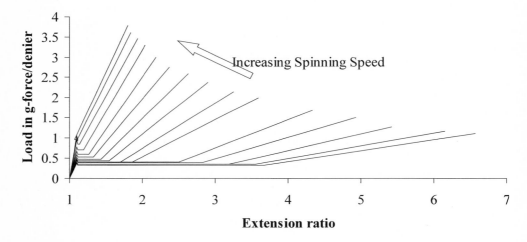

Figure 4.2   Change in PET microstructure with increased drawing speeds

Figure 4.3 shows the spun yarn load elongation curves of a series of spun yarns spun at speeds ranging from 300 m/min to 6100 m/min and birefringence value from 1.3×10^{-3} to 98.7×10^{-3}. All of these curves have a similar nature. Under load the fiber first resists extension and recovers elastically. However at a relatively low elongation, circa 5%, the fiber yields and immediately extends substantially, i.e. it "necks". The draw ratio at the neck or natural draw ratio is a characteristic of the material and the orientation. Once the entire test specimen has necked, the tension begins to increase and continues until the fiber breaks.   Beyond the yield at circa 5% elongation, the deformation is totally inelastic. Details of the physics of fiber deformation can be found in the works of Ward [5].

---

[1]Birefringence is the diameter normalized difference between the index of refraction parallel to the film axis and the index of refraction perpendicular to the fiber axis. It is essentially a measure of the number of covalent bonds aligned with the fiber axis, hence, a direct measure of molecular orientation in linear polymers.

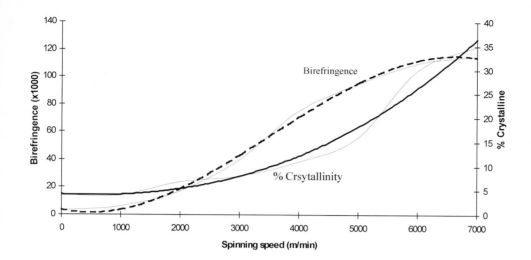

Figure 4.3 PET spun-yarn load-elongation curves for spinning speeds 300m/min to 600m/min

Solvent diffusion is much slower than the analogous heat transfer problem in melt spinning. Consequently, speeds in solution spinning are much lower than in melt spinning. In wet spinning, speeds are further limited by tension created by fluid drag in the wash bath. Therefore fibers spun from isotropic solutions generally have low molecular orientation. The molecular orientation and crystallinity are developed in subsequent (drawing and heat-treating) processes.

The structured nature of lyotropic solutions requires a modification of the typical wet spinning process to achieve both high global molecular orientation and slow enough filament coagulation to preclude void formation. (See Section 4.4.3.3 below.) This is accomplished by positioning the spinnerets above the coagulation bath, allowing the use of higher polymer temperatures and permitting greater drawdown to take place in the air gap between spinning jet and bath surface. Descriptively, the process is referred to as dry jet wet spinning.

The aforementioned capital-intensive nature of fiber production is implicitly illustrated in Figure 4.1. The single spinning position shown may produce only a few kilograms of fiber per hour. The capital costs of the cheapest commercial spinning machines, "compact spinners", start at about $2 per kilogram of fiber per year. A state of the art polyester tire cord machine could cost $6 per kilogram of fiber per year. A small wet spinning plant for producing fibers from a rigid rod polymer could cost $100 per kilogram per year. Increasing spinning speeds has been a major industrial focus since the 1970's, primarily to increase unit productivity and reduce the capital costs of spinning.

Quality control in fiber production is critical and difficult. Because the small filament diameters, extreme lengths, and high stresses in fiber processing, polymer quality is critical. The purity standards for fiber grade polymers exceed those of the pharmaceutical industry. The use of multiple positions puts additional quality demands on the fiber producer. Not only must he maintain constancy in time i.e. along the threadline, but also in space, i.e. from filament to filament within a single position and from spinning position to spinning position.

4.4.2   Modelling the spinning process

The physical parameters of a typical spinning process are extraordinary. In commercial melt spinning processes the threadline velocity after stretching and solidification or spinning ranges from <5 to >360 km per/hour. At the higher speeds, accelerations can easily exceed 100 g's - with the driving force being transmitted as tension through the liquid. Cooling rates of 10,000°C/hr are common. Actual measurements on filaments are extremely difficult and generally restricted to laboratory environments. Consequently, most fiber producers and many textile universities have developed mathematical spinning models to more fully understand the spinning process.

All spinning models start as fluid flow problems based on first principles, as discussed below. The earliest published work on spinning models [6] predates the advent of high speed spinning. In the low speed spinning regime rheological effects are dominant and the liquid-solid transition is free of the complexities of high molecular orientation and stress crystallization. At higher spinning speeds, inertial effects dominate the process and the development of micros-structure becomes quite complex. Incorporation of more sophisticated treatments of the solidification and microstructure development is the dominant areas of modern modeling research. An overview the modeling process follows. Several different models are presented in the excellent book "High Speed Fiber Spinning" by Ziabicki and Kawai [7].

4.4.2.1 Constitutive equation

The stretching of the spinning threadline is governed by its rheology, that is

$$F/A = \beta * dv/dx \qquad (1)$$

Here F is the filament tension, A is its cross-section area, v is the threadline velocity, and x is the distance from the spinneret. A and v are related through the continuity equation as discussed below. β is the extensional viscosity function as distinct from the shear velocity. The measurement of extensional viscosities is much more difficult than the measurement of shear viscosities. In fact extensional viscosity experiments closely resemble the spinning process.

In the simplest case, low speed, isothermal spinning of a shear independent polymer , β would be a constant and integration of equation 1 yields an exponentially increasing velocity. Fiber spinning does not meet the above restrictions and thus requires a more sophisticated viscosity function. A very simple and surprisingly effective approach has been to apply Troutans's rule,

$$\beta = 3 \bullet \eta \qquad T \geq T_g \qquad\qquad (1\text{-}a)$$

Here η is the Newtonian shear viscosity of the melt. Below the glass temperature, the filament is assumed to be rigid, i.e. its modulus significantly exceeds the spinning stress,

$$\beta = \infty \qquad T < T_g \qquad\qquad (1\text{-}b)$$

4.4.2.2 Conservation of mass

In differential form, the conservation of mass is expressed as

$$(\rho \partial A / \partial t)\Delta x = \rho A v - \rho (A v + v \partial A / \partial x \Delta x + A \partial v / \partial x \Delta x) \qquad (2)$$

Here, ρ is the polymer density and the quantity $\rho A v$ is instantaneous mass flow rate. The left hand side of equation (2) is the mass being accumulated in the incremental length Δx, while the two terms on the right hand side represent flow into and out of Δx receptively. For a steady state process, the left hand side =0 or $\partial A / \partial t$ =0 and equation (2) reduces to

$$v \partial A / \partial x + A \partial v / \partial x = 0 \qquad\qquad (2\text{-}a)$$

In solution spinning additional terms must be added to account for solvent mass transfer [8].

The principle of conservation of mass in fiber processing can take a very simple but useful form, i.e.

$$\text{Denier (or decitex)} \bullet \text{velocity} = \text{a constant} \qquad (2\text{-}b)$$

4.4.2.3 Conservation of momentum

The spinning process is driven by the external force supplied by the winding or take up device with a small assist from gravity, (almost all spinning is done downward). Inertia, fluid drag and surface tension effects all oppose the draw down of the filament and result in a decrease in filament tension between the winder and the spinneret.   (Both the gravitational and surface tension effects are small and can be ignored. In practice surface tension is usually ignored because it complicates the calculations and introduces additional experimental parameters. On the other hand gravity is usually included because it is simple and requires no new physics).

The change in force moving away from the spinneret and ignoring surface tension is

$$dF/dx = \rho A v dv/dx - \rho A g + \tfrac{1}{2} \rho v^2 \pi D C_d. \qquad (3)$$

New symbols used here are g for the acceleration due to gravity, D for the filament diameter and C_d for the air drag coefficient. The first term on the right hand side of equation 3-a represents the inertial effects, the second term is the gravitational contribution and the third term is air drag.

All of the terms in equation 3 are small in low speed melt spinning, and the process is dominated by viscosity. At higher speeds the acceleration and air drag become quite large. Viscosity still dominates the top section of the threadline where the polymer is a true liquid, while the inertia component dominates the liquid-solid transition region where the structure is formed. This has the practical effect that in high speed spinning, geometric uniformity (denier uniformity) may still be affected by viscosity but the microstructure is dominated by the acceleration. In melt spinning the drag component becomes significant only at the highest fiber speeds and large distances from the spinneret. In wet spinning, the viscous drag replaces the inertial component as the most important factor and limits the actual achievable spinning speeds.

4.4.2.4 Conservation of energy

Calculation of the viscosity β requires knowledge of the filament temperature T_f. The spinning threadline looses heat very rapidly via convection. The radiative contribution is small and is usually disregarded. A simple form expressing the temperature change as a function of distance from the spinneret is

$$dT_f/dx = -2\sqrt{(\pi A)} \cdot h \cdot (T_f - T_a)/(C_p \cdot \rho A v) \qquad (4)$$

Here T_f is the filament temperature, h is the fiber to air heat transfer coefficient, T_a is the quench air temperature, and C_p is the polymer specific heat. Terms reflecting radiation and the changes in internal energy and entropy related to chain alignment and solidification may be added but are small compared to bulk temperature effects.

4.4.2.5 Experimental parameters

As mentioned above, a form must be chosen for the extensional viscosity. In addition C_d, the fiber air drag coefficient and h the fiber to air heat transfer coefficient must be determined experimentally. Over the years these factors have been the subject of much theoretical and experimental work. Several independent discussions can be found in reference 6. On going improvements in measurement techniques have made it possible to more accurately measure threadline parameters on laboratory threadlines. It is now possible to do "reverse threadline modeling" that is to input the measured threadline profiles into computer models and determine the appropriate transfer coefficients.

4.4.2.6 Boundary conditions

The problem as formulated above is a boundary value problem, with conditions specified at both ends of the threadline.

At the spinneret, x=0:
- The cross sectional area = the spinneret hole area,
- The filament velocity is uniquely determined by the hole size and the mass flow rate.
- The temperature at spinneret is the melt temperature.
- The tension level at the spinneret is unknown.

The lower boundary of the process is set by application of equation (1-b),
- The filament temperature is the glass transition temperature.
- The threadline velocity at T_g is the winding or take out speed.
- The filament cross-sectional area is uniquely determined by the wind up speed and the mass flow rate.
- Neither the tension nor the distance at which the threadline reaches T_g are known. .

4.4.2.7 Solution of the equations

Computationally, it is much easier to solve an initial value problem than it is to solve a boundary value problem. If all the values are known at the start of the process, functional evolution can be computed by incrementally updating the functions away from the boundary. The usual practice in modeling spinning is to convert the mathematical problem to an initial value problem by assuming a value for the unknown stress at the spinneret. The modeler then calculates the threadline profiles. Comparing the calculated threadline velocity at T_g with the actual take up velocity determines whether the assumed stress at the spinneret was too high or too low. The estimated stress can be adjusted until a satisfactory solution is obtained.

4.4.2.8 Introduction of structure and properties

A mechanism is required to relate the solution of the mechanical problem outlined above to the resulting fiber microstructure. The most common approach is to adopt stress optical relationships from the theory of rubber networks [9]. The optical birefringence of the oriented spun fiber is assumed to be directly proportional to the stress in the threadline just above the glass transition temperature. In high speed spinning, significant crystallization can occur during spinning and more sophisticated approaches are required.

A separate mechanism is required to relate the fiber microstructure to the spun yarn physical properties. Physical properties are usually correlated with birefringence using actual process data.

4.4.2.9 Model sophistication

The outline above only hints at the sophistication of the spinning models in use today. In practice great emphasis has been placed on:

- The various mathematical algorithms that can used to solve the mathematical problem presented above. Increases in computing power have resulted in progressively more sophisticated computations
- Refinements of the transport coefficients, specifically air drag and heat transfer coefficients and wet spinning solvent diffusion. This information tends to be proprietary.
- Utilization of temperature dependent densities and specific heats.
- Enhancements to account for multiple filaments and their interactions
- Unique spinneret and quenching geometry's.
- Computation of radial structure variations within the filament, which can be very important in high stress melt spinning and in all solution processes.
- Incorporation of non round cross-sections
- Bi-component spinning – two polymers in a single filament
- Time dependence via responses to perturbations.

Models incorporating all of the above and more exist. However, much of the best spinning model work is proprietary within the large fiber producers.

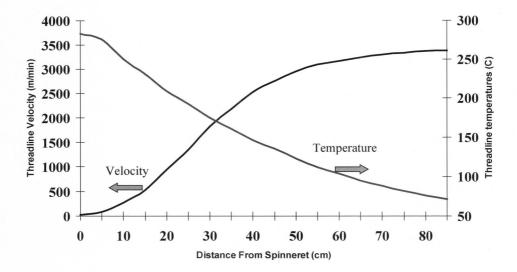

Figure 4.4   Threadline speed and temperature changes versus distance from the spinneret for PET POY spinning at 3400 m/min, predicted by a spinning model.

Spinning models greatly improve the understanding of the spinning process. As an example, Figure 4.4 shows the velocity and temperature profiles calculated for a polyester fiber spinning process operating at 3400 meters per minute.   In this process the fiber is extruded at 283 C and reaches its nominal freeze temperature, 265 C, in about 0.1 seconds and at 8 cm below the spinneret. The fiber requires only 0.03 additional seconds to reach its glass transition temperature, 70 C, about 80 cm below the spinneret. While in the "super cooled" state between T_m and T_g, the fiber is stretched approximately 23:1.

In an industrial environment, the economic benefits of spinning models can be very large. Properly used, models can significantly reduce the number experiments and the time necessary to develop new products. They can also aid in the trouble shooting of production problems.

4.4.3.   Development of structure during melt spinning

Between the spinneret and the glass transition temperature, the fiber experiences a complex thermal and stress history that determines the spun yarn microstructure. The spun yarn microstructure serves as a template, which influences all further modification to the fiber microstructure.[2]

The strain experienced during drawdown leads to an increase of molecular orientation of the fiber – generally in agreement with the affine[3] deformation calculations of Kuhn. Concurrently, entropic relaxation is occurring and some molecular orientation is being lost. At any point along the threadline, the net molecular orientation is the result of the dynamic balance between these two processes. Close to the spinneret, the filament is moving slowly and is clearly a liquid. In this region relaxation times are short compared to process times and net molecular orientation increases slowly, if at all. Further from the spinneret and below the equilibrium melting temperature strain rates are higher and relaxation times are significantly longer. As the fiber cools and relaxation times approach and then exceed characteristic process times, more and more of the imposed molecular orientation persists. Increasing spinning speed has a dramatic effect on strain rates, consequently net orientation increases dramatically as spinning speed increases. The imparted molecular orientation varies both between and within the polymer chains with the most highly oriented segments dominating the further development of morphology.

Most commercially attractive polymers are partially crystalline. Crystallinity serves to increase the thermal stability of the fiber by shifting the temperature of dimensional and mechanical stability from the molecular relaxation processes of the glass transition to the melting transition. As molecular orientation increases the rate of crystallization increases

---

[2] The development of structure in the spinning threadline is an extremely complex subject. Even a simple literature survey is far beyond the scope of this work. Instead, we shall employ a condensation model approach to develop a general understanding.
[3] An affine deformation is one in which the individual network elements deform in the same manner as the sample in general. This requires the chain segments to both rotate and stretch. An alternate deformation scheme, the pseudo affine deformation is based on rotation but does not allow the chain segments to extend.

(up to six orders of magnitude). Consequently, as net molecular orientation increases and the temperatures decrease in a spinline, it becomes increasingly probable that orientation driven crystallization will occur before entropically driven relaxation. The phenomena described above (often referred to as stress or orientation induced crystallization) dramatically reduces the entropy differences between the crystalline state and the amorphous material prior to crystallization. Nucleation and crystallization thus occur far more rapidly and a greater number of growth sites are created. The process of oriented crystallization dramatically increases the fiber's resistance to further deformation

With these concepts, a simple structure development map can be created, Figure 4.5. Structure development in the spinline may be conveniently be divided into three regimes, true liquid, the condensation/ crystal nucleation region, and the crystal growth region. Concurrent with the development of the fiber crystallinity, the topology of the molecular chains, i.e., the degree of interconnectivity of the spun yarn structure, is also determined. The most critical of these steps is the crystal nucleation, which serves as a template for all further structure formation processes within the fiber.

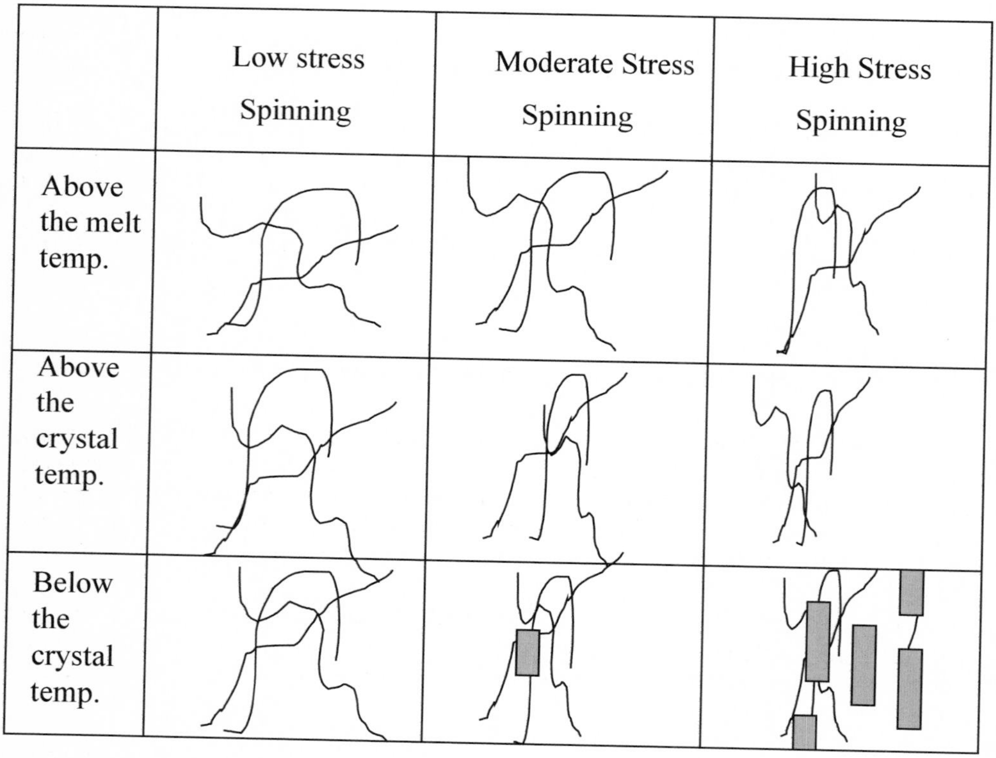

	Low stress Spinning	Moderate Stress Spinning	High Stress Spinning
Above the melt temp.			
Above the crystal temp.			
Below the crystal temp.			

Figure 4.5   Schematic of the development of structure during spinning

Since the crystallization of oriented molecular chains occurs at high temperatures, the non-crystalline or amorphous[4] material remains quite mobile. Chain segments exiting the ends of the crystals will be physically anchored and are consequently subjected to higher stresses. Chains segments in close proximity to and orientational alignment with existing crystallite surfaces may join them via epitaxy. Those parallel to but removed from the crystallite surface will be shielded from additional strain and may disorient. The net result is that stress crystallization in the spinline results in the crystallites being tightly bound to the load bearing structure, whereas the surrounding matrix material is free to relax and remains relatively unoriented. As stresses increase this segregation becomes progressively greater and the morphology of the crystals changes from point nucleated spherulites to line nucleated row structures.

The details of these complex processes have remained elusive over the past several decades and a number of models and suggestions exist in the literature. Until the details of the nucleation process in a spinning fiber are elucidated, however, description of structure formation in the spinline is deduction and conjecture. It is likely that two crystallization processes occur in the crystallization of oriented chains; the formation of an oriented fibrillar species (perhaps from an orientation induced mesogenic phase) and the decoration of this oriented fibrillar phase by folded chain lamellae growing from remaining and somewhat relaxed melt. Further definition of structure formation during spinning awaits the experimental elucidation of the nucleation mechanism.

The effects summarized above allow relative slowly crystallizing polymers such as PET and nylon to crystallize in the spinline as the spinning stress increases (usually increased by increasing spinning speed). The fibers thus produced change from unstable low orientation yarns or LOY's to progressively more stable partially oriented yarns or POY's and then to highly oriented yarns or HOY's. Note that HOY's are distinct from fully oriented yarns or FOY's produced via drawing process.

4.4.4   Development of structure during solution spinning

The solution spinning process is controlled by coagulation of the fiber as the solution concentration decreases. Solvent is continuously removed from the fiber surface, either by evaporation or desorption. This creates a solvent concentration gradient in the fiber that drives the remaining solvent outward to the fiber surface. Solvent diffusion also continually reduces the mass of the coagulating fiber.

A primary concern in solution spinning is control of radial variations in microstructure Such variations are intrinsic to solution spinning. At some point the fiber forms a skin. As additional solvent is removed the skin may become too large. The skin then wrinkles due to

---

[4] It is common practice in fiber science to apply the term "amorphous" to all noncrystalline regions of a polymer regardless of whether they are isotropic or anisotropic. This leads to the apparently redundant term "unoriented amorphous" and the contradictory term "oriented amorphous".

the decrease in the contained volume and/or internal voids may form. Dogbone shapes are common.

Control of the radial structure variations depends on balancing the rate of solvent removal from the surface with its rate of replacement from within the fiber. The make up and temperature of the wash bath control the rate at which solvent is removed from the fiber surface. The internal diffusion rate depends on the radial solvent gradient and the fiber temperature.

4.4.5   Development of structure during liquid crystalline spinning

In the case of thermotropic and lyotropic polymers, structure development during spinning is more a matter of preserving the structure already existing locally in the ordered fluid than of creating new structural entities. The major change to this structure during spinning is the translation of the local order of the nematogenic domains or micelles into the global orientation of the fiber.   While crystalline order exists in some LCP's, properties are dominated by molecular orientation alone.

Because of the very high molecular orientation of LCPs in the solid state, the tensile properties of LCPs are highly anisotropic. Typical tensile moduli are 70 to > 200 GPa, more than two orders of magnitude higher than the transverse or shear modulus. This results in poor performance in shear, which translate to poor compressive performance. Consequently, LCP fibers are almost exclusively used in tension critical applications.

Lyotropic, nematogenic polymers based on truly rod-like molecular structures (para linked aromatic heterocyclic moieties such as bis oxazoles, bis thiazoles and bis imidazoles) have been processed to the highest levels of tensile properties with both strength and modulii up to 100% higher than observed with the aramids or the thermotropic polyesters. The high cost and low shear modulii of these structures has inhibited their widespread industrial use. Toyobo, has recently commercialized poly bis oxazale fiber under the trade name Zylon.

Recently, a highly three dimensionally hydrogen bonded poly bis imidazole with compressive properties 5 times higher than other LCPs has been described in the literature [10]. Whether this specific chemistry proves commercially viable is still in doubt, but the importance and advance of stronger chain to chain interactions in rod-like molecules has been proven.

4.5   The Drawing Process

Spun fibers seldom have the level of physical properties and stability required by the user. Unoriented polymers exhibit low tenacity and modulus. In particular they are prone to high irreversible plastic deformation. A common example of this is deformation, known as necking or cold drawing, is observed when plastic packaging films or refuse bags are stretched. The standard practice in fiber processing is to eliminate this possibility by drawing or pre-stretching spun yarns. Drawing increases molecular orientation and, hence, the specific tensile properties such as modulus or strength.   Simultaneously, the fiber diameter is necessarily reduced. However, it is worth noting that to a first approximation, the fiber's true breaking stress does not markedly increase during drawing. That is, the molecular structure that ultimately carries the breaking load may be elongated and aligned but it is not fundamentally altered. Figure 4.6 shows the load elongation curves for a spun yarn and the same yarn after it has been drawn 3.0:1

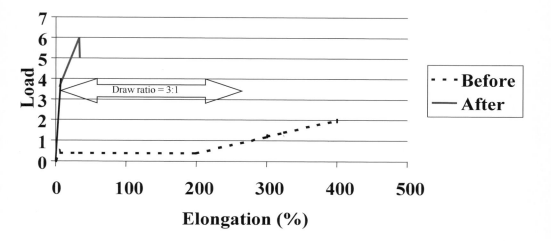

Figure 4.6: Yarn load Elongation Curves

4.5.1   Drawing technology

Industrial drawing is performed by stretching fibers continuously, passing them from a slower moving roller or rollers to one moving at a higher speed, Figure 4.7.

The speed ratios or draw ratios are typically 2:1 to 6:1 depending on the spun fiber orientation and the desired level of tensile properties. Lower or higher ratios may be used in special cases. Fibers spun from LCPs are fully oriented and cannot be drawn further.

For reason of manufacturing efficiency, drawing is generally conducted above the glass transition temperature. For spun yarns of low crystallinity or high molecular orientation it is also important to maintain the temperature below that at which spontaneous crystallization may begin. For crystalline spun yarns higher temperatures are acceptable and may be preferred.

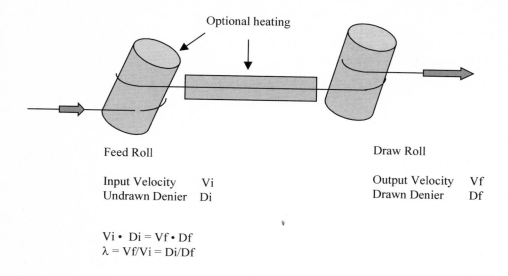

Vi • Di = Vf • Df
λ = Vf/Vi = Di/Df

Figure 4.7   Schematic of industrial fiber drawing or fiber heat treating processes

Heat may be supplied by heating the first or feed roller, or by passing the fiber through some form of heating medium between the rollers. Using heated feed rolls has the advantages of simplicity and control. However, the preheated fiber reaches the drawing temperature before it sees the drawing stress. Consequently the fiber may draw in the friction controlled stress gradient as it exits the feed roll. Stick-slip instability can lead to short term nonuniformity and threadline breaks. Preheating also increases the likelihood of the premature formation of additional crystals. When a separate heated zone is used, the fiber is under stress before it reaches the drawing temperature and drawing occurs in a temperature gradient. Variations in heat transfer lead to variations in drawing. Variations in spun yarn orientation can also alter the dynamics of temperature gradient drawing impacting process stability. Variations from filament to filament may also result in variable drawing, product irregularities and possible threadline breakage.

Choosing process draw ratios is always a compromise. The higher the draw ratio the greater the molecular alignment and the higher the strength and modulus. However, the higher the draw ratio, the lower the residual elongation and the energy to break and the higher the frequency of frequent threadline breaks. Shrinkage and dyeing difficulty also increase with draw ratio.

The lower limit of draw ratio is generally set by the natural or necking draw ratio. A continuous drawing process operating below the natural draw ratio produces yarns with alternating necked and undrawn segments such that the average draw ratio is equal to the process draw ratio. By increasing the drawing temperature. it is possible to operate below the necking draw. However, drawing still has a strong tendency to be unstable.

Many drawing processes utilize multiple steps. The advantages of multi stage processes are operational. As the structure evolves during drawing, the fibers become more robust allowing the use, for example, of higher temperatures in a second or third stage. Multistage processes also reduce the strain rates allowing the chains to be somewhat more compliant. The disadvantage to multi staging drawing is in equipment complexity and cost.

The limiting strength of the material may be determined by microstructure but, in industrial practice, it is determined by economics. Individual filaments may break due to impurities, flaws or variations in orientation. If a fiber breaks, the trailing part of the filament is no longer under tension and may adhere to the feed roller forming a filament wrap. Eventually, the wrap of even a single filament may entrap other filaments leading to a break down of the process. Increasing polymer cleanliness, product uniformity and process stability all increases achievable properties. Alternatively the fiber producer may translate improvements into economics by reducing process breaks and substandard product

The drawing process may be physically linked to the spinning process. This is ultimately an economic and engineering decision. Directly linked processes eliminate handling steps, however the final yarn speed is much greater than the spinning speed and requires much more sophisticated and expensive yarn winders. Commercial winding speeds of 6000 mpm or mach.0.3 are common. High-speed winder costs are comparable to automobile prices.

4.5.2   Modelling the drawing process

The evolution of drawing models differs significantly from that of spinning models. Spinning models start as rheological problems. Solid state structure and properties are superimposed on the fluid problem. The solid state nature of drawing and the complexity of fiber microstructure have always been appreciated. The deformations in drawing are often quite large and the changes in microstructure are usually dramatic. On the positive side, drawing experiments are much easier to perform than spinning experiments and the threadlines are much more amenable to observation and measurement, dramatically reducing industry's need for models. Modeling structural development during drawing has tended to be an academic exercise. Industrial modeling is generally based on statistical physical property models supplemented by rigorous calculations such as heat transfer.

The affine deformation of a crosslinked rubber-like network is the only polymer solid state deformation problem that has been explicitly solved. The stress-strain-temperature responses of natural rubbers can be predicted using the concepts of equilibrium thermodynamics and statistical mechanics. The key result of the rubber theory is that the stress in a stretched rubber may be described as:

$$\sigma = N_g kT \, (\lambda^2 - 1/\lambda) \tag{5}$$

where σ is the stress per unit deformed cross sectional area, N_g is the molecular weight between crosslinks, k is the Boltzman constant, T is the temperature and λ is the stretch or draw ratio. This simple approach accurately predicts both the effects of strain and temperature on rubber stress up to moderate stretch ratios of about 3:1 Above a draw ratio of three the stress begins to exceed that predicted by the model, indicating that additional processes are coming into play.

The same theory predicts that the birefringence or optical anisotropy of the stretched rubber network yields a relationship of the form

$$\Delta n = C \cdot (\lambda^2 - 1/\lambda) \tag{6}$$

Where Δn is the optical birefringence of rubber. The constant C contains the refractive index and the polarizability of the random link and the number of random links between crosslinks. Dividing equation (6) by equation (5) eliminates the draw ratio and yields the material specific stress optical coefficient.

The success of the network concept in rubbers is suggestive. The success of empirical stress optical coefficients in predicting the molecular orientation of spun yarns using the spinning models described earlier also suggests that non crosslinked polymers also have a network character. The entanglements of the interpenetrating random coil chains in linear polymers may substitute for crosslinks. After the polymer crystallizes, the crystallites may also serve as crosslinks. Of course the drawing of entangled networks is not reversible and the assumptions of the rubber model do not apply to linear polymers.

For non-crosslinked polymer deformed in the solid state, an alternate deformation process, the pseudo affine deformation, gives a better approximation of the development of orientation. In the pseudo-affine scheme, the material is assumed to deform as rod like units with fixed length rotating toward the direction of drawing. In this model the angle θ' between the unit and the direction of draw after drawing is related to the angle θ before drawing as is

$$\tan \theta' = \tan\theta / \lambda^{3/2} \tag{7}$$

Where λ is the draw ratio. In this model the birefringence of a uniaxially oriented polymer is given by

$$\Delta n = \Delta n_{max}(1 - \frac{3}{2} \overline{\sin^2\theta}), \tag{8}$$

where $\overline{\sin^2\theta}$ is the average value of $\sin^2\theta$ for the aggregate of units and Δn_{max} is the birefringence of the individual component. The pseudo-affine model gives a reasonable first order fit to the evolution of birefringence during the drawing of all of the common thermoplastic fibers.

In spite of strong micro-structural counter arguments, empirical evidence is surprisingly consistent with the assumption that all fibers have an underlying network structure and follow a well-defined deformation path. A state variable, the "network draw ratio", can be defined and used to predict fiber response to applied strains. Figure 4.8 shows a rudimentary example of how this concept can be developed. The load elongation test used in Figure 4.6 can be interrupted after the necking process is completed. The tension will immediately fall to zero, i.e. the strain is totally inelastic. When the stretching resumes, the tension rapidly rises back to the pre-stop level and then tracks the original or uninterrupted curve. The commercial objective of drawing is to eliminate the "neck" and generate a much more robust yarn. The shaded area in Figure 4.8 resembles the drawn yarn load elongation curve in Figure 4.6. If the curve is re-scaled to account for the increase gage length and reduced denier, it becomes a very close approximation to the typical drawn yarn stress-strain curve. Upon re-scaling, the elongation of the shaded curve is 33% instead of 100% and the tenacity is increased by a factor of three.

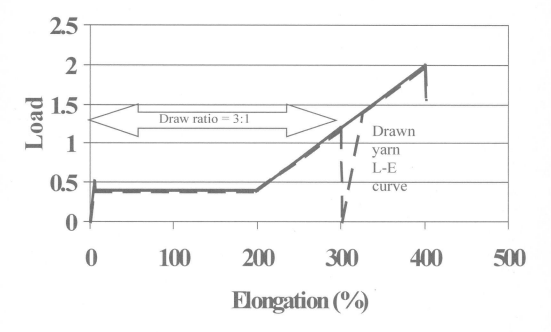

Figure 4.8   Relationship between spun and drawn yarn load-elongation curves; solid line = continuous, dashed line = interrupted.

The network model can be extended to include the molecular or network orientation introduced in the spinning process. The effect of spinning induced orientation is to reduce the total stretch in the necking region. The strain hardening region is little effected, Ward and his coworkers have shown that spun yarn stress strain curves can be restated in terms of true stress and then matched to predict the effective spinning draw ratio. Figure 4.9 illustrates the procedure. It is somewhat disappointing that the spinning draw ratio is not accurately predicted from the spun yarn birefringence by equation (5). The curve matching procedure is required to develop a relationship between the spun yarn birefringence and the network draw ratio.

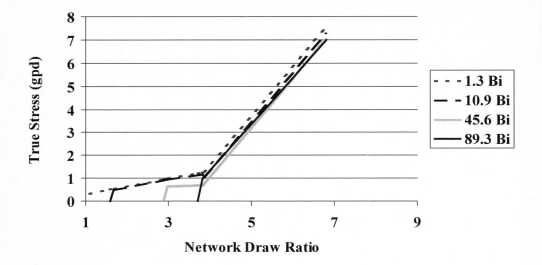

Figure 4.9   Idealised matching of PET spun yarn stress strain curves

In the simplest model, it can be assumed that:

- the true breaking stress, $\sigma_{\text{true bs,}}$ of the fiber,

- the total extensibility of the network, λ_{total}

- the strain hardening draw ratio,

- and the stress at the onset of strain hardening

are all constants.

Figure 4.10 shows experimental support for these assumptions from the series of polyester yarns spun at speeds for the set of data in Figure 4.3.[5]

The state of the network is given by network draw ratio, defined as the product of all previous draw ratios,

$$\lambda_{\text{network}} = \lambda_{\text{spin}} \cdot \lambda_{\text{draw}} \qquad (9)$$

---

[5] For this dataset the characteristic values are: breaking stress = 6.76 GPa, stress at the onset of strain hardening =0.82 GPa, the draw ratio in the strain hardening region = 1.80:1 and the total draw ratio PET at 0 orientation = 6.82:1.

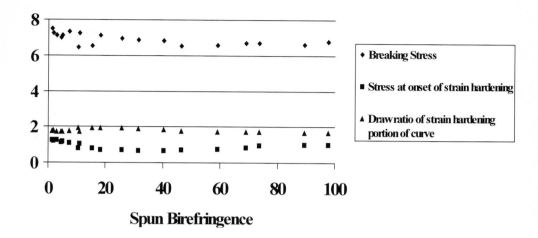

Figure 4.10 The numerical constancy of the parameters of the spun yarn PET stress strain curves, for a wide range of spun orientation levels

The residual network extensibility, after drawing, is related to the network draw or the current state as

$$\lambda_{residual} = \lambda_{total} / \lambda_{network} \qquad (10)$$

Given $\lambda_{network}$, the first two assumptions above, the tensile properties of a processed yarn as measured in a tensile test become:

$$E_b = (\lambda_{residual} - 1) \cdot 100 = (\lambda_{total}/\lambda_{network} - 1) \cdot 100 \qquad (11)$$

$$Tenacity = \sigma_{true\ bs} / \lambda_{residual} = \sigma_{true\ bs}\ \lambda_{network}/\lambda_{total} \qquad (12)$$

Here E_b is % elongation to break

This very simple model ignores all of the complexity associated with the development of the fiber microstructure but it is quite effective in predicting drawn yarn properties. The concept has been applied to various PET molecular weights, PBT, PEN, nylon, PE, PP. Note that several of the above polymers are in fact crystalline and draw by the shear breakdown and reformation of crystals.

The two assumptions regarding the strain hardening phenomena have not been used here. They can be used to predict drawing tensions, product moduli and drawing process stability. The model can be refined to include microstructure arguments, for example partitioning drawing between crystalline and amorphous components.

The predictions of drawn yarn properties in equations (11) and (12) require modification when fibers are drawn to very high levels, i.e for very high strength yarns. The experimental observation of a fixed breaking stress begins to break down when the residual elongation to break falls below approximately 30%. By carefully operating at maximum draw ratios using multiple drawing steps at progressively higher temperatures, it is possible to increase true breaking stress by about of 33%. In this region, the product of the tenacity and the square root of the residual elongation is often approximately constant. For very highly oriented fibers such as the LCP materials, it is often assumed that the work to break is a constant.

4.5.3   Development of structure during drawing of flexible chain polymers

While the network concept is surprisingly effective in predicting the effect of drawing on the mechanical properties on drawn yarns, it does not reflect the complex polyphasic yarn microstructure. The spun yarn entering the drawing process may be crystalline or non-crystalline, oriented or unoriented. The orientation may be manifested in either crystalline or non-crystalline phases or both. Each of these materials will behave differently during drawing.

4.5.3.1 Unoriented, amorphous spun fiber

These fibers are the most difficult to process, requiring the tightest control of draw temperature and stress. As a general rule, they also produce the most highly oriented, strongest and least dimensionally stable fibers from conventional melt spun yarns.

Morphologically, the orientation tends to develops as predicted by the pseudo affine deformation, equations 7 and 8. These yarns give the highest orientation of the isotropiclly spun fibers.

If the polymer is crystallizable, (the usual commercial case) it is likely that at some level of orientation, crystallization will ensue. The crystals can be viewed as being embedded in a highly aligned network. Crystallization is locally nucleated and the resulting crystals tend to be small, numerous and highly aligned. Since the crystals form after the network is extended, they have surprisingly little impact on the tensile strength or modulus. They do however stabilize the structure against entropic shrinkage.

Careful molecular characterization of draw yarns reveals a characteristic microfibrillar structure as shown in Figure 4.11. The details vary from polymer to polymer. For unoriented amorphous precursors, the intrafibrillar non crystalline regions are similar to the interfibrilar non crystalline regions. As stress induced crystallization is introduced into the spin line the intra fibrillar non crystalline regions become more oriented and the inter fibrillar non crystalline regions become less oriented, i.e. the load bearing becomes concentrated within the microfibrils.

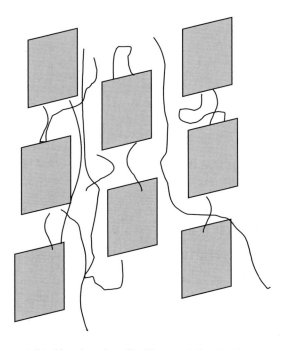

Figure 4.11 Simple microfibrillar model, after Prevorsek.

4.5.3.2 Oriented, crystalline spun fiber

The oriented crystalline spun fiber is at the opposite extreme of spun yarn structure when compared to the unoriented non crystalline fibers described above. They are the most robust precursor fibers for drawing. However the pre-existing crystallites limit the additional drawability so that they produce the weakest of the drawn yarns. Because of the already highly oriented crystalline network inherent in this fiber's microstructure, the residual draw ratio, as represented by the extreme left hand curves in Figure 4.3, is low and small orientational non-uniformities in the spun yarns are magnified in subsequent drawing. For a given level of tensile strength and modulus, however, these yarns exhibit the lowest level of residual shrinkage, making them particularly useful for industrial applications

4.5.3.3 Oriented, non-crystalline or partially crystalline spun fiber.

These fibers, typically called POYs (partially oriented yarn), are structurally intermediate to the unoriented amorphous yarns and the oriented crystalline yarns. They are more robust than unoriented yarns and can be handled more easily. They also tend to produce intermediate levels of mechanical properties and shrinkage. As a general rule the lower the spun fiber crystallinity at any given orientation, the more the fiber can be drawn and the more it tends to shrink.

Morphologically, these materials are most conveniently viewed as partially drawn fibers, i.e. the have been partially oriented in the process history prior to solid state drawing. When discussing network models above it was emphasized that partitioning the stretching between spinning and drawing has little effect on yarn tensile properties. The same is not true for total molecular orientation. The slow affine model approximates the development of molecular orientation in spinning whereas the fast pseudo affine model approximates the development of molecular orientation in drawing. Partitioning the orientation process therefore generates a family of network draw ratio, molecular orientation curves. The general rule is "the higher the spun yarn orientation, the lower the drawn yarn orientation". The total drawn yarn birefringence is, thus, decoupled from drawn yarn mechanical properties.

4.5.3.4 Unoriented crystalline spun fiber

Unoriented crystalline fibers typically show point nucleation and spherulitic morphologies. The drawing of these fibers depends on the shear breakdown of the spherulites and subsequent tilting of the reformed crystalline lamellae, Figure 4.12.

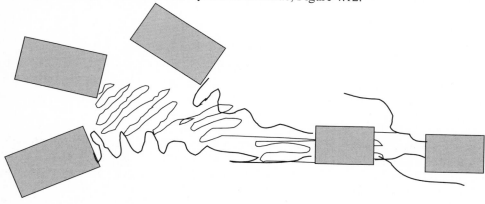

Figure 4.12 Shear break down and recrystallization of unoriented crystals during drawing

The efficacy of this process depends on the shear moduli of the crystals and the strength of the molecules. Polyethylene and nylon crystals shear before failure, above the "dispersion" temperature, whereas PET crystals do not. Consequently crystalline PE and PA can be drawn to very high draw ratios whereas unoriented, crystalline PET cannot be. Final properties and morphology of drawn oriented amorphous yarns are similar to those of the drawn unoriented non-crystalline fibers discussed above.

4.6   The Heat Treating Process

The fourth process in fiber production is heat treating or annealing. The primary purpose of heat treating is stabilize the fiber structure, hence, is to minimize the performance changes due to subsequent exposures to heat and stress. In particular, heat treatment reduces shrinkage. Heat treating temperature has limited impact on mechanical properties. The effect of heat treating temperature on diffusion properties is complex

Heat treating may follow a crimping process or a twisting process and serve to stabilize the crimp or reduce the torque liveliness of a twisted yarn.

4.6.1   Heat treating technology

In practice, a fiber heat treating process often appears similar to the drawing process illustrated in Figure 4.6 and is often done in sequence on the same machinery. The differences between drawing and heat treating are that heat treating temperatures are significantly higher than in drawing temperatures and that heat treating strains are much lower than drawing strains often involving negative strains or relaxation. As in drawing, the heat maybe applied via heated rolls, heated plates, hot air or steam chambers, etc.

The temperature of heat treating is chosen to be above the highest anticipated yarn use temperature. In general, increasing heat treating temperature increases crystallinity and lowers residual shrinkage. The upper limit on heat treating is usually set by operational considerations such as avoiding softening and sticking, chemical degradation, and ultimately, melting.

It is a common observation in fiber processing that "time at temperature" has a very small effect when compared with the temperature itself. This has significant economic ramifications because fibers are continuously moving. Increasing residence time therefore translates directly to increased machinery size and product cost.

Heat-treating may be accomplished with the fiber held at zero strain or constant length, with limited positive strain or stretch, or with limited negative strain or relax. Heat treating with out any shrinkage constraints is rare. Heat treating at controlled length or with slight stretching reduces shrinkage and may slightly increase tenacity and modulus. Strength is generally changed only in relation to % stretch or relaxation. The heat treating with a controlled overfeed termed relaxation is particularly effective for reducing residual shrinkage. In this case modulus can decrease significantly because the shrinkage takes place preferentially in the most highly stressed non-crystalline chains. The following example illustrates that while the tensile properties change very little during a small hot relax step, the shrinkage is significantly lower.

Property	Before	After	% Change
Tenacity	1 GPa	0.99 GPa	1%
Strain to Break	0.12	0.13	6%
Shrinkage	-0.08	-0.07	12.5%

4.6.2   Development of structure during the heat treating of flexible chain polymers

A variety of related structural changes can occur during heat treating. The strains experienced in manufactured fibers are generally quite high while the times associated with drawing and annealing fiber are very short. Consequently drawn yarns are seldom in their lowest energy state. The fiber usually possesses local regions where residual local stresses are quite high. During heatsetting local stresses are generally reduced via processes such as slippage of entanglements and increased crystalline perfection. If the fiber is allowed to shrink or relax during heat treating, the most highly stressed amorphous chains will tend to disorient (entropic shrinkage).

Total crystallinity, crystallite size and crystal perfection all tend to increase during heat setting, the amount of improvement measuring with increasing temperature. New crystallites may form from previously oriented but non-crystalline material. At higher temperatures smaller crystals may melt and re-crystallize. The actual number of crystallites will increase if nucleation of new units dominates or decrease if re-crystallization is significant. In practice, the total number of crystallites usually exhibits a maximum circa 50-75 °C below the melting temperature. Crystal perfection also generally increases by the expulsion of crystal defects.

Crystallization occurring in any oriented polymer is highly biased in favor of the axis orientation. Consequently, crystal orientation in any oriented semi crystalline polymer is always very high. During annealing, new crystals both the formation of new crystals and epitaxial crystallization on existing crystals will lead to higher overall orientation. Conversely, entropic relaxation of lightly constrained non-crystalline chains can lead to a decrease in overall orientation. The net chain in orientation during heat-treating will depend on which process dominates.

On the molecular level, the processes occurring during heat-treating are very similar to those that occur during high speed spinning. Key differences are that in annealing, the general arrangement of the crystalline and amorphous components is well established and the possibilities for large changes in the microstructure are severely limited.

4.6.3   Development of structure during the heat treating of rigid chain polymers

The properties of both the lyotropic aramid fibers and the thermotropic polyester fibers may be improved by annealing at elevated temperatures. In the case of Aramids, one observes a structural perfection increase leading to an increase in tensile modulus at essentially unchanged levels of tensile strength (typically about 3 GPa). This process takes place on fiber processing time scales (total treatment time on the orders of seconds or less) and correlates with an increase of the overall perfection of the fiber structure. In contrast, the annealing of thermotropic polyester fiber is performed to increase the tensile strength (typically from about 1GPa to values similar to the aramids). The process is slow, on the order of hours at temperature close to melting temperature, and correlates well with the time scale of solid state polymerization. It is felt the strength increase substantially derives from an increase in molecular weight coupled with increased chain to chain interactions.

It should be noted that while the fiber processing of LCPs is distinct from that of conventional polymers, the structure formed and the relationship of the structure to performance is similar, i.e., dominated by molecular orientation.

4.7   Fiber Structure: Multiphase Models

In most cases, fibers possess a semi-crystalline (30-80%), highly molecularly oriented microstructure. This complex structure is usually represented as a superposition of rigid crystalline phases with rubbery or glassy phases. For polyphasic models, the components can be arranged in parallel or in series, properties are then predictable with simple algebra. The modeler generally tries various parallel and series combinations until the model acceptably reflects known tensile behavior.   Extrapolation of such models beyond the boundaries of the input data has seldom proven useful.

A key feature of multiphase models is the complexity of the measurements required to characterize the model. As the number phases and the nature of their interconnectivity increases, the possible combinations also increase. Each phase must be characterized and its relative weight determined. Thus the minimum number of independent measurements is two times the number of phases minus 1. For a single phase model, a single measurement, for example optical birefringence, is sufficient.   For a two component model, three independent measurements are required. For example birefringence to determine the overall orientation, density to determine the relative amorphous and crystalline percentages, and finally orientation via X-ray scattering to determine the orientation of the crystalline component. The orientation of the amorphous component can then be calculated by subtracting the crystalline orientation from the total orientation. In order to describe the fiber's mechanical properties a specific geometric relationship, either series or parallel, must also be specified.

Further refining the model, e.g. by separating the non-crystalline component into oriented and non oriented components, or interfibrillar and intrafibrillar, or load bearing versus non load bearing all have great intellectual appeal and improve the utility of the models. Similarly, one can separate the crystalline material into structural or load bearing and non-load bearing or epitaxial components. These refinements put tremendous demands on analytical science and it is difficult to obtain unequivocal data sets.

Takayanagi and others have pursued these models extensively and shown their high utility in helping to understand the origin and manipulation of fiber properties, As noted, these models generally break down at larger deformations. A second concern with all multiphase models is that physical discontinuities arise whenever the models are strained. The network approach is much less rigorous analytically but shows surprising ability to predict responses to large-scale deformations.

4.8   General Process-Structure-Property Relationships

Far from a transition temperature, most fiber properties correlate with molecular orientation parallel to the axis of the fiber. Increasing average molecular orientation leads to increasing tensile strength and modulus and decreasing elongation to failure. In general, permeability of the fiber structure to penetrants and dyes decreases with increasing molecular orientation while all conductivities increase. Of special interest is the decrease of the linear coefficient of thermal expansion in the direction of molecular orientation, usually becoming negative at higher levels of orientation, i.e. the fiber shrinks.   The coefficient of the linear thermal expansion normal to the orientation direction increases commensurately. All these dependencies, which yield convincing monotonic plots of the property of choice versus orientation measures such a birefringence - a reflection of the much stronger covalent bonds along the chain as compared to the Van der Waals or other secondary bonds between chains. As the glass transition temperature is approached or exceeded, the molecular orientation starts to relax, manifested as modulus loss and entropic shrinkage.

Crystallization serves to introduce stable tie points into the structure, minimizing large-scale chain motions at Tg and increasing thermal stability to a morphology dependent temperature near Tm. Ultimately, the properties of a fiber reflect the level and nature of molecular chain connectivity between structural units and the overall uniformity of the microstructure along and across any given filament and between filaments in a yarn. Hence, with a given fiber chemistry and process, fiber performance may be accurately, if empirically, predicted. The next level of designing fiber microstructure for a specific balance of properties awaits increased insight into the control of structure nucleation and chain topology during spinning. Table 4.4 contains a generic set of process-structure property relationships, draw from the discussions earlier in this chapter.

As a rule it is much more difficult to characterize fiber structure than it is to measure fiber properties. Measurements of fiber microstructure utilize a vast assortment of analytical tools. A partial list includes polarized microscopy, electron microscopy, SAXS, WAXS, dynamic mechanical testing, infra red spectroscopy, thermal mechanical analysis, differential calorimetry, Raman spectroscopy and etching. It is even more difficult to make dynamic measurements of structures, which are rapidly evolving during processing. Several researchers are using synchrotrons to study model threadlines, specifically attempting to isolate and identify the initial structure defining processes in the spinline.

The value of understanding the structure component of the model lies not in testing a material but in allowing "theoretical" prediction of the results of process changes. A structurally naive engineer can run carefully designed statistical experiments and model the impact of any process change he wishes to make. The empirical approach breaks down when previously unmapped variables are introduced or when operations are conducted significantly outside the mapped regime. A clear understanding of the molecular processes is a great asset when moving beyond the established bounds. Molecular understanding is also vital in determining the limits of a material in a specific process.

Table 4.4: Generic Process-Structure-Property Trends

Process	Spinning			Drawing	Annealing
Process Variable Increased	Spinning Speed	Temperature	Mol. Wt.	Draw Ratio	Temperature
Orientation	++	-	+	+++	=
Crystallinity	+	-	+	+	++
Tenacity	++	-	+	+++	=
Elongation	--	+	Varies	---	=
Modulus	+	-	+	+	=
Shrinkage	Varies	=	+	+	--

Key
--- very strong decrease in measured property   -- a strong decrease in measured property
-    a small decrease in measured property   = a negligible change in measured property
+   a small increase in measured property   ++ a strong increase in measured property
+++ a very strong increase in measured property

4.9   Other Textile Processes

All fiber processing such as texturing, dyeing, etc. add to the stress/thermal history of the fiber and affect the fiber performance commensurate with the changes made to the fiber microstructure (usually molecular orientation and crystallinity). Some aspects specific to a given process are discussed below.

- *Texturing.* Many synthetic fibers produced for textile markets are subjected to processes designed to give them three-dimensional character. This involves bending the fiber into a zig-zag or helical path and then stabilizing the microstructure for this new configuration. The result is an orientation gradient across a filament or yarn, causing it to bend and assume the desired three dimensional trajectory when heated to above Tg and allowed to relax or shrink.
- *Dyeing.* Dye is often introduced into the fiber through a solvent or aqueous vehicle, often at elevated temperature and often under little or no controlled stress.

- *Finishing*. Textile fabrics are almost always finished, that is heat set in a stretched flat configuration, to insure dimensional stability, uniformity, and permanent press characteristics.
- *Cording*. Tire yarns are twisted into cord and then heat set to maximize the stability of the cord.

4.10   Special Processes

4.10.1 Gel spinning and superdrawing

It has been convincing shown that the draw ratio practically available in real polymer systems is limited by entanglements and other flaws (see the work of Smith, Ward). Creating precursors for drawing with a minimum of entanglements has proven effective with high molecular weight polyethylene and other polymers for allowing draw ratios of up to 100 or more, leading to extraordinary mechanical properties. If the precursor fiber is produced by placing the polymer into dilute solution such that a minimum of entanglements remain and then processing this solution into an unoriented precursor fiber, the process is known as gel spinning. The subsequent careful drawing to draw ratios above 50 is known as superdrawing. As the network tiepoints in polyethylene are crystalline, the drawing is performed above the dispersion transition temperature of polyethylene (temperature where chains can be pulled out of folded chain lamellae. This process has produced commercial polyethylene with the highest specific tensile properties of any known material. The weaknesses of these fibers are their very low compressive performance – the origin of this deficiency is the same as noted for the LCP fibers, namely, very low chain to chain interactions – and their low maximum use temperatures. Hence it may be deduced that the microstructure of fibers comprised of highly oriented flexible chains are, in principal, identical with the microstructure derived from stiff, nematogenic polymers.

4.10.2 Protein fibers

Measurement of the mechanical performance of a variety of silks, especially spider dragline silk, has shown these fibers to posses a balance of tensile and compressive performance unmatched in synthetic fibers (see the work of Kaplan, Jelinski). It was observed that the silk protein entered a lyotropic phase prior to exiting the spider's spinnerets, suggesting an analogy between the spinning of synthetic LCPs and biology of silk filament spinning. As pointed out originally by Viney, however, silk in solution is a typical globular protein with a hydrophilic shell protecting a hydrophobic care and allowing water solubility. In the absence of an interpenetrating network, it is unclear how the silk molecule would be deformed to satisfy the orientational requirements of the nematic structure. Viney suggests that the lyotropicity may originate from a linear assembly of the globular protein molecules, i.e. assemble into a string of beads of high aspect ratio. If correct, this could be the basis of entirely new concepts of oriented fiber structure formation.

4.11   Conclusions: what do you want to make - what really matters

The fiber processor never produces a final product, that is, the ultimate consumer never buys product from the fiber manufacturer. Usually there are several intermediate operations performed on the fiber after fiber manufacture and before consumption by the consumer. For example, polyester staple could be 1) blended with cotton, 2) spun into a yarn (several steps), 3) woven into a fabric, 4) dyed, 5) heat set (ironed), 6) cut and, 7) sewn into a shirt. Each of these processes depends on properties of the original fiber. In terms of importance, each processor will rank performance in his own area as the most important. In the shirt example the yarn spinner will be primarily interested in "spinnability" or how fast can he run his equipment and still produce a product acceptable to the weaver. The dyer wants ease of dyeing and uniformity, properties related directly to the fiber microstructure and its uniformity. The consumer will be interested in appearance, comfort and cost. Clearly, producing acceptable fibers is extremely complex and is loaded with pitfalls. Most of the details are closely held trade secrets.

References

1.   Web sites containing up to date listings of specialty Fibers include www.textileworld.com and www.amfa.com
2.   Flory, P. F., *Proc. R. Soc. London Ser*. Part A, 234, 73 (1956).
3.   Onsager, L., Ann. N.Y. Acad. Sci., 51, 627 (1949, 1956).
4.   Prevorsek, D. C., "Spectra: The Latest Entry in the Field of High-Performance Fibers" in High Technology Fibers Part D, Marcel Dekker, Inc New York (1996)
5.   See for example I.M. Ward, Developments in Oriented Polymers, Applied Publishers, Barking, UK (1982 and 1987); I.M. Ward and D.W. Hadley, An Introduction to the Mechanical Properties of Solid Polymers, John Wiley and Sons, Chichester, UK (1993).
6.   Kase, S. and Matsuo, T., Journal of Applied Polymer Science., 11, 251 (1967)
7.   Ziabicki, A. and H. Kawai, High-Speed Fiber Spinning, Wiley Interscience, (1985)
8.   Ziabicki, A. Fundamentals of Fiber Formation, John Wiley and Sons, London(1976)
9.   Treloar, L. R. G. "The Physics of Rubber Elasticity", Oxford Press, London (1958)
10.   Jiang Hao, Adams W W, Eby RK "Fibers from polybenzoxazoles and polybenzothiazoles" in *High Technology Fibers*, part D, Marcel Dekker Inc, New York (1996)
11.   Takayanagi, M. , Imada K., and Kajiyama T., J. Polymer Sci C 15 263 (1966) Models Ward Book
12.   Mather, P. T., Angel Romo-Uribe, Chang Dae Han and Seung Su Kim, *Macromol,*, 30, 7977-7989 (1997).
13.   Peterlin, A. "The strength and stiffness of polymers", A. Zachariades and R.S. Porter (eds.), Marcel Dekker, New York, 97 (1984).

5 High Modulus Fibres

5.1 Melt Spun Polyethylene, Polypropylene and Polyoxymethylene (Polyacetal) Fibres

I M Ward
IRC in Polymer Science & Technology, University of Leeds, UK

5.1.1 Introduction

The starting point for the high modulus melt spun polyolefine and polyacetal fibres was the cold drawing studies of Ward and co-workers [1,2] on polyethylene. Two key results which underpinned the science and technology were:

(i)   The Young's modulus depends only on the draw ratio so that the achievement of high modulus requires the determination of conditions for providing high draw ratios.

(ii)   The drawing behaviour, including the highest draw ratio achievable, depends on polymer molecular weights, in particular the weight average molecular weight $\overline{M}_w$, with low $\overline{M}_w$ being required for the easy preparation of high draw.

These key results are shown in Figures 5.1 and 5.2, and the drawing behaviour will be discussed in detail below.

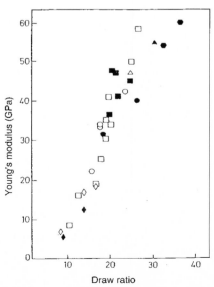

Figure 5.1   Modulus versus draw ratio for a variety of quenched (open symbols) and slow-cooled (solid symbols) of linear polyethylene samples drawn at 75°C. Reproduced with permission of Elsevier Applied Science.

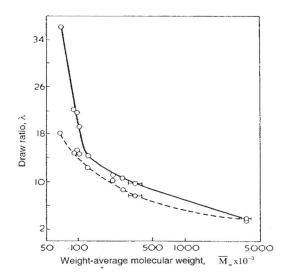

Figure 5.2   Draw ratio, λ, as a function of $\overline{M}_w$ and drawing time for quenched samples drawn at 75°C,   --- 60 sec draw   — 90 sec draw.   Reproduced by permission of Elsevier Applied Science.

Following the polyethylene result, it was soon found that similar results could be obtained for polypropylene (PP) and polyoxymethylene (POM). Cansfield et al [3] showed that low molecular weight PP could be drawn to draw ratios greater than 20 to give moduli in the range of 20GPa (Figure 5.3). This result was later confirmed by Taylor and Clark[4]

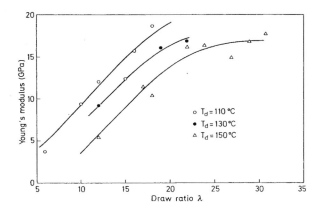

Figure 5.3   Modulus-draw   ratio   relationship   for   low   molecular   weight ($\overline{M}_w = 1.8x10^5$) polypropylene samples drawn at different temperatures. Reproduced with permission of the Society of Plastics Engineers.

High draw and high moduli were also obtained for polyoxymethylenes by Capaccio and Ward[5] and independently by Clark and Scott,[6] who regarded a two-stage drawing process as important. Although two-stage drawing may be the most viable route to a commercial process, Brew and Ward [7] showed that a single stage process produced marginally better products, and that the modulus/draw ratio relationship (Figure 5.4) is still almost unique.

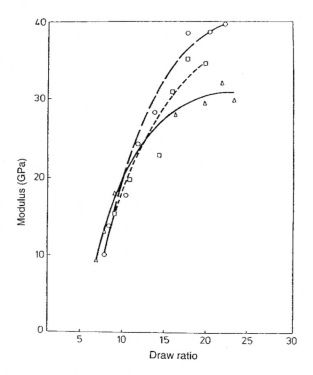

Figure 5.4   Modulus-draw ratio relationship for polyoxymethylene samples drawn at 145°C. Single-stage drawing at 10 cm/min (Δ) and 1cm/min (O). Two-stage drawing experiment (□) first for 72s at 10cm/min, then 1cm/min. Reproduced with permission of Elsevier Applied Science.

The tensile drawing behaviour of PP and POM was not studied in any great detail after these early investigations, essentially because the maximum stiffnesses achieved at 20 GPa and 40 GPa for PP and POM respectively did not appear to be outstanding compared with PE at 70 GPa. There has been, however, a revival of interest in the drawing behaviour of these polymers recently, due to the development of die-drawing processes for thick sections, monofilaments and fluted cores for wire ropes. During the 70's and 80's the scientific and technological effort was concentrated on high modulus polyethylene fibres, and this will now be described in detail.

5.1.2   The tensile drawing behaviour of polyethylene

In a comprehensive series of investigations, Capaccio, Ward and co-workers [8,9] explored the influence of three key variables.

(1)   The chemical composition of the polyethylene, including molecular weight and molecular weight distribution, and copolymer content.

(2)   The influence of initial morphology and molecular orientation.
(3)   The requirements for optimising drawing conditions, in terms of temperature and strain rate.

Two major conclusions emerged at a very early stage. First, it was necessary to optimise the draw temperature to an optimum temperature range for each grade of polymer so that there was adequate molecular mobility for a very high degree of plastic deformation without incurring void formation or fracture. It is essential that both the interlamellar shear process (which enables the deformation of the initial spherulitic structure) and the c-shear process (which permits slip of the chains through the crystalline regions) are both thermally activated. The draw temperature for practical processing must therefore be at a very high temperature, close to the melting point. In this respect there is a link between tensile drawing and the α-relaxation process in polyethylene.

Secondly, there are important relationships between the drawing behaviour and molecular weight and morphology, both of which are interrelated. The clearest indication of these effects was shown in a study where the drawing conditions were fixed at a comparatively low strain rate and a draw temperature of 75°C. A range of linear polyethylenes, varying in molecular weight and molecular weight distribution, was then produced by cooling from the melt at different cooling rates.

It is important at this stage to emphasise that, in contrast to the cold drawing of polyethylene terephthalate, discussed in Chapter 1, which draws through a neck to a constant fixed draw ratio (called the natural draw ratio) the drawing of polyethylene is time-dependent. On drawing polyethylene a neck forms at a comparatively low draw ratio in the initial neck (~10), but on further extension of the sample, there is continuous drawing of the material beyond the neck as the draw time is increased, and the draw ratio continues to increase, perhaps to as high as 30 or 40, until the whole sample reaches this draw ratio.

Figure 5.5 shows a series of results for the draw ratio, determined as the maximum value in the centre of a dumbbell sample as a function of the time of draw. The extreme cooling conditions, either quenching to ambient temperature or very slow cooling, are shown. It can be seen that for these fixed drawing conditions.

(i)   Very high draw ratios are only obtained for the lowest molecular weight samples.

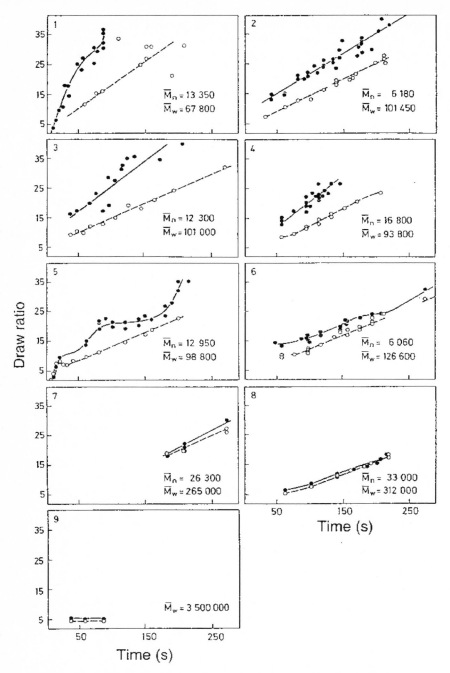

Figure 5.5 Draw ratio as a function of time of draw for quenched (o) and slow cooled (•) LPE samples drawn at 75°C. Reproduced with permission of John Wiley and Sons.

(ii)   With the exception of the ultra high molecular weight sample, the draw ratio is time-dependent and the slow cooled samples draw to high draw ratios more rapidly than the quenched samples. This difference is, however, not evident for $\overline{M}_w$ >200,000.

A further result of major significance is that for all these samples, irrespective of molecular weight or the initial thermal treatment, there is good first-order correlation between modulus and draw ratio, which has already been shown in Figure 5.1.

The interpretation of these results has been the subject of much discussion, aided by subsidiary structural studies. It is generally agreed that the over-riding consideration is that drawing involves the deformation of a molecular network. For high molecular weight polymers the junction points of this network are predominantly physical entanglements so that the initial morphology does not affect the drawing behaviour. A plausible explanation of the effect of morphology at low molecular weights is that junction points can be formed by molecular chains being incorporated in the lamellae. Slow cooling implies that the crystallisation occurs at low supercooling under conditions of low viscosity where homogeneous crystal growth occurs, as distinct from quenching where crystallisation occurs very rapidly and many tie molecules are formed to produce a more comprehensive molecular network. The ease of draw of the slow cooled low molecular weight polymers has therefore been attributed to the unfolding of a more regular lamellar texture and the reduction in the number of tie molecules. An additional factor is that in the case of slow cooling, segregation of low molecular weight material can occur, because the initial crystallisation takes place at a temperature which is higher than the melting point of this material, quite apart from its slower rate of crystallisation. It is suggested that this low molecular weight material can act as a plasticiser in the drawing process. Such effects were first proposed by Way and Atkinson[10] for polypropylene, and confirmed by Barham and Keller [11] who showed the deleterious effect on drawing of removing this low molecular weight material.

The results presented in Figure 5.5 show the effect of molecular weight on the drawing behaviour at 75°C. The next major practical issue which was examined by Ward and co-workers [9] was the effect of draw temperature, with a view to obtaining high draw ratios and hence high modulus for higher molecular weight polymers. It was found that this approach was only successful to a limited degree. Figure 5.6 shows results for a polymer with $\overline{M}_w$ =8x10^5. The maximum draw ratio achievable does increase steadily from about 10 at 75°C to 40 at 130°C. Also shown in Figure 5.6 is the plot of the room temperature Young's modulus. It can be seen that this has a peak value of about 40 GPa for drawing at 120°C, after which it falls steeply because the drawing process is then ineffective due to the very high molecular mobility. Jarecki and Meier [12,13] subsequently confirmed this result, showing that it is desirable to select as high a draw temperature as possible for each grade of polyethylene, so as to obtain high effective draw and hence high modulus.

From a practical viewpoint, these results taken together defined the boundaries of subsequent practical processes for melt spun high modulus polyethylene fibres. The Leeds

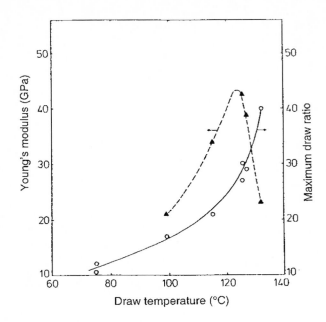

Figure 5.6 Effect of draw temperature on maximum attainable draw ratio (—) and room temperature modulus of samples drawn to the maximum at each temperature (---). (high molecular weight LPE, $(\overline{M}_w = 8x10^5)$. Reproduced by permission of John Wiley & Sons.

group, first in collaboration with ICI Fibres, and then in collaboration with SNIA (Italy) and Hoechst-Celanese (USA) devised two stage melt spinning and hot drawing processes for commercial production of these fibres. These were marketed first by SNIA as Tenfor fibre and secondly by Hoechst-Celanese as Certran fibre. Key factors in the practical process are

(i)   For satisfactory drawing behaviour it is necessary to control the spun yarn morphology by simulating the slow cooling effects shown for compression moulded samples (Figure 5.5). This requires the placing of a stroud immediately below the spinneret   It is also necessary to spin to a low birefringence i.e. to operate at a low wind up speed. Preorientation limits the subsequent draw ratio.

(ii)   The drawing process is most effectively undertaken in several stages. This reduces the heat transfer problems and enables the draw temperature to be increased to obtain higher draw ratios at higher processing rates.

An important practical extension of the research described above, which is concerned entirely with the drawing of polyethylene homopolymers, was the extension of such studies to polyethylene co-polymers, and more specifically the effect of short chain branches[13]. It was found that the incorporation of small numbers of short chain branches had a very significant effect, increasing the strain-hardening and reducing the maximum draw ratio.

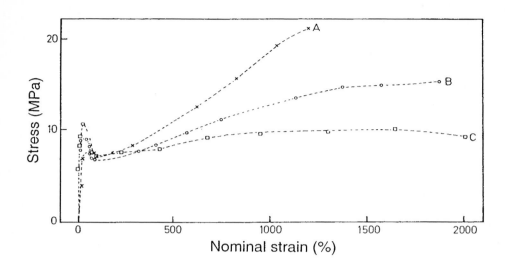

Figure 5.7   Load extension curves for ethylene-Hexene 1 copolymers. A: 4 butyl/10³C; B: 1 butyl/10³C; C: homopolymer. Reproduced by permission of John Wiley & Sons.

Figure 5.7 shows comparative results for two copolymers with n-butyl branches in concentrations of 1 and 4 per 1,000 carbon atoms respectively. For quenched samples drawn under identical conditions at 75°C the maximum draw ratio falls from 20 to 11. It was also found that large branches (n-butyl compared with ethyl) had a greater effect at similar concentrations. Following our previous exploration of the effect of molecular weight on drawing, in terms of the deformation of a molecular network, it is concluded that the incorporation of branches produces a more effective molecular network. This could be due either to the direct creation of more physical entanglements or to the effect of branching on molecular motions. De Gennes was first to indicate the effect of branching on reptation, showing that the reptation time approaches infinity in branched systems.

Finally, it was notable that the relationship between the room temperature Young's modulus and the draw ratio for these short-branch copolymers was virtually identical to that established for the homopolymers. There were, however, some differences in the morphology of the drawn polymers between the copolymers and the homopolymers. It appeared that the average crystal lengths in the copolymers were significantly less than in the homopolymers, suggesting that there may be contributions to the stiffness from taut tie-molecules in the copolymers, rather than solely from interlamellar crystalline bridges, as in the homopolymers

5.1.3   Tensile drawing of polypropylene and polyoxymethylene

It has been shown that polypropylene (PP) [15] and polyoxymethylene (POM) [7] can also be drawn to high draw ratios and high modulus, and in the case of POM, high strength, fibres. In the case of PP, there has been only moderate interest, because the maximum room temperature modulus which has been obtained is only about 20GPa.   Although this compares favourably with the maximum theoretical chain modulus of ~45GPa, 16] it is significantly less than that obtained for PE, where values of 40-60GPa are readily achieved. In the case of POM, modulus values of 30-40GPa can be obtained, but there has been no commercial interest, in spite of the excellent creep behaviour and high temperature performance.

The key factors to achieving high draw and high modulus in PP and POM, are identical to those established for PE, although much less research has been undertaken.

In PP, Cansfield et al [15] showed that a high draw temperature was essential, and that the best results were obtained with low molecular weight samples. The low molecular weights were achieved by degradation of higher molecular weight polymer. This gives rise to a narrow molecular weight distribution, which probably explains why Cansfield et al saw no effect of initial morphology (i.e. initial thermal treatment). Typical modulus - draw - ratio data are shown in Figure 5.3. Taylor and Clark[17] subsequently confirmed that high modulus PP fibres can be produced by tensile drawing.

Capaccio and Ward [5], and independently, Clark and Scott[6] showed that high draw can lead to high modulus POM fibres. Clark and Scott considered that this must be a two stage drawing process, but subsequent studies by Brew and Ward showed that single stage drawing actually produced slightly superior fibres, because of the deleterious effects of annealing the first stage product of a two stage drawing process. Brew and Ward[7] also found that the processing window in terms of temperature and strain rate, was extremely narrow, which does support the idea that multi-stage drawing could be required for any cost-effective production route.

5.1.4   The structure of ultra high modulus polymers

Plastic deformation of crystalline polymers gives rise to the breakdown of the initial spherulitic structure and the eventual formation of a fibrillar structure. At modest draw ratios ($\lambda\sim10$) this fibrillar structure shows a characteristic two point SAXS pattern which is interpreted as arising from alternating blocks of crystalline and amorphous material (Figure 5.8(a)). Peterlin[18] proposed that the increased modulus of oriented polymers arises from fully extended molecular chains (which he called tie molecules) connecting the crystal blocks in the aligned structure (Figure 5.8(b)). Although this proposition has never been experimentally verified, it seems likely that it is correct, and relevant to the very high modulus polymers where much higher draw ratios are concerned. Detailed structural studies on ultra high modulus polyethylenes, however, have shown that in this case the

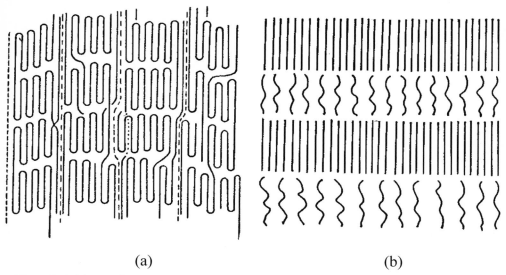

(a)   (b)

Figure5. 8 (a) Peterlin's macrofibrillar model of fibrous structure. (b)   Schematic
              diagram of simple parallel lamellar structure of alternative crystalline and
              amorphous regions.   Reproduced by permission of John Wiley & Sons.

structure is also reinforced by the crystalline regions, rather than individual extended molecules, as envisaged by Peterlin.

A combination of structural techniques, including quantitative WAXS measurements of average crystal lengths in the draw direction[19], dark field transmission electron microscopy[20], and gel permeation chromatography of nitric acid etched samples[21, 22], have shown that the average crystal lengths can reach ~ 500Å. This compares with a long period of ~ 200 Å obtained from the two point SAXS pattern. It was therefore suggested by Gibson et al.[23] that the structure of the ultra high modulus PE is that shown schematically in Figure 5.9. This can be regarded either as the original lamellar structure of Figure 5.8(a) now linked by what Gibson et al. called "intercrystalline bridges", or as akin to a fibre composite, where the structure is reinforced by long crystalline sequences which act like the fibres in the matrix formed by the original lamellar structure.

An alternative model was also proposed by Arridge, Barham and Keller[24], who proposed a different fibre composite model where needle-like crystals were embedded in a matrix of the remaining material.   On the basis of detailed neutron diffraction measurements, Barham[25] recently showed that the intercrystalline bridge model of Gibson et al. was mostly likely correct, and rejected both the Peterlin model and his own fibre-composite model as predicting neutron scattering behaviour at variance with the experimental results.

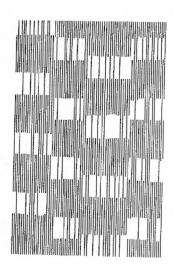

Figure 5.9   Schematic structure for high modulus polyethylene showing lamellae linked by intercrystalline bridges.   Reproduced by permission of Elsevier Applied Science.

Structural studies on die-drawn polypropylene samples by Taraiya et al.[26] suggested that the high stiffness arose from a combination of taut tie molecules and extended chain crystals, which could act as intercrystalline bridges. The intercrystalline bridges in PP are not the same as those in PE, in that the crystal structure of the bridges shows a different crystal structure from that of the lamellae, hence no increase in crystal length in the draw direction is observed. These crystal bridges were associated with a second melting peak in the DSC endotherm.

In the case of POM, it has also been concluded that the crystalline bridge model is not applicable. Yungnitz[27] found from WAXS and SAXS measurements that even for the highest modulus samples, the average crystal lengths were only comparable with the long period.

5.1.5   Fibre strength

Much of the initial interest in the achievement of high mechanical properties in polyethylene and aramid fibres stemmed from the recognition that modulus values in the range of 100GPa come very close to the theoretical estimates i.e. the so called crystal chain modulus. It was, however, very soon appreciated that for practical applications the tensile strength is usually as important, if not more important than modulus, and that it is important to achieve values significantly in excess of 1 GPa.

It is well known that fibre strength relates to two principal factors

(i)   Molecular weight

(ii)   Fibre diameter

It is part of conventional wisdom in fibre technology that tensile strength relates primarily to number average molecular weight , $\overline{M}_n$. In a key publication[28] based on the analysis of the tensile strength of cellulose acetate fractions, Flory proposed, first that the strength σ could be related to $\overline{M}_n$ by the relationship

$$\sigma = A - \frac{B}{M_n} \tag{1}$$

and consistent with this relationship, that the tensile strength of a blend of different molecular weight fractions relates to the weight average summation of the strengths of the individual components.

In more recent years, several groups have addressed the molecular weight dependence of strength for polyethylene fibres, including both gel-spun and melt spun fibres. Here, the discussion will focus on melt spun fibres and review results described by Smith, Lemstra and co-workers[29, 30], Wu and Black[31] and Ward[32-34] and co-workers.

Smith, Lemstra and Pijpers[30] found no difference between the strength/modulus relationship for melt spun and gel spun fibres. By implication, it can be concluded that ignoring size effects (to be discussed) the principal reason for the greater strength of gel spun fibres compared with melt spun fibre is the much higher molecular weight of the former ($\sim 10^6$ compared with 10^4).

Smith et al[30] and Wu and Black[31] determined the strength for a somewhat limited range of molecular weights in terms of $\overline{M}_w$ and $\overline{M}$. Wu and Black[31] included that $\overline{M}$ was the predominant factor in determining tensile strength and could find no significantly positive trends with $\overline{M}_w$ or polydispersity. Smith et al[30], on the other hand, concluded that a reduction in polydispersity from 8 to 1.1 gave an increase in strength by a factor of about 2 for a fibre with $\overline{M}_w \sim 10^5$. In retrospect, this result can equally well be interpreted as arising from an increase in $\overline{M}$.

These results are consistent with, and indeed subsumed by more recent extensive studies by Ward and co-workers[32, 33]. They are also consistent with a theoretical model for the tensile strength of oriented fibres of binary mixtures of high and low molecular weight polyethylene, by Termonia et al.[35].

Ward et al[32] made a particular point of measuring the tensile strength at -55°C of fibres drawn to fixed draw ratios (and hence fixed modulus). The low temperature tests were undertaken to eliminate any complications due to ductility i.e. in all cases brittle fracture was observed. The fixed draw ratio was in an attempt to compare similar structure so that only effects of molecular weight are significant. The results for draw ratios 15 and 20 are shown in Figures 5.10(a) and 5.10(b), presented in the form of the dependence of tensile

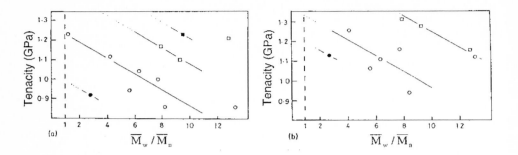

Figure 5.10 (a) Dependence of tensile strength of LPE fibres at -55°C on molecular weight for draw ratio of 15. $\overline{M}_w$ values are 60,000 (•), 100,000 (o), 220,000 (□), 330,000 (■). (b) as for (a) but with $\lambda = 20$. Reproduced by permission of Kluwer Academic Publishers.

strength on $\overline{M}_w / \overline{M}$ for different $\overline{M}_w$. It can be seen that the Flory relationship of equation (1) can be regarded as a reasonable representation of the data. In a subsequent publication Hallam et al[33] carried the analysis one stage further by extrapolating the strength values to those corresponding to a polydispersity of unity, as indicated by the dotted lines in Figures 5.10(a) and (b). This procedure gives a method for determining the strength of a monodisperse sample which defines the intrinsic molecular weight dependence of the polymer for each draw ratio. Plotting these results on a logarithmic scale for the two draw ratios tested showed a good correlation from which the line of best fit gave the relationship

$$\sigma = KM^{0.25} \qquad\qquad (2)$$

where K had the value 68 MPa for draw ratio 15 and 78 MPa for draw ratio 20.

From their collected data on tensile strength measurements Hallam et al[33] were able to test the Flory relationship[28] for a range of polyethylenes differing very significantly in respect of molecular weight and molecular weight distribution. The tensile strength using equation (2) were then predicted on the basis of a simple weight average distribution from the molecular weight distributions of the different polymers, as obtained from GPC measurements. In the case of the draw ratio 15 samples, the correlation coefficient between the experimental and calculated results was found to have the value of 0.69; the value at the higher draw ratio was 0.79. It is also informative to consider the graphical representation of the relationship between calculated and experimental data which is shown in Figure 5.11. It can be seen that although there is a good overall correlation the experimental results tend to fall below the calculated results particularly at the higher draw ratio. This is consistent with the observation that tests at these high draw ratio often result in premature failure. It can,

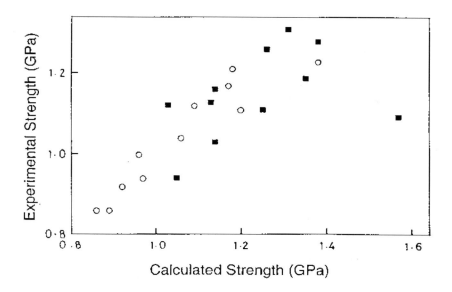

Figure5.11   Comparison of experimental and calculated tensile strength values (o) λ=15; (■) λ =20.   Reproduced by permission of Kluwer Academic Publishers.

however, be concluded that these results provide good support for the theoretical proposals of Flory[28] and Termonia et al.[35]

It is generally accepted that the factors which determine the tensile strength of fibres can be divided into two groups, intrinsic factors such as molecular weight or molecular orientation and extrinsic factors such as flaws or fibre diameter. The discussion so far has concentrated on intrinsic factors, and shown that molecular orientation and structure (as reflected by draw ratio) and molecular weight are major key parameters.   The intrinsic factors can be identified by attempting to keep extrinsic factors constant or as nearly similar as possible. We will now consider the effect of extrinsic factors, and here the key variable will be fibre diameter, but we also wish to attempt to identify the role of flaws. In this respect Weibull analysis is the principal statistical method which is widely used to study the failure mechanism of brittle materials such as glass or ceramic fibres. Provided that the tensile strength data are determined at -55°C, where polyethylene fibres show clear brittle fracture, it is therefore to be anticipated that Weibull analysis may be a valuable tool in exploring the nature of the tensile failure. Weibull analysis is based on two assumptions (a) the material is statistically homogeneous, which implies that the probability of finding a flaw of a given severity within either an arbitrary small value of the fibre or along its length is the same throughout all the fibre, and (b) failure at the most critical flaw leads to total failure (the concept of the weakest link of a chain).

If $P_f(\sigma)$ is the probability of failure of a specimen containing n links at a stress σ two key relationships can be derived.

(1) Assuming that the threshold stress for failure for any sample is zero

$$\ln \ln \left[1 / (1 - P_f) \right] = \text{constant} + m \ln \sigma \tag{3}$$

where m is called the Weibull modulus and defines the nature of the distribution of failure stress.

(2)   More specifically, it follows from the initial assumption that $P_f(\sigma)$ can be written as

$$P_f(\sigma) = 1 - \exp\left[-n\phi(\sigma) \right] \tag{4}$$

Weibull assumed that $\phi(\sigma)$ takes the form

$$\phi(\sigma) = \left[\frac{\sigma - \sigma_n}{\sigma_n} \right]^{m} \quad \text{for } \sigma > \sigma_n$$

and $\phi(\sigma) = 0$ for $\sigma \leq \sigma_n$
where σ_n is the minimum fracture stress and σ_n is a normalising factor

from which it follows that the mean strength of the fibre $\overline{\sigma}$ is given by

$$\overline{\sigma} = \left(\frac{\sigma_0}{n^{1/m}} \right) \Gamma(1 + 1 / m) \tag{5}$$

where Γ is a Gamma function.

This equation can be developed to test whether the number of flaws is either proportional to the surface area of the fibre or alternatively the volume.

For the surface model $\overline{\sigma} = \left[\frac{\sigma_0}{(\Pi d\ell)^{1/m}} \right] \Gamma(1 + 1 / m)$ $\tag{5a}$

and for the volume model $\overline{\sigma} = \left[\frac{\sigma_0}{\Pi d^2 \ell)^{1/m}} \right] \Gamma(1 + 1 / m)$ $\tag{5b}$

Amornsakchai et al[34] applied these Weibull equations to a study of melt spun polyethylene fibres. Figure 5.12 shows a typical Weibull plot which was obtained. A result of overall significance was that low draw ratio fibres showed a high value of m, which indicates a narrow distribution of strengths and is consistent with the conclusion of the previous studies described above, that the strengths of these fibres are determined by the intrinsic parameters of molecular weight and draw ratio rather than extrinsic parameters such as fibre diameter and flaws.

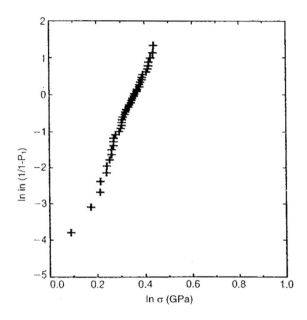

Figure 5.12   Typical Weibull plot for an oriented polyethylene fibre. Reproduced by permission of Kluwer Academic Publishers.

Typical results for high modulus/low molecular weight fibres had a lower value of m from the Weibull plot, in good agreement with the value obtained from the gradient of the ln(tensile strength) versus ln(diameter) line assuming a volume distribution of flaws (Equation 5b).

In several instances, the results of the Weibull analyses were less clear, even suggesting the possibility that the fracture of higher molecular weight melt-spin fibres may be related to surface flaws rather from volume flaws. It could, however, be firmly concluded that there are significant effects of fibre diameter on the strength of high draw melt spun fibres. (For an example see Figure 5.13).

Smook et al[29] have proposed that the strength/diameter effect in gel spun fibres can be explained as the basis of the Griffith theory of fracture. Figure 5.14 shows typical results for melt spun fibres plotted in the form $1/\sigma$ versus $d^{1/2}$. In theory, by extrapolating to zero

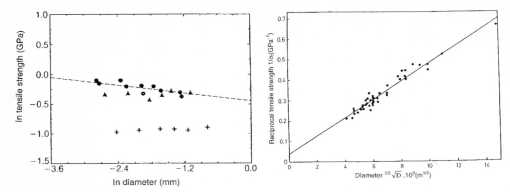

Figure 5.13 L_u tensile strength versus lndiameter for oriented polyethylene fibres. (+) E = 10 GPa; (▲) E = 20 GPa; (•)   E = 30 GPa.   Reproduced by permission of Kluwer Academic Publishers.

Figure 5.14   Linear strength diameter relationship   observed   for   high molecular weight high modulus gel spun and drawn polyethylene fibres. Reproduced by permission of Kluwer Academic Publishers.

diameter the intrinsic strength of a fibre without flaws is obtained. Although there is a good correlation between σ and $d^{1/2}$, the extrapolated values for strength obtained in this way are in the range 1-3 GPa which is much lower than the value of 19GPa estimated for the theoretical strength of oriented polyethylene and are much lower than those obtained by Smook et al[29] from similar plots for gel-spun fibres. On the other hand, these values are not too far from the maximum value for tensile strength obtained by Smith and Lemstra[29] by extrapolation of the strength/modulus relationship to the maximum theoretical modulus. It can only be concluded that this is an area where further research would be valuable in terms of both science and technology.

5.1.6   Other mechanical properties

5.1.6.1 Dynamic mechanical behaviour

The dynamic mechanical behaviour of the high modulus PE fibres[23] (Figure 5.15) shows the α and γ relaxations typical of linear polyethylene, and the storage modulus E′ rises at low temperatures to a value of 160GPa (for λ ~ 30) which compares favourably with the chain modulus value of ~ 300 GPa estimated theoretically or measured by neutron diffraction and X-ray measurements under stress. The magnitude of the γ relaxation is significantly reduced for high draw ratio samples, but the α relaxation is still very significant (and consistent with this, there is appreciable creep at high temperatures).

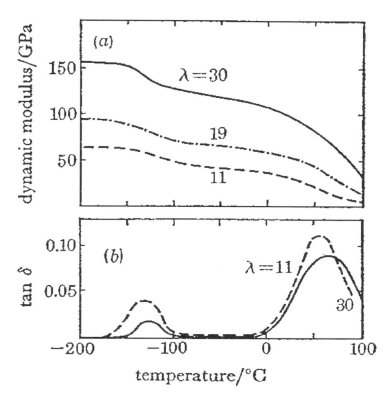

Figure 5.15   (a) Dynamic modulus and (b) tan δ plotted against temperature for drawn
                 LPE at indicated draw ratio λ. Reproduced by permission of The Royal
                 Society.

A quantitative interpretation of the dynamic modulus and loss has been given by Gibson et
al.[23], on the basis of the intercrystalline bridge model, and a fibre composite model,
where the reinforcing elements (the fibres) are identified with clumps of crystalline regions
of diameter ~ Å. Using the SAXS and WAXS measurements to estimate the long period
and the average crystal lengths respectively, it was shown that quantitative predictions for
the fall in the tensile storage modulus with temperature and the magnitude of tan δ could be
obtained.

The dynamic mechanical behaviour of the ultra high modulus PP[37] and POM[38] is
qualitatively similar. In both cases the storage modulus at high draw ratios (λ~20) rises to
about half the theoretically estimated value. This is 25 GPa compared with 42 GPa for PP
and 65 GPa compared with 106 GPa for POM. In PP the β relaxation is very markedly

reduced with increasing draw ratio and a similar result is observed for the γ relaxation in POM. In both cases this reduction has been attributed to the reduced mobility of the non-crystalline regions due to their increased orientation and the production of taut tie molecules.

5.1.6.2 Creep behaviour of polyethylene fibres

It was recognised very soon after the discovery of the melt-spun high modulus polyethylene fibres in 1973, that a major disadvantage was their very poor creep resistance, and this was particularly the case for the very low molecular weight polymers which were selected for initial commercial exploitation. It was quite quickly discovered that the creep behaviour improves with increasing molecular weight, even more dramatically than the strength. However, despite the invention of the gel-spun fibres in 1979, creep is still a limitation of any polyethylene fibre, and a working rule is that these fibres can only be expected to sustain without permanent deformation, a stress equivalent to about one tenth of their breaking stress.

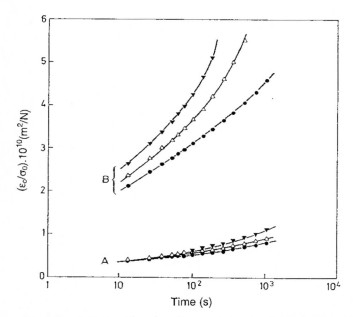

Figure 5.16   Creep compliance $(\varepsilon_c / \sigma_0)$ for drawn polyethylene monofilaments at 0.1 (•), 0.15 (Δ) and 0.2 GPa (▼) applied stress σ_0 as a function of time; — are least squares fit to the mechanical model of Figure 5.17(a), A, λ = 30; B, λ = 10.   Reproduced by permission of Elsevier Applied Science.

In a series of publications Ward, Wilding and several co-workers[39-43] described the creep and recovery behaviour of a range of melt spun drawn polyethylene fibres, where the key variables of molecular weight and draw ratio were examined, together with other factors, notably copolymer content and radiation cross-linking were also considered. A primary result shown in Figure 5.16, is the reduction in creep compliance with increasing draw ratio and a corresponding reduction in the degree of non-linear behaviour, so that the creep compliance becomes closer to being independent of the stress level. It was also found that to a first approximation there are two independent components to the creep response, a recoverable part which has a comparatively short time constant, and a non-recoverable part which is markedly dependent on stress, so that for low stress levels the creep strain is totally recoverable to a very good approximation[39]. This behaviour can be well described by the simple model representation of Figure 5.17(a).

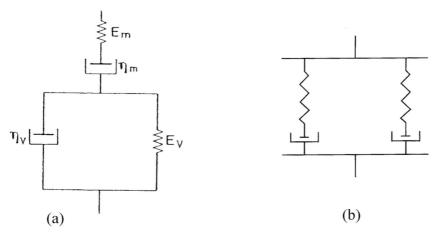

(a)   (b)

Figure 5.17   Schematic representation of the four element models.

The creep response ε_c at a constant applied stress σ_o can therefore be described by the equation

$$\varepsilon_c = \frac{\sigma_o}{E_m} + \frac{\sigma_o}{E_v}\left(1 - e^{-t/\tau}\right) + \frac{\sigma_o t}{\eta_m} \qquad (6)$$

where E_m, E_v, η_m and η_v correspond to the springs and dashpots of Figure 5.17(a) and $\tau = \eta_v/E_v$ is the retardation time of the Voigt element. The validity of this simple representation is illustrated in Figure 5.18 which shows creep curves fitted to best values for the model parameters E_m, E_v, η_m and η_v at each stress level and predicted recovery curves based on those parameters. The values obtained for E_m and E_v increase with draw ratio but are comparatively insensitive to stress level. This is consistent with the dependence of the modulus (the short term response, say, a 10 sec isochronal value) on the draw ratio. Although both η_v and η_m depend on stress level, this is much more marked in the case of

$\eta_m$where the stress dependence was shown to correspond to that expected for a thermally activated process.

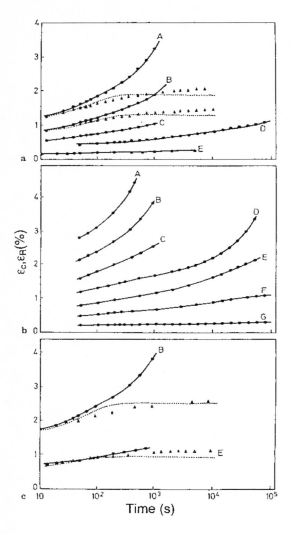

Figure 5.18   Creep ε_c (•) and recovery strain ε_R (▼) versus time for (a) Rigidex 50 monofilaments, λ=20:   A, 0.2; B, 0.15; C, 0.1; D, 0.05; E, 0.025GPa and (b) and (c) H0-20-54 λ=20:   A, 0.5; B, 0.4; C, 0.3; D, 0.2; E, 0.15; F, 0.1; G, 0.05GPa.
(——) in (a) and (c) are least squares fit to the mechanical model of Figure 5.17(a) and the dotted lines are the respective recovery curve; (——) in (b) are visual fits to the data. Reproduced by permission of Elsevier Applied Science.

Further work showed that, after an initial period of rapid creep response which incorporates the time dependent Voigt element response, the creep rate at moderate and high stress levels reaches a constant value. This has been termed the plateau creep rate $\dot{\varepsilon}_p$ and is very well represented by the Eyring equation (Eq. 7):

$$\dot{\varepsilon}_p = \dot{\varepsilon}_o \exp-\frac{\Delta H}{KT}\sinh\frac{\sigma v}{KT} \qquad (7)$$

where ΔH is the activation energy and v is the activation volume for the thermally activated process.

For high molecular weight fibres, copolymer fibres and radiation cross-linked fibres there is a continuous fall in the creep rate with time for low and moderate stress levels i.e. there is no permanent deformation represented by the plateau creep rate and the strain is totally recovered when the stress is removed. This is clearly of some practical importance as it enables the definition of a "critical stress" below which fibres under load creep to an equilibrium strain rather than extend continuously to failure.

For this reason it was suggested by Ward that a useful representation of the data is to produce a plot of strain rate versus strain. Ward called this representation a Sherby-Dorn plot, after their research paper on creep in polymethylmethacrylate. There is, however, no general similarity with Sherby and Dorn's results, in that first they did not observe a plateau creep and secondly their initial creep response (the Voigt element component in PE fibres) could be represented by thermally activated creep (equation 7) i.e. permanent deformation, which is not true for PE fibres.

Sherby Dorn plots for a typical fibre are shown in Figure 5.19 where the total creep rates $\dot{\varepsilon}_c$ are plotted as a function of strain ε_c for different stress levels. From this plot a second plot is produced showing the plateau creep rate on a logarithmic scale as a function of stress σ. It can be seen that this plot is not linear, as would be required if $\dot{\varepsilon}_p$ could be described by the simple thermally activated process of equation (7). The simplest explanation of this result[40] is that $\dot{\varepsilon}_p$ is the sum of two thermally activated processes acting in parallel, one process with a large activation volume where σv is large compared with KT, so that we can write

$$\sigma = \frac{2.3KT}{v_1}\left[\log\dot{\varepsilon}_p - \log\frac{[\dot{\varepsilon}_o]_1}{2}\right] + \frac{\Delta H}{KT} + \frac{KT}{v_2}\sinh^{-1}\left[\frac{\dot{\varepsilon}_p}{[\dot{\varepsilon}_o]_2}\exp\frac{\Delta H_2}{KT}\right] \qquad (8)$$

where the subscripts identify the quantities similar to those of equation (7) referring to the two processes. This representation is shown schematically in Figure 5.17(b).

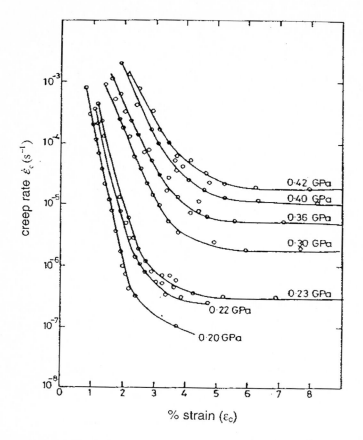

Figure5.19 Sherby-Dorn plot for a typical sample (Linear polyethylene H020, $\lambda = 20$) at indicated stress levels. Reproduced by permission of Elsevier Applied Science.

Ward and co-workers[40-42] have used this representation to compare creep data for a wide range of PE fibres to explore the effects of draw ratio, molecular weight, copolymer content and cross-linking. Process 2 is most affected by draw ratio with the activation volume falling from about 400 Å^3 to about 100 Å^3 as the draw ratio is increased from 10 to 30. This process dominates the creep behaviour at high stress and its temperature dependence could be readily determined. For all samples values in the range of 30 Kcals/mole were obtained similar to that for the α-relaxation process. Ward and Wilding[42] suggested that it is the c-shear process, including the movement of a defect such as a Reneker defect, through the crystalline regions. The fall in ν_2 with increasing draw ratio, suggests that the activated event becomes more localised as the structure reaches greater perfection.

Process 1 is much less well defined than Process 2, because it dominates the behaviour at low stress levels, where accurate measurements of creep are difficult to achieve. Process 1 does contribute more to the creep response with increasing molecular weight and copolymer content, and also for cross-linked samples. It has therefore been tentatively suggested by Ward and co-workers[42] that it relates to a molecular network, which acts in parallel with the crystalline structure responsible for the high modulus. This suggestion is consistent with the quantitative interpretation of the large negative thermal expansion of the PE fibres, which can be explained on the basis of the contraction forces of a rubber-like molecular network which become more important as the crystalline structure softens with increasing temperature. (See below).

From a practical engineering viewpoint any change in structure which increases the contribution of Process 1 is valuable, because it leads to an increase in the 'critical stress'.

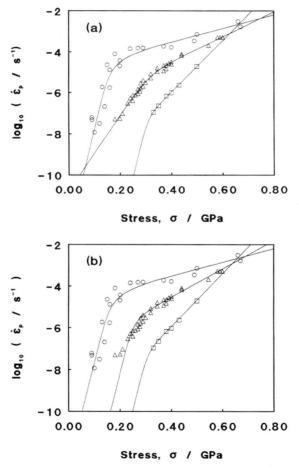

Figure 5.20 Log plateau strain rate versus stress for drawn polyethylene monofilaments: (○) Alathon 7030; (Δ)Rigidex 002-55; (□) Rigidex 002-47; (a) Free fits to data (b) constrained fits. Reproduced by permission of John Wiley and Sons.

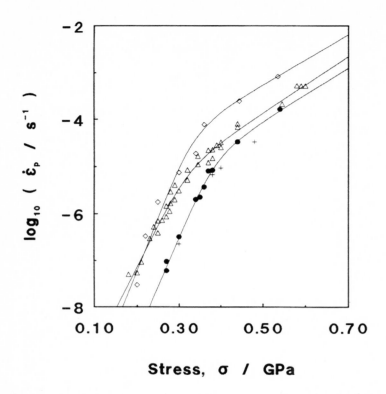

Figure 5.21   Log plateau strain rate versus stress for Rigidex 002-55 drawn
monofilaments: (Δ) Unirradiated; (◊) 10MRad in nitrogen; (•) 0.5MRad
in acetylene; (+) 1MRad in acetylene. Reproduced by permission of
John Wiley and Sons.

Figure 5.20 shows the effect of branch content and Figure 5.21 the effect of cross-linking.
It can be seen that in terms of an analysis based on equation (7) increasing branch content
affects both Process 1 and Process 2, giving rise to an increasing contribution from Process
1 and increasing the activation volume of Process 2, by limiting the size and perfection of
the crystalline regions, but also more importantly reducing the contribution of this process,
perhaps by restricting the c-shear process. The effect of cross-linking appears to be more
straightforward, giving an increasing contribution from Process 1, without much affecting
Process 2. The effect of increasing molecular weight is similar.

The improvements in creep performance obtained by increasing molecular weight and
branch content are very significant, but there is a concomitant reduction in the maximum
draw ratio, which can be a severe limitation for the melt processing/drawing route as
distinct from the gel spinning/drawing route when high molecular weight polymers (but not
branched polymers) can be successfully drawn to high draw ratios. Moreover, the creep
performance, especially at high temperatures, is still only moderate compared with that of

higher melting point polymers (polyesters, polypropylene, polyoxymethylenes). For these reasons, there has been a substantial research program at Leeds University on radiation cross-linking of PE fibres, which will now be reviewed[44-51].

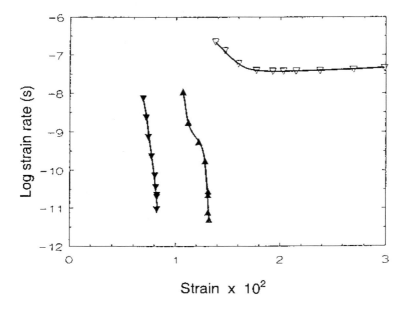

Figure 5.22   Log creep strain rate versus strain for melt spun and drawn polyethylene fibres cross-linked by γ-irradiation in acetylene and annealed in acetylene: − 0.17GPa stress applied at 20°C; (∇) untreated 0% gel; (▲) 3MRad dose, 75% gel; (▼) 6MRad dose, 82% gel. Reproduced by permission of Elsevier Applied Science.

The pioneering research of Charlesby and co-workers showed that PE could be cross-linked by either electron beam or γ-irradiation, both producing identical effects for a given radiation dose in Gy (or Mrads), providing that heating effects are not present. The practical issue is that the irradiation treatment leads to both cross-linking which is advantageous in terms of creep and high temperature performance, and chain scission which is disadvantageous because it leads to a reduction in overall strength (and creep). A significant breakthrough in this respect was made by the discovery that irradiation in an atmosphere of acetylene promoted cross-linking by initiating chain reactions which could produce a given degree of crosslinking for a much lower radiation dose[44]. The chain scission reactions, on the other hand, depended only on the total radiation dose, hence, high degrees of cross-linking and major improvements in creep performance can be obtained without significant loss of fibre strength. Figure 5.22 illustrates the dramatic improvement in creep behaviour which can be obtained, and Figure 5.23 shows that for the dose levels required there is only a small strength loss. It is important to note that for practical purposes it is essentially to irradiate the fibre after drawing. Irradiation before drawing has been shown to lead to improved creep, but the draw ratio which can be achieved is severely limited because of the formation of too many cross-links.

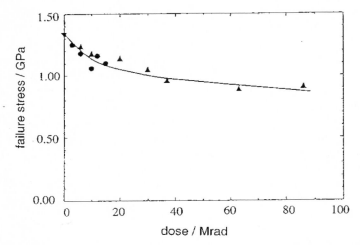

Figure 5.23   Brittle failure stress versus dose for cross-linked melt spun and drawn polyethylene fibres, E – beam irradiation, annealed 6h at 110°C:
(▼) untreated; (▲) irradiated in nitrogen, annealed in nitrogen; (•) rradiated in acetylene, annealed in acetylene.   Reproduced by permission of Elsevier Applied Science.

In a series of papers, Jones, Ward and co-workers[48-53] have explored the reasons for the effectiveness of the use of an acetylene atmosphere during irradiation.   Electron spin resonance and ultraviolet visible spectroscopy were used to show the formation of diene, triene and tetraene following irradiation in acetylene[48].   It was concluded that polymer bridges, predominantly diene, form cross-links between radical pairs[49].   Computer modelling[50] subsequently confirmed that the formation of the diene bridges is consistant with the anticipated radical pair distances on adjacent PE chains of ~ 5.6Å.

Recent studies have involved the time-dependent recovery behaviour[53], which somewhat surprisingly, was found to be closely similar for a range of samples differing in molecular weight and/or draw ratio.   The recovery behaviour was very well described in terms of the model where two thermally activated processes are acting in parallel.   In recovery, however, only two elements of the initial model, the spring E_2 and the dashpot V_1 are involved.

5.1.7   Thermal properties

5.1.7.1 Melting behaviour

It was initially anticipated that the high modulus melt spun PE fibres would show the very high melting point of chain extended PE. Low molecular weight grades of LPE drawn at 75°C showed melting temperatures of 138-139°C at the highest draw ratios, whereas higher molecular weight polymers drawn at 115°C showed melting temperatures as high as 141.5°C. It is clear from morphological studies that the higher melting temperatures of 139°C and higher are not to be associated with extended chain material. Detailed DSC measurements were undertaken by Clements et al.[54,55] to explore the effects of molecular weight and draw ratio. All the samples showed a dependence of melting temperature on heating rate, an effect which is generally known as superheating. For high molecular weight highly drawn PE, this could lead to melting temperatures as high as 145°C. The superheating effects were attributed to configurational constraints on the amorphous regions due to the very high degrees of molecular alignment being achieved.

5.1.7.2 Shrinkage and shrinkage force

Measurements of irreversible shrinkage and the magnitude of the shrinkage force as a function of draw ratio[56] are consistent with the stretching of a molecular network, which is central to understanding the drawing behaviour, as discussed above. On heating the drawn samples to a temperature above the melting point, complete retraction is observed, suggesting the continuity of at least part of the initial molecular network during the whole drawing process. This conclusion is supported by recent neutron scattering results of Barham[28].

At temperatures significantly below the melting point (120°C compared with 135°C) the shrinkage of the highly drawn fibres is quite low (~5%) and this is attributed to the degree of crystal continuity provided by the intercrystalline bridges (or long crystals). There is, however, a significant development of shrinkage force at comparatively low temperatures, generally just above the initial draw temperature. The shrinkage forces are ~ 10-20MPa, and their development with temperature and rapid decay as the temperature is increased into the melting range at 120°C, is reminiscent of similar results for amorphous polymers such as PET.   These observations confirm the existence of a molecular network which has importance consequences for the thermal expansion behaviour now to be discussed.

5.1.7.3 Thermal expansion

It is well known from X-ray measurements that the c-axis of the crystalline regions in PE shows contraction with increasing temperature.   In the highly drawn PE fibres, the surprising result was obtained that the thermal coefficient of expansion in the draw direction was significantly more negative than the X-ray diffraction results suggest[57]. Values of -

30 x $10^{-6}K^{-1}$ were obtained, compared with a value of -12 x $10^{-6}K^{-1}$ for the c-axis contraction.

Orchard et al.[57] showed that this surprising result was consistent with a model in which the structure consists of two components acting mechanically in parallel. The first component is responsible for the polymer stiffening and can be considered as the intercrystalline bridge structure of Figure 5.9c. The second component is a large scale molecular network which we have seen to be responsible for the recoverable shrinkage and the shrinkage force, and it gives rises to an internal stress. Orchard et al showed that their model provided a quantitative correlation between the observed increase in the magnitude of the negative axial thermal expansion coefficient and the change in stiffness with temperature. Moreover the model predicted values for the internal shrinkage stress which agreed well with the measured values of 10-30 MPa at 76°C.

5.1.8   Surface treatment

The chemical inertness of polyethylene is an advantage in many respects, especially for biomedical applications and where the material is exposed to acids, alkalis and sunlight. It also means that the fibres possess poor intrinsic adhesion to other materials which gives rise to very low interlamellar shear strengths (ILSS) for fibre/resin composites. This has led to extensive studies of surface treatment for both melt spun and gel spun PE fibres, predominantly using plasma treatment in an atmosphere of a chosen gas.

The earliest research in this area was by Ladizesky and Ward[58] on the Leeds melt spun fibres but there has been much further research, with particular emphasis on the chemical application of PE fibres to the reinforcement of acrylic resins for dental plates.

Ladizesky and Ward[58] studied the pull-out adhesion of both chemically treated and plasma treated melt spun PE monofilaments. For low draw ratio monofilaments, chromic acid etching was equally effective to plasma treatment in an atmosphere of oxygen, but for high draw ratio monofilaments plasma treatment was very much the more effective. For this reason, subsequent research by Ward and co-workers[59,60] focussed on plasma treatment with different gases (oxygen, helium, argon and CF_4). Although significant improvements in pull-out adhesion were obtained with argon and helium gas, the best results were obtained with oxygen gas, so this was adopted in all future Leeds research. A combination of pull-out tests and physico-chemical studies (gel formation, ESCA, contact angle measurements) by Tissington et al.[61] showed that the improved fibre-resin adhesion arises from three factors.

(1)   General oxidation of the surface, shown by determination of surface energies by contact angle measurements[60] to give improved surface melting.

(2)   Cross-linking of the surface. A significant gel fraction was obtained, suggesting cross-linking of the surface which removes a weak boundary layer.

(3)   Surface-pitting which is revealed SEM photographs. The oxygen treatment produces a remarkable cellular structure (Figure 5.24). The resin penetrates the surface pits and gives rise to mechanical keying. Ladizesky and Ward[59]

presented a theoretical analysis for the two stage failure of samples with pitted surfaces.

The ILSS values for polyethylene composites are increased from about 15 MPa to about 30 MPa by the optimum plasma treatment in oxygen gas. This value of 30 MPa has been shown to be close to the intrinsic shear strength of oriented polyethylene[62]. Woods and Ward[63] developed a continuous process for plasma treatment and also obtained results for PE fibres cross-linked using γ irradiation in the presence of acetylene. The cross-linked fibres shows a small increase in ILSS, consistent with removal of a weak boundary layer, but required similar plasma treatment to unirradiated fibre to reach a satisfactory level of ILSS. Woods and Ward[63] also showed that both melt spun and gel spun fibres are similarly affected by oxygen plasma treatment. The plasma treatment of gel spun fibres has been extensively reported by other workers, notably Hild and Schwartz[64] and Li and Netravali[65], and a recent review of plasma treatment of both melt spun and gel spun fibre has been presented by Ladizesky and Ward[66].

5.1.9   Applications of melt spun PE fibres

The key features of the melt spun PE fibres include high tenacity and modulus, good abrasion resistance and excellent resistance to sunlight and most chemicals and solvents. Although the absolute values of tenacity and modulus are significantly lower than for the gel spun PE fibres, the melt spinning process is intrinsically less costly, so that the melt-spun fibres can be seen as complementary, providing that the comparison can be made on the basis of comparable production volume.

The primary applications for the melt spun fibres[67] include safety belts and ropes, dental floss and protection clothing where the cut resistance is comparable to that of the Spectra gel spun fibre and Kevlar. Other applications include fishing nets, sail cloth and sewing thread.

5.1.10 Composites

Ladizesky and Ward followed their adhesion studies with an assessment of the behaviour of the epoxy resin composites[68] incorporating the melt-spun PE fibres, either as a continuous filament yarn or woven fabric. The composites showed very satisfactory values of stiffness and strength, and high energy absorption in Charpy impact tests. As anticipated, the interlaminar shear strength of the composites could be significantly increased by plasma etching of the fibres in oxygen gas. Although the plasma treatment reduced resin cracking in flexural and impact tests, the impact energies were not greatly affected, because these relate primarily to deformation of the fibres. Similar results were obtained for polyester resin/fibre composites[69].

It was of particular interest to explore hybrid composites[70], where the PE fibres are combined with either glass or carbon fibres. These hybrid composites showed unusual and potentially very useful combinations of mechanical properties, in terms of combining high tensile modulus and strength, adequate compressive strength and high impact energy, including the capability of withstanding high levels of deformation without disintegrating. It is particularly important that the hybrid composites, especially in the case of the PE/carbon fibre hybrids do not shatter on impact and show high resistance to disintegrating in bending and compression.

Recent studies of the effect of plasma surface treatment on the impact behaviour of PE fibre/epoxy resin composites have been undertaken for both melt-spun and gel spun fibres[71,72]. It was shown that in both cases the improvement in fibre/resin adhesion due to plasma treatment reduced delamination during impact, but there was a corresponding reduction in energy absorption.

Further studies of hybrid composites reinforced with melt-spun PE fibres and glass fibres[73], showed that although 100% untreated PE fibre composites had the maximum impact energy absorption, when other properties such as flexural strength are included, a plasma treated PE fibre composite with a few layers of glass at the centre offers the best compromise of properties.

5.1.11 Hot compaction

A new development of scientific and technological interest, is the discovery that solid sections of highly oriented melt-spun PE fibres can be produced by a simple hot compaction process[74]. In the initial research a unidirectional array of fibres was heated in a matched metal mould to a temperature on the range 136-140°C which is just below the peak melting temperature of the fibres, under a comparatively low pressure (100ps $\cong$ 0.7MPa) and allowed to soak for a few minutes. A high pressure (typically 3000psi $\cong$ 21MPa) was then applied for a few seconds, before removing the mould from the hot press and cooling to room temperature.

Transmission electron microscopy on compacted samples etched with permanganic reagent[75], showed that successful hot compaction involves the melting and recrystallisation of a thin layer of material on the surface of the fibres. In this case only 10-20% of the fibres are melted, and all the melting occurs on the fibre surface so that the recrystallised material forms the matrix of a polyethylene fibre/polyethylene composite. The electron microscopy studies show that at optimal compaction temperature (~138°C) the lamellae recrystallised from the molten polymer share the c-axis orientation of the fibres on which they nucleate, whereas at higher compaction temperatures twisting of the lamellae is observed and a corresponding reduction in composite stiffness. The recrystallised material is also clearly identified in DSC measurements, where a lower melting point fraction can be observed. The hot compaction of gel spun PE fibre (SPECTRA 1000, Allied Signal Corporation) has also been studied[76]. In this case it is necessary to heat the fibres to a temperature where substantial fibre melting occurs and consequent loss of fibre modulus and strength. The mechanism of fibre to fibre bonding appears to involve a combination of mechanical interlocking and fibre to fibre fusion akin to spot welding.

Table 5.1.1 shows a comparison of the stiffness constants of the compacted PE fibre sheets with die-drawn sheet, theoretical estimates based on molecular modelling and a PE fibre/ epoxy composite.   The elastic constants were obtained in all cases by ultrasonic sound velocity measurements[77].   The extrapolated fibre constants were obtained by hot compaction at different temperatures to produce different properties of melted material determined from DSC measurements.

Table 5.1.1 Comparison of the stiffness constants for compacted PE fibres, etc

Elastic constant: /Process	C_{33}	C_{11}	C_{13}	C_{12}	C_{44}
Compacted PE fibre	62.3	7.16	5.09	4.15	1.63
Die Drawn Sheet (draw ration 20)	66.0	6.90	4.40	3.90	1.60
PE fibre/epoxy composite	54.8	7.62	5.89	4.39	1.71
Extrapolated fibre constants	68.5	7.17	5.16	4.02	1.59
Theoretical values for PE fibre elastic constants	290	9.15	5.15	3.95	2.86

The results shown in Table 5.1.1 confirm the effectiveness of the hot compaction process, especially with regard to the conventional PE/epoxy composite.   The deformation of a complete set of elastic constants enables a valuable comparison with theoretical estimates[78]. The measured pattern of anisotropy compare well with theoretical estimates. As expected, the experimental value of C_{33} is lower than the predicted value due to the imperfect structure of the melt-spun and drawn fibres.

For commercial applications of the hot compaction process a woven fabric gives a better balance of properties, and it has been shown that single step process at comparatively low pressures (100-400psi) in a hot press or an autoclave gives satisfactory compaction. It has also been shown that cross-linked fibres show a wider temperature window for satisfactory compaction, as well as improved creep and high temperature behaviour[79]. The very low absorption of high frequency radiation in 100% PE composites makes the hot compacted PE a viable material for radomes and medical imaging screens. Protective sports goods such as shin pads and shoulder guards have also been made from hot compacted PE.   These applications use the additional quality of the hot compacted sheets of post formability i.e. the ability to postform flat sheet into a given shape.

References

1.   J.M. Andrews and I.M. Ward, J. Mater. Sci., **5**, 411 (1970).
2.   G. Capaccio and I.M. Ward, Nature Physical Science, **243**, 143 (1973) Brit. Patent Appl. 10746/73 (filed 6.3.93).
3.   D.L.M. Cansfield, G. Capaccio and I.M. Ward, Polym. Eng. Sci., **16**, 721, (1976).
4.   W.N. Taylor Jr., and E.S. Clark, Polym. Eng. Sci., **18**, 518 (1978).
5.   G. Capaccio and I.M. Ward, Brit. Pat. Appl., 52644/74 filed 3 Oct 1973.
6.   E.S. Clark and L.S. Scott, Polym. Eng. Sci., **14**, 682, (1974).
7.   B. Brew and I.M. Ward, Polymer, **19**, 1338 (1978).
8.   G. Capaccio, T.A. Crompton and I.M. Ward, J. Polym. Sci., Polym. Phys. Edn., **14**, 1641 (1976).
9.   G. Capaccio, T.A. Crompton and I.M. Ward, J. Polym. Sci., Polym. Phys. Edn., **18**, 301 (1980).
10.   J.L. Way, J.R. Atkinson and J. Nutting, J. Mater. Sci., **9**, 293, (1974).
11.   P.J. Barham and A. Keller, J. Mater. Sci., **11**, 27, (1976).
12.   L. Jarecki and D.J. Meier, Polymer **20**, 1078, (1979).
13.   L. Jarecki and D.J. Meier, J. Polym. Sci., Polym. Phys Edn., **16**, 2015 (1978).
14.   G. Capaccio and I.M. Ward, J. Polym. Sci., Polym. Phys. Edn., **22**, 475 (1984).
15.   D.L.M. Cansfield, G. Capaccio and I.M. Ward, Polym. Eng. Sci., **16**, 721 (1976).
16.   K. Tashiro, M. Kobayashi and H. Tadakoro, Polymer J., **24**, 899 (1992).
17.   W.N. Taylor and E.S. Clark, Polymer Eng. Sci., **18**, 518 (1978).
18.   A. Peterlin, Ultra High modulus Polymers, A. Ciferri and I.M. Ward (Eds) London, Applied Science Publishers 1979, Chapter 10.
19.   J. Clements, R. Jakeways and I.M. Ward, Polymer **19**, 639 (1978).
20.   C.J. Frye, M.G. Dobb, D.J. Johnson and I.M. Ward, J. Polym. Sci., Polym. Phys. Edn., **20**, 1677 (1982).
21.   G. Capaccio and I.M. Ward, J. Polym. Sci., Polym. Phys Edn., **19**, 667 (1981).
22.   G. Capaccio and I.M. Ward, J. Polym. Sci., Polym. Phys. Edn, **20**, 1107 (1982).
23   A.G. Gibson, G.R. Davies and I.M. Ward, Polymer **19**, 683 (1978).
24.   R.G.C. Arridge, B.J. Barham and A. Keller, J. Polym. Sci., Polym. Phys. Edn, **15**, 389 (1977).
25.   D.M. Sadler and P.J. Barham, Polymer **31**, 46 (1990).
26.   A.K. Taraiya, A.P. Unwin and I.M. Ward, J. Polym. Sci., Polym. Phys. Edn. **26**, 817 (1988).
27.   S. Jungnitz, Ph.D. Thesis, Leeds University 1983.
28.   P.J. Flory, J. Amer, Chem. Soc., **67**, 2048 (1945).
29.   P. Smith and P.J. Lemstra, J. Polym. Sci., Polym. Phys. Edn., **19**, 1007 (1981).
30.   P. Smith, P.J. Lemstra and J.P.L. Pijpers, J. Polym. Sci., Polym. Phys. Edn., **20**, 2229 (1982).
31.   W. Wu and W.B. Black, Polym. Eng. Sci., **19**, 1163 (1979).
32.   M. Hallam, D.L.M. Cansfield, I.M. Ward and G. Pollard, J. Mater. Sci., **21**, 4119 (1986).
33   M. Hallam, G. Pollard and I.M. Ward, J. Mater. Sci., Letters, **6**, 975 (1987).

34.   J. Amornsakchai, D.C.M. Cansfield, S.A. Jawad, G. Pollard and I.M. Ward, J. Mater. Sci., **28**, (1993).

35.   Y. Termonia, W.R. Greene and P. Smith, Polymer Commun. **27**, 295 (1986).

36.   J. Smook, W. Hamersma and A.J. Pennings, J. Mater. Sci., **19**, 1359 (1984).

37.   A.J. Wills, G. Capaccio and I.M. Ward, J. Polym. Sci., Polym. Phys. Edn., **18**, 493 (1980).

38.   B. Brew, J. Clements, G.R. Davies, R. Jakeways and I.M. Ward, J. Polym. Sci., Polym. Phys. Edn., **17**, 351 (1979).

39.   M.A. Wilding and I.M. Ward, Polymer **19**, 969 (1978).

40.   M.A. Wilding and I.M. Ward, Polymer **22**, 870 (1981).

41.   M.A. Wilding and I.M. Ward, Plastics Rubber Proc. Appl. **1**, 167 (1981).

42.   I.M. Ward and M.A. Wilding, J. Polym. Sci., Polym. Phys. Edn., **22**, 561 (1984).

43.   M.A. Wilding and I.M. Ward, J. Mater. Sci., **19**, 629 (1984).

44.   D.W. Woods, W.K. Busfield and I.M. Ward, Plast. and Rubb. Process. Applic., **5**, 157 (1985).

45.   P.G. Klein, D.W. Woods and I.M. Ward, J. Polym. Sci., Polymer Phys. Edn., **25**, 1359 (1987).

46.   D.W. Woods and I.M. Ward, Plastics Rubber and Comp. Processes and Applic., **18**, 255 (1992).

47.   J. Rasburn, P.G. Klein and I.M. Ward, J. Polym. Sci., **B**, Polym. Phys **32**, 1329 (1994).

48.   R..A. Jones, G.A. Salmon and I.M. Ward, J. Polymer Sci., **B**, Polym. Phys., **31**, 807 (1993).

49.   R.A. Jones, G.A. Salmon and I.M. Ward, J. Polymer Sci., **B**, Polym. Phys., **32**, 469 (1994).

50.   R.A. Jones, D.J.R. Taylor, R.E.T. Stepto and I.M. Ward, J. Polym. Sci., **B**, Polymer Phys., **34**, 901 (1996).

51.   R.A. Jones, I.M. Ward, D.J.R. Taylor and R.F.T. Stepto, Polymer **37**, 3643 (1996).

52.   R.A. Jones, D.J. Groves and I.M. Ward, Polym. International **44**, 300 (1997).

53.   S.A. Gordeyev and I.M. Ward, J Mater Sci, **34**, 4767 (1999).

54.   J. Clements, G. Capaccio and I.M. Ward, J. Polym. Sci., Polym. Phys Edn., **17**, 693 (1979).

55.   J. Clements and I.M. Ward, Polymer **23**, 935 (1982).

56.   G. Capaccio and I.M. Ward, Colloid Polym. Sci., **260**, 46 (1982).

57.   G.A.J. Orchard, G.R. Davies and I.M. Ward, Polymer **25**, 1203 (1984).

58.   N.H. Ladizesky and I.M. Ward, J. Mater. Sci., **18**, 533 (1983).

59.   N.H. Ladizesky and I.M. Ward, J. Mater. Sci., **24**, 3763 (1989).

60.   M. Nardin and I.M. Ward, Mater. Sci. Technol., **3**, 814 (1987).

61.   B. Tissington, G. Pollard and I.M. Ward, J. Mater. Sci., **26**, 82 (1991).

62.   N.H. Ladizesky and I.M. Ward, J. Mater. Sci., 23, 72 (1988).

63.   D.W. Woods and I.M. Ward, Surfaces Interface Analysis **20**, 385 (1993).

64.   D.N. Hild and P. Schwartz, J. Adhesion Sci. Technol., **26**, 897 (1992).

65.   Z. F. Li and A.N. Netravali, J. Appl. Polym. Sci., **44**, 333 (1992).

66.   N.H. Ladizesky and I.M. Ward, J. Mater. Sci., Materials in Medicine, **6**, 497 (1995).

67.   G. Farrow, Processing Properties and Applications of the High Performance Melt Spun Polyethylene Fibres, ACS Meeting, New Orleans, March 1996.

68.   N.H. Ladizesky and I.M. Ward, Composites Science & Technology, **26**, 129 (1986).

69.   N.H. Ladizesky and I.M. Ward, Composites Science & Technology, **26**, 169 (1986).

70.   N.H. Ladizesky and I.M. Ward, Composites Science & Technology, **26**, 199 (1986).

71.   B. Tissington, G. Pollard, I.M. Ward, Composites Science & Technology, **44**, 197 (1992).

72.   D.W. Woods, P.J. Hine, R.A. Duckett and I.M. Ward, J. Adhesion, **45**, 173 (1994).

73.   D.W. Woods, P.J. Hine and I.M. Ward, Composites Science & Technology, **52**, 397 (1994).

74.   P.J. Hine, I.M. Ward, R.A. Olley and D.C. Bassett, J. Mater. Sci., **28**, 316 (1993).

75.   R.A. Olley, D.C. Bassett, P.J. Hine and I.M. Ward, J. Mater. Sci., **28**, 1107 (1993).

76.   R.J. Yan, P.J. Hine, I.M. Ward, R.H. Olley and D.C. Bassett, J. Mater. Sci., **32**, 4821 (1997).

77.   P.J. Hine and I.M. Ward, J. Mater. Sci., **31**, 371 (1996).

78.   D.J. Lacks & G.C. Rutledge, J. Phys. Chem., **98**, 1222 (1994).

79.   M.J. Bonner, P.J. Hine, and I.M. Ward, Plast. Rubber Comp. Proc. and Apps., **27**, 258, (1998).

5.2   Aramid Fibres

J.E. McIntyre
School of Textile Industries, University of Leeds, UK

5.2.1   Historical introduction

In terms of scale of production, the most important step-growth synthetic polymers used in fibre manufacture have been, from their first commercialisation until the present day, two aliphatic polyamides (nylon 6 and nylon 6,6) and an aromatic polyester (PET). A major factor in their success is their melting behaviour. All three melt at temperatures high enough to make them useful over a range of temperatures that approaches 200°C, yet sufficiently below their decomposition temperatures to permit them to be made by melt polymerisation and converted into fibres by a melt extrusion process.

It was clear from an early stage that in the polyamides the association of the amide groups, and in the aromatic polyesters the conformational rigidity of the aromatic *para*-phenylene groups, were factors that led to higher melting temperatures and to increased chain stiffness. If these properties were enhanced, the products could be expected to open up new, more specialised, and therefore perhaps more valuable uses. Consequently there was an incentive to evaluate polyamides containing either para-phenylene groups or groups that possessed similarly higher conformational rigidity. It was soon obvious that most, if not all, of the polymers based on such structures did have very considerably higher melting temperatures, but they were not melt-processable at either the polymerisation or the extrusion stage. It was therefore necessary to use solvents for both stages. Interfacial polymerisation and interfacial spinning, two solvent based processes that would obviate the need to find solvents that dissolved the polymers themselves, did not provide the basis for viable processes. Solvents for the polymers themselves that were suitable for one stage or both were sought and gradually found.

The first aromatic polyamide fibre to be commercialised was poly(*m*-phenylene isophthalamide)(MPIA) (Figure 5.24 (1)), under the trade-name Nomex, by the U.S. firm DuPont. The basic patents that describe its production were filed in 1957 and the fibre itself became available, at first on a relatively small commercial basis, in 1962. In 1972 the Japanese firm Teijin introduced a fibre named Conex (or Teijinconex) that had the same chemical stucture but a somewhat different production route. A third product of this chemical structure, Apyeil, from the Japanese firm Unitika, was launched commercially in 1986. A fourth such product, Fenilon, was produced on a small commercial scale in the former USSR. All of these products were aimed particularly at markets requiring fibrous structures possessing high resistance to combustion coupled with good dimensional stability at very high temperatures, which remain their main outlets.

A different chemical structure and production route were adopted for the same outlets by the French firm Rhône-Poulenc, who developed an aromatic polyamide-imide fibre named

Kermel (Figure 5.24 (2)), launched commercially in 1971. Both MPIA and Kermel possess tensile properties that are within the range of other fibres already available. Although in the case of modulus their properties lie towards the top end of that range, they cannot nowadays be classed as high-modulus fibres and are only briefly described here for comparative purposes.

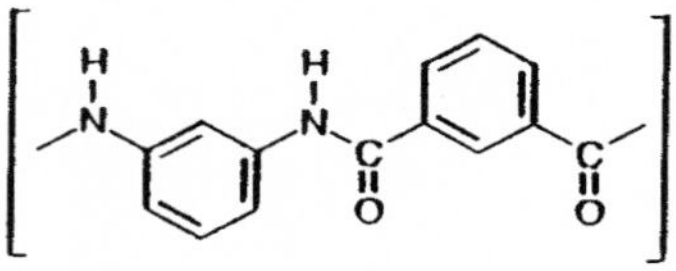

(1)

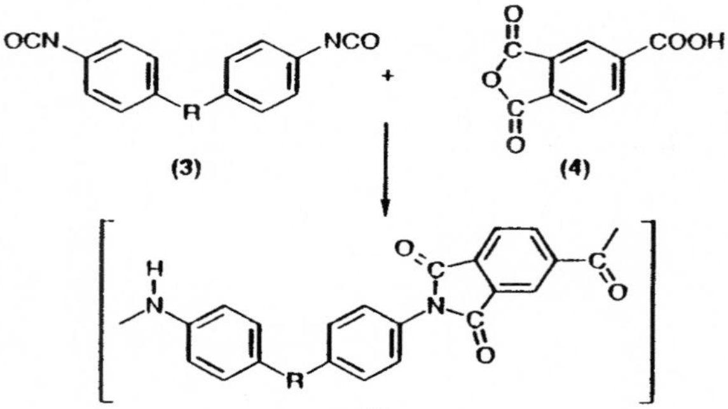

(3)   +   (4)

(2)

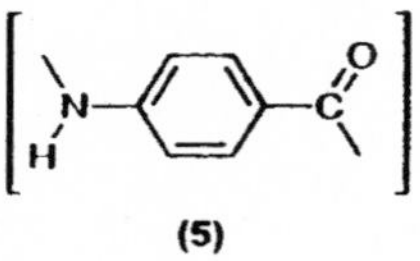

(5) 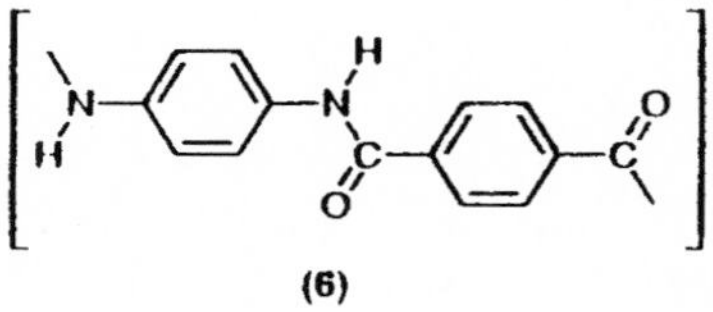   (6)

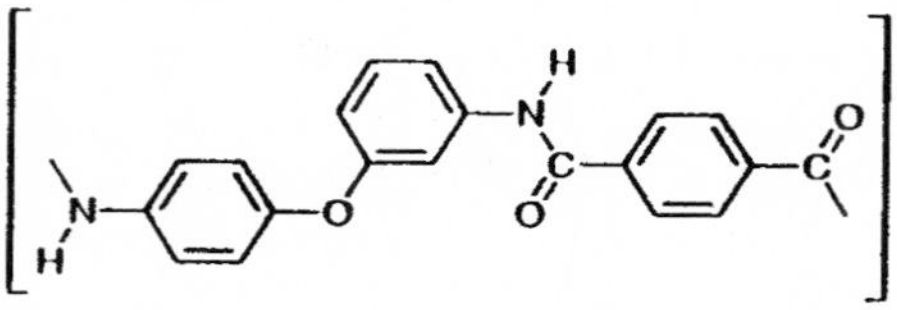

(7)

Figure 5.24   Formulae

In 1970, DuPont announced a new high-strength, high-modulus fibre called Fiber B, which at that stage probably consisted of poly-1,4-benzamide (Figure 5.24 (3)), their patents for which had been filed in 1966. When the new fibre became available commercially under the name Kevlar it consisted instead of poly(*p*-phenylene terephthalamide) (PPTA) (Figure 5.24 (4)) and appeared as two variants, both of very high tenacity but one of higher elongation to break and the other of considerably higher modulus. DuPont's patents covering the polymer synthesis were filed in 1968, and those covering the fibre spinning and heat treatment were filed in 1971. Akzo began production on a full plant scale in 1987 of a fibre of the same chemical structure, called Twaron, which was later transferred to Twaron Products, part of Acordis.

Teijin introduced a high-strength, high-modulus fibre called Technora (at first HM-50) in about 1974. It was made from a copolymer containing 50 moles % each of the repeating unit of PPTA, Figure 5.24 (4), and the repeating unit, Figure 5.24 (5).

Since all these fibres were very different in their properties from previous polyamide fibres, the existing generic names nylon and polyamide were deemed inappropriate and a new generic name, aramid, was coined for them. However, the internationally agreed meaning of this term and that current in the U.S.A. are different. In 1974, the U.S. Federal Trade Commission (FTC) defined aramid fibres as those having in the polymer chain recurring amide groups, at least 85% of which are joined directly to two aromatic groups, whereas the International Standards Organisation (ISO) definition further permits imide groups to be substituted for up to 50% of the amide groups, thereby including Kermel, which is not included in the FTC definition.

5.2.2   Heat- and flame-resistant meta-aramid fibres

5.2.2.1 Poly(metaphenylene isophthalamide) fibres

When Nomex and Conex were launched commercially, it became apparent that the two products, although both essentially MPIA, were being made by different processes. In the case of Nomex, the polymer is made from *m*-phenylenediamine and isophthaloyl chloride by a solution polymerisation technique using dimethyl acetamide (DMA) as the solvent. An acid acceptor, calcium oxide or hydroxide, is added to take up the hydrochloric acid produced by the polymerisation reaction, and it is thereby converted into calcium chloride. Fibres are then dry spun from a spinning solution that contains about 20% by weight of polymer, along with about 9% of calcium chloride. The washed fibres are drawn at a high temperature of about 300°C, which results in crystallisation [1].

More recently, DuPont have described a process in which the dry-spun filaments are incompletely washed and then simultaneously thoroughly washed and drawn, typically by 4:1 at 90°C, in a counter-current extraction system. Tow formation and associated processes are carried out at similarly low temperatures, so that a water-swollen amorphous tow is obtained [2]. While in the water-swollen state, these fibres can be dyed with basic dyes by padding a dye solution on to a tow then heating the tow with steam, first to about

120°C to diffuse the dye into the fibres, and then to about 165°C to crystallise the fibres and improve the dye fastness [3].

In the case of Conex, the early patents describe a more complex polymerisation process in which the *m*-phenylenediamine and isophthaloyl chloride are dissolved in tetrahydrofuran, which is not a solvent for the polymer. Consequently a prepolymer of modest molecular weight precipitates, forming a slurry. This slurry is then contacted with an aqueous solution of an acid acceptor such as sodium carbonate or triethylamine [4]. The product is separated from the other constituents, dried, and dissolved at about 20% concentration in N-methylpyrrolidinone (NMP) by slurrying it in the cold solvent and then heating [5]. Fibres are wet spun from this solution into a coagulant consisting of concentrated (about 40%) aqueous calcium chloride at 80-100°C [6]. Upward spinning with co-current coagulant flow has been described. The fibre is washed and then drawn in two stages, first by about 3:1 in boiling water, then, after drying, by about 1.4:1, at 310°C [7]. A later patent describes dry-jet wet spinning of a hot spinning solution with two-stage coagulation, first in cold water, then in hot aqueous calcium chloride [8].

Although the solutions containing calcium chloride that have been dry spun by DuPont exhibit higher solution stability than solutions without the salt, they have not previously been found suitable for wet spinning, which results in fibres containing large voids and requiring hot stretching for improvement of fibre properties. Solutions of MPIA for wet spinning have normally been based on removal of acid without salt formation, or on formation of insoluble and therefore removable salts, or on stagewise removal of soluble salts. However, by extruding a hot spinning solution in DMA containing calcium chloride, of the type normally dry spun, into a hot aqueous coagulant containing about 20% DMA and 40% calcium chloride at about 110°C, then spraying the fibres just after they leave the coagulant bath with an aqueous conditioning solution containing, for example, 40% of DMA and 9% of calcium chloride, the solvent content of the fibres is kept high enough to plasticise them sufficiently for them to be drawn to a draw ratio above 4:1 in a single stage in an aqueous solution of DMA. Without the conditioning step, efficient drawing is not possible. The fibres made in this way are non-crystalline, and therefore give much higher dye uptake, using carrier at 130°C, than hot-stretched wet-spun yarns, which are crystalline [9].

5.2.2.2 Polyamide-imide fibres

Up to 50% of the connecting groups in aramid fibres that are formed during polymerisation may be imide groups instead of amide, based on the ISO standard that defines the term *aramid*. One such structure is used to produce the commercial fibre Kermel, produced originally by Rhône-Poulenc. The name Kermel has been used to describe fibres from a range of polymers, the essential structure of which is that they are made from an aromatic di-isocyanate and trimellitic anhydride (TMA). Initially two different commercial products were made by using two different di-isocyanates, diphenyl ether-4,4'-di-isocyanate and diphenylmethane-4,4' di-isocyanate (MDI).[10,11,12] According to the Kermel company's literature, the di-isocyanate in use when a 'third-generation', higher-tenacity Kermel was introduced in 1993 was MDI, and at that time an improved polymerisation process led to a much narrower molecular weight distribution than previously and hence to improved fibre properties.

The polymerisation is carried out in an aprotic, dipolar solvent, DMA or NMP being preferred initially.   Recent patents use different di-isocyanates, notably phenylene-1,3-di-isocyanate or tolylene-di-isocyanate, have described the incorporation of minor amounts of terephthalic acid, isophthalic acid, or sulphoisophthalic acid in place of part of the TMA, thus increasing the proportion of amide present, and have used other solvents, notably γ-butyrolactone or N,N'-dimethylethyleneurea.   Fibres can be produced directly from the solutions of these polymers by either wet or dry spinning, but wet spinning has been, and probably still is, the preferred process for Kermel production [13].

A particular structural consideration with these aramids based on trimellitic anhydride is that the amide-imide (AI) sequences can be either head-to tail (i.e. AIAI or IAIA) or head-to-head (AIIA or IAAI) in any two successive repeating units, probably in a random order. This should lead to a reduced tendency to crystallise.

5.2.3   High-tenacity high-modulus fibres from anisotropic solution

5.2.3.1 Production and mechanical properties

The development of poly(p-phenylene terephthalamide) (PPTA) fibre followed closely upon that of poly(p-benzamide), which was dry-spun from an anisotropic solution in the solvent, dimethyl acetamide containing about 2% of lithium chloride, in which it was polymerised. The fibre so obtained had high tenacity and modulus,e.g. tenacity 0.77 N tex^{-1}, elongation to break 3 %; modulus 42 N tex^{-1}, and could be made still stronger (e.g. 1.77 N tex^{-1}) and stiffer (e.g. 106 N tex^{-1}) by annealing briefly at a temperature of about 550°C [14].   In the case of PPTA the solvent initially used for polymerising to a high molecular weight was a mixture of two aprotic dipolar solvents, N-methylpyrrolidinone (NMP) and hexamethylphosphoric triamide (HMPA), neither of which was adequate on its own. PPTA polymer could be made from p-phenylenediamine and terephthaloyl chloride in this solvent mixture at inherent viscosity (logarithmic viscosity number) values up to about 6 dLg^{-1}[15]. An alternative polymerisation solvent, NMP containing 10 to 20% of calcium chloride, was discovered later [16] and is now preferred because of concern about the possible toxicity of HMPA, and a few further solvents also capable of giving sufficiently high molecular weight have since been identified. However, these organic solvents are not suitable for spinning fibres from PPTA. It is necessary to isolate the polymer from the gel-like solution initially obtained and to re-dissolve it in sulphuric acid, which cannot be used directly for preparation of the polymer because of its reaction with the p-phenylenediamine component.

PPTA dissolves in sulphuric acid to form an anisotropic solution provided (i) that the molecular weight of the PPTA is above a certain limit, (ii) that the sulphuric acid is of a strength close to 100% (ordinary concentrated acid is not suitable), (iii) that the polymer concentration is above a certain limit but below a further, higher limit, and (iv) that the temperature is within a specific range [17]. Typical values for solutions used to extrude fibres are (i) a molecular weight corresponding with an inherent viscosity of about 6 dL g^{-1}; (ii) 100.1%; (iii) 19.4%; and (iv) 70-90°C. The solution can, however, conveniently be made at low temperatures, such as 0°C, at which 100% sulphuric acid is a solid, by thoroughly mixing the polymer and solid solvent [18]. The components react to form a crystalline complex of PPTA and sulphuric acid, which is suitable for storage and melts at about 70°C to give the anisotropic spinning solution.

The process used to manufacture high-strength, high-modulus fibres from these solutions of PPTA is dry-jet (air-gap) wet spinning into water or very dilute sulphuric acid [19]. This process has several advantages over conventional immersed-jet wet spinning, particularly when it is used in conjunction with co-current flow of the coagulant together with the nascent fibres through a vertical spinning tube positioned below the air gap and directly under the spinneret (Figure 5.25). It makes it possible to apply high elongational shear in the air gap and thus to orient the nematic domains in the extrudate before it reaches the coagulant, so that the fibres produced are highly oriented. It also makes it possible to establish a stable temperature differential between the extrudate, which is at a temperature in the range 70-90°C, and the coagulant, which is at a temperature in the range 0-5°C. This process leads to PPTA fibres of high tenacity, such as Kevlar 29, and of high modulus, but a further significant increase in modulus can be obtained by treatment of the filaments under tension at a very high temperature, up to 550°C, for only a few seconds [20]. This further process leads to PPTA fibres of very high modulus, such as Kevlar 49, but they have somewhat reduced elongation to break and work to break.

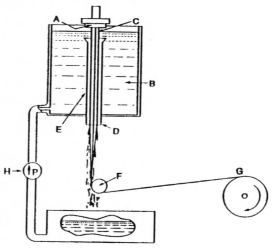

Figure 5.25 Air-gap (dry-jet) wet spinning with a spinning tube. A = spinneret; B = coagulant; C = filaments in air gap; D = spinning tube; E = filaments in spinning tube; F = haul-off roll; G = wind-up; H = coagulant recycle pump [adapted from Ref. 20]

In order to obtain still higher tenacity in the as-spun PPTA yarn, it is necessary to improve the uniformity of the fibres and the degree of orientation without reducing the molecular weight of the polymer. Higher tenacity achieved by increasing the degree of orientation is commonly accompanied by some reduction in elongation to break and increase in initial modulus. Methods adopted and used in conjunction by DuPont to increase the tenacity of Kevlar are based on arranging non-turbulent flow of coagulant into the top of the spinning tube, thus improving the uniformity of the fibres [21]; on increasing the spinning speed (i.e. the speed of the haul-off rolls) from about 400 m min^{-1} in the original process to above 800 m min^{-1} by injecting additional coagulant downwards into the spinning tube just after the filaments enter it [22]; on using spinneret holes of lower diameter, such as 50 μm [23]; and on applying a closely controlled fibre tension during the washing and drying stages [23].

These two last methods lead to high tenacity together with high elongation to break. If the drying tension is high, for example 3.5 gpd (0.31 N tex^{-1}), tenacities of above 28 gpd (2.47 N tex^{-1}) and as high as 31.3 gpd (2.76 N tex^{-1}) can be obtained [24]. In order to obtain still higher modulus without the reduction in tenacity associated with the original high-temperature process, the spun yarn is thoroughly washed and any residual acid content is neutralised. The yarn is not dried before being subjected to treatment at a high constant tension, typically 80% of the breaking tension, at a low temperature, below 50°C, to produce orientational drawing, and then at a higher temperature, such as 175°C, and a lower tension to dry it [25].

An alternative process for higher modulus is based on a short treatment of the undried spun yarn, swollen with water that is as neutral as possible, at a much higher temperature, about 600°C, for a short time, less than 3 s, in a hot turbulent air jet, which is an excellent heat-transfer device [26].   This process gives a high value (e.g. 79%) for an empirical crystallinity index and high apparent crystallite size (>7 5 nm) based on the width of the X-ray diffraction peak near to θ = 23°, and is capable of raising the modulus to values above 170 GPa (116 N tex^{-1}). It is also said to give an increase in molecular weight, whereas unwashed fibres give a reduction in molecular weight and tenacity when heated to high temperatures.

In order to obtain increased fatigue resistance, particularly for use in tyre cords, a coagulating bath temperature of 20-40°C is used, then the yarn is washed and neutralised under a defined low tension and dried at 100-200°C under a still lower tension. These yarns have relatively high elongation to break of 5% or above [27].

A greater range of PPTA fibre variants is now available as a result of process improvements such as those described above. In the case of Kevlar, there are products with higher tenacity (Ht; Kevlar 129), higher extension to break (He; Kevlar 119), higher modulus (Hm; Kevlar 149), and an intermediate modulus product that retains much of the higher work to break of Kevlar 29 (Hp; Kevlar 68). The properties of these fibres given in Table 5.2.1 are those provided by DuPont [28], and Figure 5.26 shows how they correspond with properties given in some of the relevant patents.

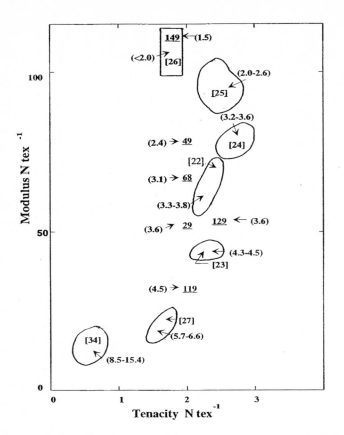

Figure 5.26 Modulus and tenacity (both N tex^{-1}) for various types of Kevlar and for PPTA
fibres made by various patented procedures. Kevlar types underlined; data
from patent examples lie within enclosed areas; patent references between
square brackets; range of elongations to break between round brackets.

Table 5.2.1. Physical properties of some Kevlar variants [28]

Type	Tensile Strength (GPa)	Tenacity (N tex^{-1})	Elongation to break (%)	Initial Modulus (GPa)	Moisture Regain (%)	Density (g.cm^{-3})
RI and 29	2.9	2.05	3.6	60	7	1.44
Ht (129)	3.3	2.35	2.9	75	7	1.44
He (119)	2.9	2.05	4.5	45	7	1.44
Hp (68)	2.9	2.05	3.1	90	4.2	1.44
49	2.9	2.05	1.9	120	3.5	1.45
Hm (149)	2.4	1.70	1.5	160	1.2	1.47

Twaron, originally available in HT and HM types, also appears in specialised versions such as Twaron 2300 for reinforcement of rubber hoses, with increased tensile and loop strength and flexibility, Twaron 1099 pulp fibres with increased specific surface area, and Twaron CT Microfilament.

Microfilament yarns are highly desirable for certain uses of PPTA continuous-filament yarns, notably in ballistic protection and hose reinforcement, where it is possible to obtain either improved performance in terms, for example, of loop strength, at the same yarn linear density, or equivalent performance at a lower yarn linear density, by reducing the diameter of the individual filaments. In addition, in agreement with Griffith fracture theory, filaments of lower linear density exhibit higher tenacity than otherwise equivalent filaments of higher linear density. It is calculated that values of linear density as low as about 0.01 dtex are required to attain half the ultimate tenacity of a flawless filament (16 GPa; 11 N tex^{-1}) calculated by extrapolating to zero filament linear density the tenacities of yarns made from PPTA of the usual commercial molecular weight. A theoretical value obtained by calculation from bond strengths and chain cross-sectional areas for fully oriented crystals is 29 GPa (20 N tex^{-1} [29].

In order to make fibres of lower linear densities, several approaches can be envisaged. One, use of lower polymer concentration in an otherwise unmodified process, is ineffective because it results in a lower degree of orientation in the fibes and hence gives lower tenacity and modulus. A second approach, increasing the spin-stretch ratio in the air gap, is limited by the onset of draw resonance, and hence of high variability of fibre diameter, at a critical value of spin-stretch ratio. Although it is possible to increase the critical spin-stretch ratio very substantially by cooling to very low temperatures in the air gap [30], such a process is not readily adaptable to commercial practice. A third approach, reducing the diameter of the spinneret holes, does not affect the diameter of the filaments if the feed rate per hole and the haul-off speed are unchanged. However, if the feed rate per hole is reduced proportionately to the reduction in cross-sectional area of the hole, the critical spin-stretch ratio above which draw resonance occurs remains approximately the same, so that filaments of lower linear density can be produced. Linear densities as low as 0.36 dtex are obtainable using spinneret holes of diameter 40 μm.

In defining a filament yarn composed of such fine filaments, Akzo-Nobel use, among other parameters, the g-value, which is a measure of the internal shear modulus [31]. In calculating this value, the Northolt version of an aggregate model [32],

$$\frac{1}{E_{son}} = \frac{1}{E_c} + \frac{<\sin^2 \phi>_E}{2g}$$

where E_{son} is the sonic modulus, E_c is the crystal modulus in the chain direction, and $<\sin^2\phi>_E$ is the orientation parameter averaged over the orientation distribution of the angle ϕ between the chain axis and the fibre axis, is modified to allow for the dependence of the orientation parameter on the tensile stress, σ_f, to

$$\frac{1}{E_{son}} = \frac{1}{E_c} + \frac{< \sin^2 \phi >_E}{2g + \sigma_f}$$

The value of g, which for the purposes of the patent must exceed 2.5 GPa, is calculated from data obtained by measuring the sonic modulus and the tensile stress at two strain levels, assuming a value for E_c of 220 GPA.

Fibrous pulps consisting of short, stalk-free fibres, up to about 10 mm in length and of diameter 0.1 to 50 µm can be produced without any extrusion step from PPTA blended with about 20% of poly(N-vinylpyrrolidinone) (PVP). A PVP of high molecular weight, e.g. 500,000, is dissolved in N-methylpyrrolidinone (NMP) containing calcium chloride, then PPTA is made in the usual way but using this mixture as the solvent. The polymer solution forms a very tough anisotropic gel, which can be broken up by agitation into highly fibrous particles. The gel is washed with water to remove the NMP, the salt and the HCl produced during the polymerisation, leaving a pulp consisting of a heterogeneous mixture of the two polymers. The PVP is believed to cause the polymerising PPTA to form oriented domains which ultimately result in pulp fibre formation, with the PVP associated with but surrounding the PPTA [33].

Although its earliest uses were in the high-strength, high-modulus field, PPTA also has excellent flame-retardant properties, to the extent that small proportions, such as 5%, and in some products as much as 23%, are blended with MPIA fibres to reduce the thermal shrinkage and increase the break-open resistance of the fabrics. For incorporation into conventional textile fabrics, the high modulus and low elongation to break of the PPTA fibres made by the established processes can create problems, since they make the fibres less suitable for processes such as textile spinning and give the fabrics a harsher handle. Fibres that are more suitable for such processes and products, with tenacity from 5 to 8 gpd (0.44-0.71 N tex^{-1}), elongation to break from 7% to 15%, and initial modulus 100-200 gpd (8.8-17.6 N tex^{-1}), can be made by using PPTA polymer of lower inherent viscosity (< 4 dL g^{-1}), more dilute, but still anisotropic, solutions in sulphuric acid (10-14%), an air-gap spinning process with the coagulant at a higher temperature (40-80°C), and drying at relatively low temperatures under zero tension. All of these conditions are designed to produce and retain lower orientation and crystallinity in the fibres [34].

Sulphonated PPTA fibres that are readily dyeable with basic dyes at the boil without the use of carriers can be made by dissolving the polymer in sulphuric acid of a higher strength than that normally used for making the spinning solution [35]. Sulphuric acid of 101% strength can be used, with a dissolution temperature of up to about 110°C and a holding time at 70°C of about 2 hours before spinning. Both sulphonation and chain scission occur, so that the final degree of sulphonation is about 15-18 moles % and the inherent viscosity falls from about 6.3 dL g^{-1} to about 3 dL g^{-1}. The solution strength used is about 12%, and the fibre is formed by wet spinning into dilute sulphuric acid without any air gap. The product, after washing, neutralisation, and drying, has tensile properties such as tenacity 3 to 5 gpd (0.26-0.44 N tex^{-1}), elongation at break about 9%, and initial modulus 90 to 140 gpd (7.9-12.4 N tex^{-1}). The fibres made from these isotropic solutions give an equatorial wide-angle X-ray diffraction pattern exhibiting one peak at 2θ about 23°, typical of the Haraguchi crystal form [36] of PPTA, instead of the two peaks (21° and 23°) characteristic of PPTA fibres spun by the normal production route [37].

This dyeable product has tensile properties typical of ordinary textile fibres and unsuitable for coloured high-tenacity or high-modulus products. However, it is likely to retain very useful flame-retardant properties and is much more suitable for conventional textile processing and use than the unsulphonated, air-gap wet-spun products.

5.3.2.2 Crystalline and supramolecular organisation

There is a distinct skin-core differentiation in many types of PPTA fibre. The skin is usually characterised by a very high concentration of microvoids, which are largely responsible for the high moisture uptakes, about 7%, of the high-tenacity and high-extensibility variants of Kevlar (Table 5.2.1). Kevlar 981, one of the Ht types of Kevlar, exhibits a particularly high skin content of 60-70%. The moisture uptakes are progressively lower for the higher modulus variants, falling to about 1.2% for Kevlar 149, which has the lowest skin content [38].

Compressive strengths of fibres made from rigid-rod liquid-crystalline polymers such as PPTA are generally low in relation to their tensile strengths. Axial compression of PPTA fibres and bending to low radii of curvature both lead to deformation of the compressed areas with the formation in the fibre, through localised compressive failure, of kink bands oriented at about 60° to the fibre axis [39]. This irreversible deformation does not lead directly to fibre breakage, but does affect the subsequent mechanical performance of the fibres. Different methods of measurement of compressive strength give different values. Elastica loop tests [39] and single-filament composite bending tests [40] give considerably higher values than single-filament recoil tests [41]. The recoil tests give values of about 360 MPa for Kevlar 49, which are close to, but slightly below, those derived from unidirectional multifilament composite tests that simulate in-use behaviour. Raman spectroscopy has been used to follow the molecular deformation of PPTA fibres in both tension and compression. In conjunction with optical microscopy it has been used to locate the variation of strength along the fibres and hence the regions of localised compressive failure manifested as kink bands. Values of strain at compressive failure based on kink band formation agree well with those derived from a modified series model of the Northolt type [42].

Staining of Kevlar 981 by silver sulphide, as well as highlighting the microvoid structure, reveals unstained lateral bands that highlight axial periodicity of 500-600 nm, coinciding with turning points of the pleat structure. The silver sulphide forms epitaxially oriented crystals within these microvoids, creating partially composite fibres that have a reduced tensile strength (2.71 GPa v. 3.65 GPa). However, they exhibit a significantly higher compressive strength (430 MPa v. 300 MPa, using the recoil test) than the original voided fibre [38]. Still greater improvements in compressive strength have been reported for PPTA fibres reinforced by dispersing 25 wt. % (10 vol. %) of silicon carbide whiskers in the spinning solution [43]. These microcomposite fibres, which contain whiskers oriented parallel to the fibre axis, no longer form kink bands when subjected to compressive loading.

The crystal structure of all the commercial high-tenacity high-modulus PPTA fibres is that determined by Northolt and van Aartsen [37]. The Haraguchi structure [36] has similar unit cell dimensions to the Northolt structure, but has a different packing that results from a lateral displacement, by $b/2$, of the chains along every other (2 0 0) plane, so that the unit cell is no longer centred (Figure 5.27). It is formed from unmodified PPTA by spinning from solutions of relatively low concentration, and in such cases can be converted into the more stable Northolt form by annealing.

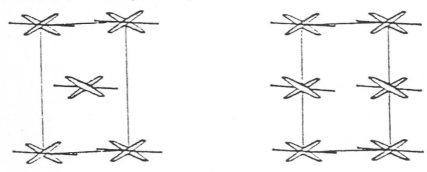

Figure 5.26 Crystal structure of PPTA. Left : Northolt crystal lattice [37]; Right : Haraguchi crystal lattice [36]

The degree of crystallinity of PPTA fibres is exceptionally high. Based on X-ray data, the value for Kevlar 29, the original high-tenacity product, has been estimated to be 80-85%. and that for Kevlar 49, the original high-modulus type, to be 90-95% [38]. Although other authors find somewhat lower values, even these far exceed those for any of the conventional textile fibres. Crystallite sizes are also high, and again higher for the high-modulus variants.

Table 5.2.2. Comparison of lateral crystallite sizes and orientation angles for a series of PPTA fibres [45]

Fibre	Lateral Crystallite Size (nm)		Overall Orientation (deg)
	110	200	
Kevlar 29	4.51	3.96	25.0
Twaron	5.41	4.48	22.2
Kevlar 49	5.08	4.06	14.4
Kevlar 981	4.52	4.04	19.7
Kevlar 149	9.14	5.85	13.5

Table 5.2.2 [45] gives comparative values for a set of PPTA fibres, and illustrates the higher values of the high-modulus products Kevlar 49 relative to the lower modulus Kevlar 49 and 149. The crystallites become progressively more asymmetric as the modulus rises, due to preferential crystal growth along the b-axis, which is the direction in which the

hydrogen bonds between neighbouring chains lie, and this asymmetry is particularly evident in Kevlar 149 (Figure 5.28) [46]. The orientation angle between the polymer chains and the fibre axis is also very much lower in the high-modulus products [45], and particularly in Kevlar 149.

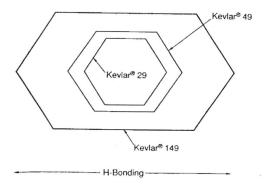

Figure 5.28   Relative crystallite sizes in Kevlar 29, 49 and 149 [46]

Various models for the internal fine structure of PPTA fibres have been proposed. A summary has been given by Krause [47]. They can be categorised as being based upon radial pleated sheet [39], concentric pleated sheet [48], container [44; 49], and periodically disorientated lattice [50] models. All of these, of course, are idealised representations. The radial pleated sheet model (Figure 5.29)[39] is the most frequently cited. It consists of a series of radially oriented, axially pleated sheets characterised by axial banding at a spacing of about 500 nm, together with a further set of narrow bands, about 30 nm in width, spaced at intervals of about 250 nm. The alternating component pleats of each sheet are arranged at approximately equal but opposite angles to the plane of a longitudinal fibre section. The angle between adjacent components is about 170°. There is a short transitional band between these components in which the molecules are briefly parallel to the plane of the section.

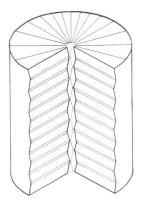

Figure 5.29 Schematic model of a PPTA fibre showing radially oriented pleated sheets [39]

This pleated structure is a useful and readily comprehensible model for interpreting, for example, changes in behaviour due to application of tension. When fibres are placed under tension there is a reduction in visibility of the banded structure that can be interpreted as pleat opening. The increase in instantaneous modulus with increasing strain exhibited by Kevlar 49 and the good elastic recovery correlate with reversible increase and decrease of the angle between the pleats [51]. Kevlar 149, the fibre type possessing the highest modulus, has a much smaller angle between the pleats and the fibre axis than other types [45].

5.3.3   High-tenacity   high-modulus   aramid   fibres   from   isotropic solutions

Technora, originally known as HM-50, which has been available commercially from the Japanese firm Teijin since about 1974, is an aramid fibre based upon a copolymeric structure. The usual composition is that of a 1:1 copolymer, probably random, formed from terephthaloyl chloride reacted with equimolar amounts of p-phenylenediamine (giving the PPTA repeating unit (Figure 5.24 (4)) and 3,4'-diaminodiphenyl ether (giving the MOPTA repeating unit (Figure 5.24 (5)), with subsequent addition of calcium hydroxide to neutralise the hydrochloric acid produced as a by-product [52]. The presence in the chain of a 120° angle due to the meta-disubstituted ring and a slightly smaller, almost compensatory, angle between the bonds at the ether oxygen atom leads to enough rotational flexibility to prevent formation of a nematic mesophase in solution, but leaves open the possibility of mutual compensation to form a highly extended chain in the solid phase. Fibres made from alternative compositions containing smaller proportions of other repeating units derived from, for example, m-phenylenediamine or naphthalene-2,6-dicarbonyl chloride have also been described by Teijin [53]. In addition, a very large number of fibres based on terpolymers and quaterpolymers from terephthaloyl chloride reacted with p-phenylenediamine plus two or three other aromatic diamines were described in patents and papers from Hoechst during a period of about 10 years from 1986. Although a trade name, Supara, was given by Hoechst to a fibre from this class, it has not emerged as a commercial product.

These polymers are all capable of producing high-strength, high-modulus fibres, but since they do not form nematic solutions they differ markedly from PPTA in the route by which the fibres are made. Technora is dry-jet wet spun from isotropic, relatively dilute (about 6%) solutions in the polymerisation solvent, NMP, into 30/70 NMP/water at 50°C, and drawn to a high draw ratio, such as 12:1, at about 500°C [52; 54; 55]. Fine inorganic particles are applied to the surface before drawing to prevent fibres adhering to one another at these high temperatures, and become embedded in the fibre surface [56]. The tensile strength (3.4 GPa; 2.5 N tex^{-1}) and elongation to break (3.6%) of Technora T221 reported by Teijin are slightly higher than those of Kevlar 29, but lower than those of Kevlar 129; the tensile modulus is reported as 74 GPa (530 N tex^{-1}). Areas in which Technora exhibits advantages over most Kevlar types are higher fatigue resistance, higher bending and frictional wear resistance, lower fibrillation tendency, and higher hydrolytic stability.[52]

The density is also lower at 1.39 g cm^{-3}, against 1.44 to 1.47 g cm^{-3} for various Kevlar variants.

According to Teijin, PPTA is less suitable for use in bristles than the Technora copolymer because of its inferior resistance to acids and alkalis. In making Technora bristles, it is necessary to overcome the general problem of making relatively thick round fibres by a wet-spinning process, which normally gives insufficient solvent extraction from fibres of these higher diameters, particularly as here when the spun fibre diameter must be high enough to allow the high draw ratio required in producing Technora fibres. This problem is solved by increasing the dope temperature to, e.g., 110°C, and the coagulation bath temperature to, e.g., 70°C, and lowering the concentration of the spinning solvent, NMP, in the coagulation bath to, e.g., 10-20% [57]. Bristles of up to 200 denier can be made in this way. They have tensile strength 15 gpd (1.35 N tex^{-1}) or more, elongation at break 4% or less, initial modulus 500 gpd (45 N tex^{-1}) or more, and flatness ratios (ratio of major transverse axis to perpendicular minor transverse axis) of 3 or less.

Technora T221 exhibits a similar type of elastic deformation to Kevlar 29, with significant modulus increase from a first modulus to a second at about 1.5% extension, although with somewhat lower modulus values than the Kevlar at high strain rates. Fatigue testing at 20cN stress on 40 mm lengths of 1.5 denier single filaments at a strain rate of 0.5 mm min^{-1} shows that the two modulus values for the fatigued Technora fibres remain nearly constant over at least 80 cycles, whereas those for PPTA fibres rise markedly during the first 20 cycles. This rise is accompanied by development of a third modulus region, similar in value to the first, at relatively high extension of PPTA, but Technora does not develop such a region [58].

There is some disagreement about the crystallinity of Technora. Ozawa of Teijin reported originally that the copolymer and fibre contain two distinct crystal structures, due to the PPTA and MOPTA repeating units in the chain [52]. A complication here is that the MOPTA repeating units, like the repeating units of Kermel, contain a structure, the MOP part derived from the diamine, that has no plane of symmetry normal to the chain axis. Consequently the MOPTA repeating units will exhibit both MOP and POM arrangements randomly relative to a given chain direction, making MOPTA itself effectively a copolymer with reduced ability to crystallise. Blackwell [59] found that the fibre exhibits a series of aperiodic diffraction maxima along the meridian, suggesting that it contains oriented chains with a completely random sequence of repeating units. Tashiro et al. [60] postulated a para-crystalline extended chain structure and found that the apparent crystallite modulus, based on meridional reflections at 2θ = 27° and 42°, for the copolymer fibres was about half that for PPTA. According to Ferreira et al. [58] Technora is ordered but not crystalline, and exhibits a substantial exotherm peaking at about 508°C. The implication is that this exotherm is due to a mesophase, and that the drawing process represents mesophase orientation.

References

1. DuPont, Brit. Pat. 871,578 (1961); Brit. Pat. 871,579 (1961); Brit. Pat. 871,580 (1961); Brit. Pat. 871,581 (1961); Brit. Pat. 877,885 (1961)
2. DuPont (H.M. Ghorashi), US Pat. 4,755,335 (1988)
3. DuPont (T.D. Zatkulak, D.J. Rodini and J.D. Hodge), US Pat. 4,919,869 (1990)
4. Teijin, Brit. Pat. 1,265,732 (1972); US Pat. 3,640,970
5. Teijin, Brit. Pat. 1,376,218 (1974)
6. Teijin, Brit. Pat. 1,423,441 (1976)
7. Teijin (K. Kouzai *et al.*), US Pat. 4,073,837 (1978)
8. Teijin, US Pat. 4,342,715 (1982)
9. DuPont (T.-M. Tai *et al.*). USP 5,667,743 (1997)
10. Societé Rhodiaceta, Brit. Pat. 1,225,735 (1971)
11. R. Pigeon, *Appl. Polym. Symp.*, 1973, (21), 33
12. R. Pigeon and P. Allard, *Makromol. Chem.*, 1974, **40/41**, 139
13. G. Desitter and R. Cassat, *Thermostable and fire-resistant fibres*, Chapter 11 in *Synthetic Fibre Materials*, ed. H. Brody, Longman Scientific and Technical, Harlow, 1994
14. S.L. Kwolek et al., *Macromolecules*, 1977, **10**, 1390
15. DuPont (S.L. Kwolek), Brit. Pat. 1,283,064 {1972)
16. Akzo, Brit. Pat. 1,547,802 (1979)
17. DuPont (S.L. Kwolek), US Pat. 3,671,542 (1972); Brit. Pat. 1,2834,064 (1972)
18. Akzo (H.T. Lammers), US Pat. 4,320,081 (1982)
19. DuPont (H. Blades), Brit. Pat. 1,393,011 (1975)
20. DuPont (H. Blades), Brit. Pat. 1,391,501 (1975)
21. DuPont (H.H. Yang), US Pat. 4,340,559 (1982)
22. DuPont (M.J. Chiou), US Pat. 4,965,033 (1990); Europ. Pat. 0 449 197 (1991)
23. DuPont (H.H. Yang), US Pat. 5,173,236 (1992)
24. DuPont & DuPont Toray (T. Furumai *et al.*), US Pat. 5,853,640 (1998)
25. DuPont (S.R. Allen), US Pat. 4,985,193 (1991)
26. DuPont (T.S. Chern and J.E. Van Trump), US Pat. 4,883,634 (1989); DuPont (T.S. Chern et al.), US Pat. 5,001,219 (1991)
27. DuPont (H.H. Yang), US Pat. 4,859,393 (1989)
28. DuPont, *Kevlar,* Dupont Engineering Fibres, Geneva, Brochure H-22999, Dec. 1993, p. 26.
29. H. van der Werff and M.H. Hofman, *Chemical Fibers Intl.*, 1996, **46**, Dec., 435-441
30. V.N. Kiya-Oglu *et al.*, *Khim. Volokna*, 1993, No.2, 17; *Fibre Chem.*, 1993, 25, 86
31. Akzo-Nobel (H. van der Werf, M.H. Hofman and J.J.M. Baltussen), Europ. Pat. 0 823 499 (1998)
32. M.G. Northolt and R. van der Hout, *Polymer*, 1985, **26**, 310
33. DuPont (K.S. Lee), Intl. Pat. Pubn. WO 96/10105; Europ. Pat. 0 783 604 (1997)
34. DuPont (S.R. Allen and D.M. Harriss), US Pat. 5,330,698 (1994)
35. DuPont (M.W. Bowen, H.M. Ghorashi and H.H. Yang), US Pat. 5,660,779 (1997)
36. K. Haraguchi, T. Kajiyama and M. Takayanagi, *J. Appl. Polym. Sci.*, 1979, **23**, 903 & 915
37. M.G. Northolt and J.J. van Aartsen, *J. Polym. Sci., Polym. Lett. Edn.*, 1973, **11**, 333

38. M.G. Dobb, C.R. Park and R.M. Robson, *J. Mater. Sci.*, 1990, **27**, 3876
39. M.G. Dobb, D.J. Johnson and B.P. Saville, *J. Polym. Sci., Polym. Phys. Edn.*, 1977, **15**, 2201
40. S.J. De Teresa et al., *J. Mater. Sci.*, 1984, **19**, 57
41. S.R. Allen, *J. Mater. Sci.*, 1987, **22**, 853
42. M.C. Andrews, D. Lu and R.J. Young, *Polymer*, 1997, **38**, 2379
43. M.A. Harmer and B.R. Phillips, *J. Mater. Sci. Letters*, 1994, **13**, 930
44. M. Panar et al., *J. Polym. Sci., Polym. Phys. Edn.*, 1983, **21**, 1955
45. M.G. Dobb and R.M. Robson, *J. Mater. Sci.*, 1990, **27**, 459
46. C.L. Jackson et al., *Polymer*, 1994, **35**, 123
47. S.J. Krause, D.L. Vezie and W.W. Adams, *Polym. Commun.*, 1989, **30**, 10
48. M. Horio et al., *Sen-i Gakkaishi*, 1984, **40**, T-285
49. L.S. Li, L.F. Allard and W.C. Bigelow, *J. Macromol. Sci., Phys. Edn.*, 1983, **B22**, 269
50. R.J. Morgan, C.O. Pruneda and W.J. Steele, *J. Polym. Sci., Polym. Phys. Edn.*, 1983, **21**, 1757
51. S.R. Allen and E.J. Roche, *Polymer*, 1989, **30**, 996
52. S. Ozawa, *Polymer J. (Tokyo)*, 1987, **19**, 119-125
53. Teijin, US Pat. 4 355 151 (1982); US Pat. 4 413 151 (1983); US Pat. 5 177 175 (1993).
54. Teijin, US Pat. 4 075 172 (1978)
55. H. Imuro and N. Yoshida, *Chemiefasern*, 1987, **37/89**, T4 & E29
56. Teijin (K. Shimada *et al.*), US Pat. 4 413 114 (1983)
57. Teijin (R. Kakihara and T. Noma), Intl. Pat. Pubn. WO 97/44510; Europ. Pat. 0 846 794 (1998)
58. M. Ferreira, T.M. Lam, P. Labache, and Y. Delvael, *Text. Res. J.*, 1999, **69**, 30-37
59. J. Blackwell, R.A. Cageao and A. Biswas, *Macromolecules*, 1987, **20**, 667
60. K. Tashiro et al., *Sen-i Gakkaishi*, 1987, **43**, 627; 1988, **44**, 7

5.3 Fibres Based on Ultra-High Molecular Weight Polyethylene - Processing and Applications

P.J. Lemstra[1], C.W.M. Bastiaansen[1], T. Peijs[1,2] and M.J.N. Jacobs[3]

1 Dutch Polymer Institute/Eindhoven University of Technology, Netherlands
2 Department of Materials, Queen Mary and Westfield College, UK
3 DSM High-Performance Fibers, Geleen, Netherlands

5.3.1 Introduction

5.3.1.1 Rigid versus flexible chains

In the last three decades of the 20th century, significant progress has been made in exploiting the intrinsic properties of the macromolecular chain, especially in the field of 1-dimensional objects such as fibres.

Two major routes can be discerned which are completely different in respect to the starting (base) materials, namely rigid as opposed to flexible macromolecules [1].

The prime examples of rigid chain polymers are the aromatic polyamides (aramids), notably poly(p-phenylene terephthalamide), PPTA, currently produced under the trade names Kevlar® (Du Pont) and Twaron® (Akzo Nobel). More recent developments include the PBO (poly-phenylene benzobisoxazole) fibre from Toyobo, Zylon®, and the experimental fibre M-5 from Akzo Nobel based on PIPD (poly-diimidazo pyridynylene dihydroxy phenylene). The latter fibre is claimed to possess an enhanced compressive strength compared with the aramids [2].

The *primus inter pares* of a high-performance fibre based on flexible macromolecules is polyethylene, currently produced by DSM (Dyneema®) and its licensee Allied Signal (Spectra®), see Table 5.3.1.

The main difference between rigid and flexible chains is the necessity to force chain extension in the case of flexible polymer molecules in order to exploit the intrinsic possibilities of the chain concerning ultimate mechanical properties, while for rigid rod polymers chain-extension has been built in by the chemist, such as is the case in PBO. Poly (p-phenylene terephthalamide), the building block of the aramid fibre, is not strictly a rigid (rod) chain, the ratio of the contour length over the persistence length is about 4 in dilute solutions [3], but during spinning, coagulation and heat setting the chains align in an extended chain conformation.

In the case of conventional flexible and (stereo) regular polymer molecules, the chains tend to fold upon solidification/crystallization and in order to exploit the intrinsic possibilities in

1-D structures, routes have been developed to transform folded-chain crystals into chain-extended structures as will be discussed in this chapter.

Table 5.3.1 Stiffness (E-Modulus) of various materials (at ambient temperature)

Material	E- Modulus [GPa]
Rubbers	< 0.1
Amorphous thermoplasts, T < Tg	2 - 4
Semi-crystalline thermoplasts	0.1 - 3
Wood (fibre direction)	15
Bone	20
Aluminum	70
Glass	70
Steel	200
Ceramics	500
Carbon fibre	500-800
Diamond	1200
Polyethylene Fibre (Dyneema®))	100 - 150
Aramid Fibres (Kevlar®, Twaron®)	80 - 130
PBO (Zylon® /Toyobo)	180 – 280
"M-5 (Akzo Nobel)	300

Figure 5.30 shows the development of the tensile modulus of oriented polyethylene structures in the 20th century. In the following sections the various routes will be reviewed and basic aspects of chain extension will be discussed for the case of polyethylene fibres.

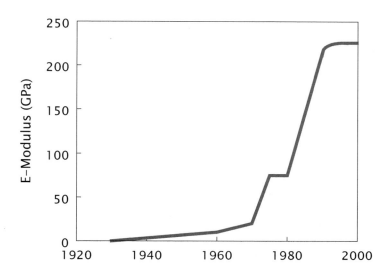

Figure 5.30   The development of the tensile (E-) modulus of oriented polyethylene
                    structures in the 20th century

5.3.1.2 Chain-folding versus chain-extension

Since the first scientific routes for the synthesis of high molar mass polymers were discovered by Carothers in the 30s of this century, polymer scientists have attempted to improve the mechanical properties by orienting the chain molecules. In fact, the prerequisites for actually producing 'useful fibres', viz. high modulus and high strength fibres, were already formulated by Hill and Carothers [4] in the early 1930s including the necessity of long chain molecules which should be in an extended chain conformation and in a parallel (crystalline) register with the fibre axis.

The concept of a chain in an extended conformation was not so obvious during the time Carothers made his pioneering investigations. Intrinsic rigid chains were synthesized much later, in the sixties, whereas (stereo)regular flexible chains tend to fold upon solidification (crystallization) which is in fact the opposite from chain-extension.

During the time that Carothers made his pioneering work on the synthesis of polyamides in the 30s, crystallization of synthetic polymers was hardly studied. Abitz, Gerngross and Hermann [5] introduced in 1930 the so-called 'fringed micelle' model for polymer crystals. Bunn published in 1939 that polyethylene crystallizes in an orthorhombic crystal structure [6] and Storks [7] mentioned in 1938 the possibility of chain-folding during crystallization based on electron-diffraction studies of thin films of gutta-percha, trans-polybutadiene. The concept of Storks concerning chain folded crystallization remained unnoticed until the mid-fifties when Keller [8], Jaccodine [9], Fischer [10] and Till [11] published independently that the newly developed linear polyethylenes by Ziegler (1953) form platelet, so-called lamellar crystals, upon precipitation from dilute solutions. Since the contour length of the polyethylene chains, which were investigated, is much longer than the thickness of the platelet crystals, typically 10 nm, Keller concluded [12] that the chains are folded in these single crystals.

The number of papers concerning chain-folding during crystallization overshadowed by far the number of papers concerning chain-extension and exploiting the intrinsic mechanical properties of the polymeric chain in the period 1930-1960. Meijer and Lotmar (13) already mentioned in 1936 the high stiffness of an extended polymer chain. Treloar [14] published in 1960 a seminal paper, at least to the authors view, concerning the ultimate stiffness of an extended polymer (polyethylene and polyamide) chain. Upon loading an extended polyethylene chain, both the C-C bonds and the bond angles in the main chain increase. Taking the force constants from low molar mass compounds, Treloar calculated the tensile (Young's) modulus of a single, extended polyethylene chain to be 182 GPa, to be compared with a tensile modulus < 2 GPa, see Table 5.3.1, for an isotropic PE sample! These simple calculations triggered studies to pursue chain orientation/extension in order to improve the mechanical and physical properties of polymer systems, viz. fibres and tapes.

5.1.3.1 Chain extension in the melt

In the literature various processes have been described to orient the chains directly in the molten state. The problem of chain-orientation and extension in the melt is that extensive relaxation processes occur, the chains resist deformation and retract back to a random coil conformation. Lowering the extrusion-spinning temperature is not a real solution for this problem. It was shown already in 1967 by van der Vegt and Smit [15] that on lowering the extrusion temperature of polyethylene, and other crystallizable polymers, elongational flow-induced crystallization will occur and the solidified polymer will block the flow. This topic has been studied extensively by Porter et al. [16] and they observed that the 'plugs' which are formed in the die, as a result of crystallization during blocked flow, are highly oriented and possess a high stiffness. Young moduli of 70 GPA were reported [17] but, of course, blocked capillaries are rather incompatible with technological demands concerning high production speeds. For more details on capillary blockage and related phenomena, the reader is referred to references 17 and 21, and Chapter 8 of this volume.

The conclusion is that the ultimate fate of experiments concerning chain extension directly in the melt is, or relaxation and lack of orientation/chain-extension at elevated temperatures, or flow-induced crystallization in the processing equipment at temperatures close to the melting/crystallization temperature. Consequently, in order to obtain a high degree of chain-extension, drawing should be performed in a separate step, after processing/shaping and below the melting point, viz. in the (semi)-solid state.

5.3.1.4 Solid-state drawing

In the 70s, Ward et al. [18 - 21] started systematic studies concerning the drawability of linear polyethylenes in the solid state and developed a technological route for optimised melt-spinning and subsequent solid-state drawing of linear polyethylenes. Typically the spinning of fibres and their subsequent orientation is decoupled with respect to the temperature regime. By optimising the polymer composition and process conditions, PE fibres could be produced possessing tensile (Young) moduli up to 75 GPa and a strength level up to 1.5 GPa, see also Chapter 5.1.

The process of melt-spinning/drawing is limited with respect to the molar mass of the polyethylenes. With increasing molar mass, both the spinnability (a strong increase of melt-viscosity causes difficulties to produce homogeneous filaments) and the drawability in the solid-state decrease, which sets an upper limit to melt-spinning of polyethylenes of typically 500 kg/mole. The limited drawability of semi-crystalline polymers in the solid state will be discussed below in paragraph 5.3.1, and is often referred to as the natural draw ratio.

In conclusion, melt-spinning followed by drawing in the solid-state, encounters two major limitations:

a) *with increasing molar mass, melt-spinning/extrusion becomes more difficult related to the strong increase in melt-viscosity; the zero-shear viscosity η_0 scales with $M_w^{3.4}$, and*

b) *with increasing molar mass the drawability in the solid-state decreases, viz. the chains in the extruded and solidified filaments become more difficult to extend.*

5.3.1.5 Solution-processing

Solution-spinning
An obvious route to increase the spinnability of high molar mass polyethylenes is to use solvents to lower the viscosity. Jürgenleit filed [22] a patent in 1956 concerning solution-spinning and subsequently drawing of linear polyethylene but the results were not impressive, a strength level of < 1.2 GPa was obtained, to be compared with ~1.5 GPa in the case of optimised melt-spinning. Solution-spinning of ultra-high molecular weight (UHMW)-polyethylene, M_w typically > 10^3 kg/mole, was performed by Zwick but no post-drawing nor fibre properties were mentioned in his patent application [23]. Blades and White (Du Pont) introduced their so-called flash spinning [24] technique of pressurised solutions of linear polyethylenes. The fibrillated strands were subjected to slow drawing. Maximum values for the tenacity (tensile strength) were 1.4 GPa and 20 GPa, respectively.

In retrospect, the early attempts in the sixties of the 20th century to employ solution-spinning of high molar mass polyethylenes, to obtain high strength/high modulus fibres failed since no proper understanding was available at that time concerning optimised drawing of solution-spun fibres. It took two decades, 1960 – 1980, in order to properly understand the mechanisms for ultra-drawing via a tortuous path involving chain extension in dilute solution to solution(gel)-spinning.

Chain-extension in dilute solutions
Mitsuhashi [25] was probably the first to attempt inducing chain extension in solution, using a Couette type apparatus, and reported in 1963 the formation of fibrous 'string-like' polyethylene structures upon stirring. His work remained unnoticed until ~ 10 years later Pennings et al., using a similar apparatus, reported the so-called 'shish kebab' type morphology of polyethylene crystals [26].

Stirring polymer solutions to induce chain-extension is less obvious than might be anticipated at first sight. Simple shear flow is inadequate and in order to obtain full chain-extension, the flow has to possess elongational components. The importance of elongational flow fields on the transformation from a random coil into an extended chain conformation in dilute solutions has been originally recognised by Frank [27] and experimentally investigated by Peterlin [28] and addressed theoretically by de Gennes [29]. The conclusion is that an isolated chain will fully stretch out beyond a certain critical strain rate $\dot{\epsilon}_{cr}$ which scales with $M^{-1.5}$ as determined experimentally for monodisperse samples by Odell and Keller [30]. This relationship implies that longer chains are more readily extensible.

Chain-extension in dilute solutions can be made permanent if extension is followed by crystallization. Taking into account that with increasing molar mass the chains become more readily extensible, and given the fact that polymers such as polyethylene are usually polydisperse, one can easily envisage, in retrospect, that in an elongational flow field only the high molar mass fraction becomes extended and crystallizes into a fibrous structure ('shish'). The remaining part will stay in solution as random coils and upon subsequent cooling, nucleates and crystallizes as folded-chain crystals, nucleating onto the fibrous structures ('kebab').

The structure of shish-kebab type fibrous polyethylene is far from the ideal arrangement of PE macromolecules for optimum stiffness and strength. Due to the presence of lamellar overgrowth, the moduli of precipitated fibrous PE 'shish-kebabs' were limited to up to about 25 GPa [31], to be compared with tensile moduli > 50 GPa in the case of direct melt-spinning/drawing, as performed by Ward et al. In fact, the 'shish-kebab' structure is only the intermediate between the folded-chain crystal and the extended chain crystal.

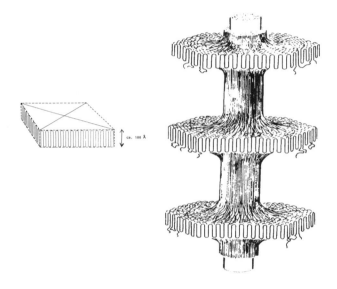

Figure 5.31   Structure of folded-chain crystal vs. 'shish-kebab'

Fibrous structures without lamellar overgrowth were obtained by Zwijnenburg and Pennings [32,33] using their so-called surface technique, see Figure 5.32. A seed fibre (polyethylene or even cotton) is immersed in a dilute solution of UHMW-PE and from the surface of the rotating inner-cylinder fibrous, tape-like, polyethylene structures could be withdrawn at low speeds. This pulling of fibres from the rotor is due, as was found out later after the discovery of the solution(gel)-spinning route, to the formation of a thin gel-layer on the rotor surface. Upon stirring, the chains in the dilute solutions, especially near the wall of the rotor, become elongated to some extent, they phase separate and adhere to the surface of the rotor in the form of a gel-layer. Under optimised conditions, with respect to solution concentrations, temperatures and take- up speeds, oriented UHMW-PE structures could be obtained possessing tensile moduli over 100 GPa and strength values above 3GPa!   In general, with increasing solution temperature, the lamellar overgrowth decreases and finally rather smooth oriented UHMW-PE structures could be obtained. The surface growth technique was another milestone on the route to high-performance UHMW-PE fibres and, in fact, the first experimental proof that high-modulus/high-strength structures could be made. The technique, however, possesses intrinsic draw backs such as very low production speed, a non-uniform thickness of the tape-like structures which were pulled of from the rotor and the problem of scaling up this process. Attempts have been made to develop technologies for continuous production of UHMW-PE tapes, such as the rotor technique by M. Mackley [34], see Figure 5.32. Supercooled UHMW-PE solutions were sheared and tape-like PE structures could be produced possessing stiffness values of ~ 60 GPa at take-up/roll-off speeds of several meters/min. The belt, see Figure 5.32, increases the local shear rate and drags the produced tape from the surface of the rotor. Although the linear take-up speed is still not impressive, the width of the tape is 'unlimited' and hence the mass of tape per unit of time is much higher than for the surface growth technique. However, these and related attempts became of less interest for industrial applications, after the discovery of the simple and straightforward solution(gel)-spinning technique.

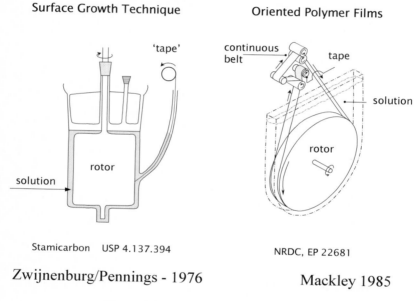

Figure 5.32 Surface growth techniques

Solution(gel)-spinning

At the end of the seventies, solution(gel)-spinning of UHMW-PE was discovered at DSM [35,36,37,38]. In the solution (gel)-spinning technique, semi-dilute solutions are employed during spinning but the elongation of chains is performed by drawing in the semi-solid state, i.e. below the melting c.q. dissolution temperature   Figure 5.33 shows schematically this process, now often referred to as solution(gel)-spinning. A solution of UHMW-PE with a low polymer concentration, typically of 1-2 %, was spun into water. Upon cooling a gelly filament is obtained consisting of a physical network, obtained by thermoreversible gelation, containing a large amount of solvent. The as-spun/quenched filaments are mechanically sufficient strong (gel-fibres) to be transported into an oven in which drawing is performed. At first glance, the ultra-drawability of these gel-fibres seems not too surprising in view of the large amount of solvent which could act as a plasticizer during draw. The remarkable feature, however, is that ultra-drawing is still possible after *complete removal* of the solvent *prior* to the drawing process. The solvent is necessary to facilitate processing of the rather intractable polymer UHMW-PE (melt-processing is impossible due to the excessive high melt-viscosity) and induces a favourable structure/morphology for ultra-drawing but the *solvent is not essential* during the drawing process.

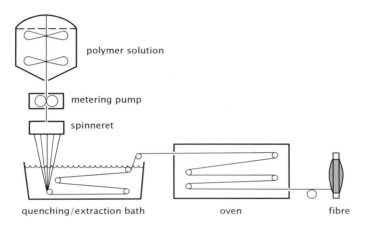

Figure 5.33 Solution (gel)-spinning of UHMW-PE (batch process)

Before discussing the actual drawing mechanisms, chapter 5.3.3, involved in ultra-drawing UHMW-PE structures, we address first some fundamental aspects concerning stiffness and strength as documented in literature in order to comprehend the following sections concerning drawability 5.3.3 and fibre properties 5.3.4.

5.3.2   The ultimate stiffness and strength of flexible polymers

5.3.2.1 The ultimate stiffness

In the previous section it was mentioned that Treloar calculated the tensile (Young's) modulus of an extended polyethylene chain to be 182 GPa. Figure 5.30 shows in fact that the tensile modulus of (experimental) polyethylene fibre grades surpasses the limit of Treloar calculations. Using modern force field calculations, the ultimate moduli are estimated in the range of 180 – 340 GPa [39,40].

Estimates of the ultimate tensile moduli of polyethylene and other polymer systems, can also be obtained from X-ray diffraction measurements on oriented fibres during mechanical loading. For the interpretation of the X-ray data a homogeneous stress distribution [41,42] has to be assumed. Table 5.3.2 shows some representative data from literature [42-48].

Table 5.3.2 Tensile Moduli derived from X-Ray studies

Material	X-Ray Modulus(GPa)	ref
Polyethylene (PE)	235	[43]
Poly(vinyl alcohol) (PVAL)	250	[44]
Poly(ethylene terephthalate) (PETP)	110	[45]
Polyamide-6 (PAM-6)	165	[46]
Polypropylene (i-PP)	46	[47]
Polyoxymethylene (POM)	70	[48]

Generally, the tensile moduli derived from X-Ray data are lower in comparison with data derived from theoretical calculations. Nevertheless, all literature data show that the E-Modulus of polyethylene in the chain direction is extremely high, viz. > 200 GPa. This is due to the small cross-sectional area of the chain, the absence of side groups, and the planar zig-zag conformation in the orthorhombic crystal lattice. Polyethylene is not unique and other linear polymers with similar crystal structures possess similar theoretical moduli, for example poly(vinyl alcohol). Isotactic polypropylene, on the other hand, has a lower ultimate modulus in the chain direction related to the 3_1 helix conformation in the crystal lattice.

5.3.2.2 The ultimate tensile strength

In the past, a variety of studies has been devoted to the theoretical tensile strength of oriented and chain extended structures, i.e the breaking of chains upon loading [49,50]. The theoretical tensile strength of a single, extended, polymer chain can be calculated directly from the C-C bond energy. These calculations show that the theoretical tensile strength's are extremely high, in the order of 20-60 GPa! These values for the theoretical tensile strength are, in general, considered to represent the absolute upper limit of the theoretical tensile strength. The theoretical value of the tensile strength is calculated as the product of the E- Modulus and the strain for which the energy of the bonds is at a maximum. The values thus obtained are valid at absolute temperature (or infinite loading rate). Taking thermal vibrations into account, the strength levels decrease by 20 –65% at ambient temperature [51]. Moreover, in an array of chain extended polyethylene macromolecules, these theoretical values are approached only if all C-C bonds fracture simultaneously. This requires a defect-free, chain-extended structure and *infinite* polymer chains. In practice, however, we are dealing with *finite* chains and a completely different situation is encountered as will be addressed in the next section.

5.3.2.3 Infinite versus finite chains

The theoretical estimates in sections 5.3.2.1. and 5.3.2.2 concerning the ultimate stiffness and strength of (extended) polymer chains were based on loading infinite chains or, alternatively, infinite chains in perfect crystals. In practice, however, we are dealing with finite chains and, consequently, notably the tensile strength is determined not only by the primary bonds but equally well by the intermolecular secondary bonds. Upon loading an array of perfectly aligned and extended finite polymer chains, the stress transfer in the system occurs via secondary, intermolecular, bonds. Chain overlap is needed in order to be able to transfer the load through the system, see Figure 5.34.

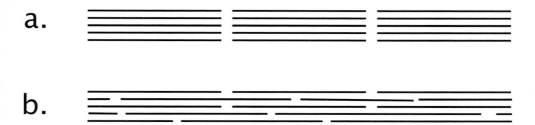

Figure 5.34   Hypothetical extended-chain PE structures; a) no chain overlap, b) chain overlap

Qualitatively, one can easily envisage that the bonds in the main chains are only activated when the sum of the small secondary interactions, $\Sigma\varepsilon_l$, approaches E_i, the bond energy in the main chain. In this respect, one can distinguish between weak Van der Waals interactions, as is the case in polyethylene, or specific hydrogen bonds as encountered in the case of the polyamide or aramid fibres. Intuitively, one expects that in order to obtain high-strength structures in the case of polyethylene, a high molar mass is needed, in combination with a high degree of chain-extension, to build up sufficient intermolecular interactions along the chains.

Termonia and Smith [52,53] used a kinetic model to simulate the fracture behaviour of an array of aligned and extended *finite* polymer chains. Both chain slippage and chain rupture were considered by introducing a stress dependent activation barrier for rupture of both inter- and intramolecular bonds. It was found that the molecular weight (or number of chain ends) has a profound influence both on the fracture mechanism and on the theoretical tensile strength of these hypothetical structures. It was shown that chain slippage prevails at a low molecular weight, as expected. Figure 5.35(a) shows the calculated stress-strain behaviour of polyethylene as a function of the molecular weight. In Figure 5.35(b), polyethylene is compared with PPTA. Figure 5.35(b) clearly demonstrates the influence of secondary interactions, viz. Van der Waals vs. hydrogen bonds. In order to obtain a strength level of 5 GPa, a molar mass of $> 10^5$ Dalton is needed for polyethylene whereas 5×10^3 Dalton is sufficient for PPTA!   The conclusion is, polymers possessing strong secondary bonds require a smaller overlap length to obtain a high tenacity. This conclusion does not imply that any flexible polymer possessing hydrogen bonds, for example the conventional polyamides, is automatically an ideal candidate for obtaining high tenacity fibres. On the contrary, the hydrogen bonds also exist in the folded-chain crystals which are formed upon solidification of the melt or from solution. These hydrogen bonds provide a barrier for ultra-drawing [54], see also section 5.3.5.

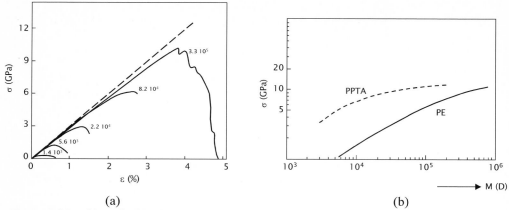

(a)   (b)

Figure 5.35   (a) Calculated stress-extension behaviour of polyethylene; and (b) breaking strength as a function of molecular weight, M for polyethylene and PPTA.

5.3.3   Chain-extension, on the borderline between solid and melt

5.3.3.1 Solid-state drawing of polyethylenes

Traditionally in the fibre/textile industry, chain orientation and extension is generated in melt- and solution-spun fibres by two different methods: (i) applying a draw-down to the fibres during or immediately after spinning (in the molten state or super-cooled melt) and (ii) drawing of fibres at temperatures close to but below the melting- or dissolution temperature. Drawing in the (semi-)solid state, i.e. below the melting and/or dissolution temperature is usually much more effective, in terms of the development of the tensile modulus as a function of draw ratio, since relaxation processes are   restricted since the chains are trapped into crystals which act as physical network junctions.

In the case of polyethylenes, a well-known observation made by Ward et al. [20,55] based on numerous *isothermal* drawing experiments, is that with increasing molar mass the maximum draw ratio decreases towards a limiting value of 4-5 at M_w values over 10^6 Dalton, see the shaded area in Figure 5.36.   A limited drawability in the solid state is not unique for polyethylenes. Many other polymers demonstrate a limited drawability, for example polyamides, often referred to as the natural draw ratio.

To understand the drawing behaviour of polyethylenes in the solid-state, one automatically focuses on the role of crystallites, viz. the folded-chain crystals which are organised in more or less well-developed spherulites in melt-crystallised samples. There is, however, no direct correlation between crystal size/crystallinity or morphology in general vs. the maximum draw ratio as shown in Figure 5.36.   Depending on the crystallization conditions, the drawability can change dramatically for one specific sample. For example, slow cooling from the melt can promote the drawability in the solid state [21] but also can cause identical polyethylene samples to become brittle [56] in the case of very slow cooling. Crystallization from solutions at low supercoolings, to promote the formation of more well-developed single crystals, will enhance the drawability of the single crystal mats [57] after filtration and solvent evaporation/extraction.

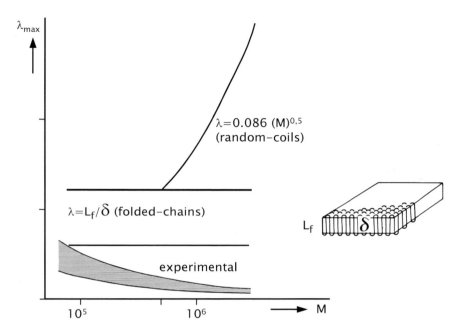

Figure 5.36 Maximum draw ratio λ_{max} as a function of molar mass M: comparison between random-coils, folded chain crystals and experimental values.

In order to understand the possible role of the morphology on the drawing behaviour in the solid-state, we take a first order approach. The topology and arrangement of molecules in melt-crystallized polymers is dependent on many parameters such as molar mass, crystallization temperature, degree of supercooling etc. and is still a matter of debate. Let us take two extreme, and therefore not realistic, cases, respectively:

a)   upon crystallization from the melt perfectly folded (single) crystals are formed,
b)   the chains retain their folded-chain conformation upon solidification.

For case a), the maximum draw ratio λ_{max} is simply given by the ratio of the fold length L_f and the diameter of the chain δ:

$$\lambda_{max} = L_f/\delta \tag{1}$$

Taking typical values for the fold length L_f , 20 - 30 nm, and δ, 0.5 - 0.7 nm, respectively, the maximum draw ratio for the case of well-stacked folded-chain lamellar crystals is between 30 - 60, *independent of the molar mass.* This simple model was proposed by Peterlin already in the early seventies [58].

For case b), the chains retain their random coil conformation upon solidification from the melt, the maximum draw ratio is given by the ratio between the fully extended contour length L and the average unperturbed end-to-end distance, the square root of $<r^2>o$:

$$\lambda_{max} = Nl_b \cos(\theta)/(C_\infty Nl_b^2)^{0.5} = 0.086(M)^{0.5} \qquad (2)$$

In equation (2), N is the number of C-C bonds (M/14), θ is the bond angle, (112^0 in the case of polyethylene) and C_∞ the characteristic ration (6.7 for polyethylene). The maxium draw ratio scales with the square root of M, see Figure 5.36. For example, taking a molar mass M of 10^6 Dalton, the maximum draw ratio is equal to 86.

It is clear that both simple, first approximation calculations, predict a totally different maximum draw ratio than observed experimentally, see shaded area in Figure 5.36.

5.3.3.2 Solution(gel)-crystallized polyethylenes

As discussed in section 5.1, solution(gel)-spinning of UHMW-PE rendered as–spun/cast structures which are still ultra-drawable after *complete* removal of the solvent prior to the drawing process. The solvent is necessary to facilitate processing of the rather intractable polymer UHMW-PE (melt-processing is impossible due to the excessive high melt-viscosity) and induces a *favourable structure/morphology* for ultra-drawing but the *solvent is not essential* during the drawing process.

A very simple model for this enhanced drawability of solution-spun/cast UHMW-PE was put forward by Smith et al. [36] based on a network approach, ignoring completely the morphology and crystal structure. This model is derived from classical rubber elasticity theory and assumes that entanglements are trapped in polyethylene upon crystallization and act as semi-permanent crosslinks in a physical network upon solid state drawing. The crystals as such do not resist deformation in the temperature domain, close to but below the melting temperature, which is employed for (ultra-) drawing. The maximum draw ratio scales with the ratio between a fully stretched strand, scaling with M_e, between two entanglement points, and the average unperturbed end-to-end distance which scales with the square root of M_e, hence λ_{max} scales with $M_e^{0.5}$. In fact, the entanglement model is similar to case b) discussed above, but focuses not on the whole chain but on a part of the chain, the strand between adjacent entanglement loci.

Upon dissolution, the entanglement density decreases and the molar mass between entanglement loci increases, in good solvents about proportional to the inverse of the volume fraction of polymer in solution,: M_e (sol.) = M_e(melt)/ φ. The maximum draw ratio, λ_{max}, the ratio between a fully stretched strand, scaling with M_e, between two entanglement points and the original distance which is, based on Gaussian statistics, proportional to $M_e^{0.5}$, hence $\lambda_{max} \propto M_e^{0.5}$.

Assuming that upon solidification (quenching) M_e remains constant (ignoring reeling-in of chains onto the crystal surface which promotes additional disentangling), the maximum draw ratio of solution-cast/spun polyethylenes is given by:

$$\lambda_{max} \propto \varphi^{-0.5} \tag{3}$$

Equation 3) teaches that the maximum attainable draw ratio in solution-crystallized PE samples is enhanced compared with melt-crystallized polyethylenes, which is related to a looser physical network structure.

The entanglement model is remarkable versatile and can explain various phenomena, such as:

a) *the limited drawability of melt-crystallised UHMW-PE, λ_{max} 4-5, see Figure 5.36, since the molar mass between entanglements, Me, of polyethylenes is ~2kg/mole,*
and
b) *the dependence of the maximum draw ratio of solution-cast/spun UHMW-PE on the initial polymer concentration in solution, $\lambda_{max} \propto \varphi^{-0.5}$,*
and
c) *the increase of the drawability in the solid-state upon isothermal crystallization at low supercoolings of UHMW-PE solutions (53), or from the melt in general, due to the fact that the chains are reeled in, viz. pulled out their entanglement network.*

One should notice that the simple entanglement network model, relating the maximum draw ratio solely to a single parameter, the initial polymer concentration in solution, should be used and applied with care. In the model it is tacitly assumed that entanglement slippage does not occur and that the chain elements between entanglement loci are fully stretched out. This assumption might well be true for high-molar mass polyethylenes under optimised drawing conditions. However, with increasing drawing temperatures and with decreasing molar mass, chain slippage will occur and drawing becomes less effective.

In the literature [59,60] a unique relationship is presented between the draw ratio, the degree of chain extension/orientation and the resulting tensile modulus as function of draw ratio for various polymers, including PE [59,60], PP [61] and PETP. Moreover, it was shown that the birefringence and the tensile (Young) modulus of drawn polyethylene fibres, under optimised drawing conditions, solely depend on the macroscopic draw ratio, independent of the molar mass, the molar mass distributions and the type and content of branches and the initial morphology. Detailed neutron scattering experiments, using partly deuterated polyethylene samples, confirm the above view on affine deformation and these experiments favour all the simple 'entanglement model'.

Last but not least, the proposed 'entanglement network' model is not universally valid. It can be applied to apolar polymers such as polyethylenes and polypropylenes but not to polymers possessing relatively strong secondary interactions, such as hydrogen bonds. In the case of polyamids, the folded chain crystal resists deformation [54], see also chapter 5.3.5.

5.3.3.3 Solvent-free processing of UHMW-PE; nascent reactor powders

The 'entanglement model' explains qualitatively the influence of the initial polymer concentration on the maximum draw ratio and also teaches that a relatively large amount of solvent is needed to remove entanglements prior to ultra-drawing. Especially in the beginning of the solution (gel)-spinning technique, only very low UHMW-PE concentrations could be handled, typically below 5%. Due to extensive development efforts and the use of efficient mixing equipment, such as twin screw extruders combined with temperature-gradient drawing processes, makes it nowadays feasible to handle more concentrated solutions but, nevertheless, solution (gel)-spinning requires a major amount of solvent which has to be recycled completely.

Solvent-free routes have been a challenge ever since the invention of the solution (gel)-spinning process and numerous attempts have been made to obtain disentangled precursors via different routes. The rationale behind this approach is that once disentangled UHMW-PE structures are obtained via some route, subsequent melt-processing should become feasible, at least one would expect a time-temperature window in which disentangled UHMW-PE should possess a lower initial melt-viscosity in comparison with a standard equilibrium melt. Processing disentangled UHMW-PE and preventing re-entangling as much as possible with the help of specific shear fields viz. shear-refining conditions [62,63,64], should open the possibility to make disentangled structures possessing favourable drawing characteristics.

Additional arguments to this approach are the experimental observations that relaxation times in UHMW-PE melts over 10^4 seconds are present [65], even at 180°C. Moreover, it is well-established nowadays that it is virtually impossible to obtain homogeneous products by compression-moulding UHMW-PE powders [66,67], even at very long moulding times (> 24 hrs). The very long chains do not cross boundaries between the powder particles. Consequently, chain diffusion/mobility in UHMW-PE melts is seemingly extremely slow and one expects a certain time scale for the transformation from a disentangled structure into a 'equilibrium' melt which could be used favourably.

To prepare disentangled UHMW-PE structures is feasible and rather straightforward. A rather obvious, but not very practical approach, is to collect precipitated single crystals grown from dilute solutions. This does not solve the problem of solvent recovery, the only advantage could be that solvent recovery is performed before the spinning/drawing operation in a different unit operation.

A much more elegant method is to make disentangled UHMW-PE directly in the reactor. Polymerisation conditions are known, viz. low temperature and rather low catalyst activity, which promote the formation of folded-chain crystals directly on the surface of the (supported) catalyst [68,69]. During low temperature polymerisation on (supported) Ziegler/Natta and/or Metallocene-based catalysts, the growing chain on the catalyst surface will crystallize since the temperature of the surrounding medium is below the dissolution temperature. In the limit of a low concentration of active sites on the catalyst (surface), one could expect that the individual growing chain will form his own monomolecular crystal. Summarising, the polymerisation technology is available to provide disentangled UHMW-PE directly form the reactor and can even be optimised to provide UHMW-PE powder particles possessing long polymer chains which *have never "embraced" each other before the processing step*, viz. an extreme case of disentangling prior to processing .

Despite all efforts made to prepare specific disentangled UHMW-PE precursors for subsequent melt-spinning, the ultimate conclusion at this point in time is that processing disentangled UHMW-PE with the aim to benefit from an initial lower melt-viscosity and to preserve the disentangled state to some extent during processing and prior to drawing, is not feasible at all. The salient feature is that disentangled UHMW-PE, either obtained by collecting precipitated single crystals or via specific low-temperature polymerisations shows [69]:

a)   *the same high melt-viscosity (in shear) upon heating above the melting temperature as standard 'equilibrium' UHMW-PE melts. No memory effect from any previous polymerisation/crystallization history can be depicted,* and moreover,

b)   *upon re-crystallization from the melt, the favourable drawing characteristics of disentangled UHMW-PE are lost completely and the drawing behaviour is indistinguishable from a standard melt-crystallized UHMW-PE sample.*

These experimental observation seem at first sight incompatible with the 'entanglement network' model and current theories on polymer melts where long chain molecules reptate in virtual tubes with characteristic tube renewal times scaling with M^{3+x} where $x = 0.4 - 0.8$ [70,71].   The simple fact that no memory effects could be found, implying that the entanglement network is fully restored instantaneously, is rather contradictory to the experiments mentioned above, concerning very long relaxation times in UHMW-PE melts and the impossibility to produce homogeneous UHMW-PE products by compression-moulding.

This intriguing problem of a) no memory effect and b) immediate loss of drawability upon re-crystallization, has been addressed by Barham and Sadler following the fate of a polyethylene chain using neutron scattering upon melting [72]. They showed that the radius of gyration, R_g, which is rather low in solution-crystallized samples due to chain-folding, immediately jumps to its 'equilibrium' random coil dimensions upon surpassing the melting temperature. The authors introduced the concept of '*chain explosion*' for this instantaneous increase in the radius of gyration. In this fast chain expansion process, the individual chain seemingly does not notice its neighbouring chains. De Gennes pointed out in a recent note that if a chain starts to melt, the free dangling end of the molten chain will create its own tube and moves much faster than anticipated from reptation theory, and in fact rather independent from the molar mass. The driving force is related to the gradient of the ratio of gauche to trans conformers, between the free (molten) chain, and the chain segments in the crystal [73].

The experiments concerning the fast decay in drawability upon re-crstallization, related to the phenomenon of '*coil explosion*', point out that rather local changes, involving chain segments, can dominate the drawing behaviour and these results prompted one of us (P.J.L.) to modify the 'entanglement model' to some extent [74], without altering the consequences concerning e.g. the relationship between drawability and initial polymer solvent concentration, by focussing on chain segments and stem arrangement in crystals rather than taking complete in a 'entanglement network' into account. As will be discussed in chapter 5.5, the shear moduli of polyethylene crystals, perpendicular to the chain direction, are low. Upon crystallization from solution, the molecules fold usually along the {110} plane and the stems of a test chain (heavy dots) are shown in Figure 5.37(a) without indicating the folds.

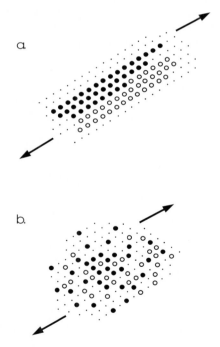

Figure 5.37   Stem arrangement in PE crystals: (a) adjacent; (b) random re-entry. Solid
               dots = test chain stems (folds not shown).

For the sake of simplicity, we assume that adjacent re-entry occurs during crystallization and that the chain is located within one crystal plane. Shearing and unfolding in the direction perpendicular to the chain and along the {110} plane is rather easy in view of the low shear moduli. Upon melting these crystals the chains will immediately adopt a random coil conformation as discussed before and stems of different molecules will interpenetrate in the 'coil explosion' process. Upon re-crystallization, the stems of the test chain are now crystallized in a more random order within the crystal and shearing (slip) is more difficult since the chains can not cross mutually during deformation, see Figure 5.37(b).

The schematic representation of stems within the crystals is, of course, an oversimplification. In actual practice, superfolding will occur and crossover of stems belonging to one chain [75]. The presented model, however, only serves the purpose to demonstrate that adjacent re-entry and locality of molecules within a crystal facilitates the process of ultra-drawing which comprises fragmentation of lamellar crystals via shearing, tilting and subsequent unfolding of clusters. The instantaneous loss in drawability upon melting and re-crystallization is due to re-arrangement and intermixing of stems involving only local chain motions rather than movement of the complete chains chain as proposed for self-diffusion in polymer melts.

The driving force for the mixing of stems and coil-explosion upon melting is the gain in entropy (which is absent in the diffusion/reptation processes in quiescent polymer melts). The polymer chain which is folded and rather compact in the lamellar crystal gains entropy upon melting and its corresponding transformation into a random coil.

Summarizing, the drawing behaviour of linear polyethylenes can be understood in terms of disentangling long chain molecules. The drawability is related to the degree of disentangling, either by crystallization from (dilute) solutions of by slow crystallization from the melt. The crystals fixate the disentangled state but do not resist deformation (unfolding). Whether entanglements, located outside the crystals, or the arrangement of stems within the crystals determine the (maximum) drawability is a matter of fine tuning which can not be confirmed by current experimental techniques. The entanglement model predicts, at least for high molar masses, the dependence of the maximum draw ratio on the initial polymer concentration in solution semi-quantitatively, (λ_{max} scales with $\varphi^{-0.5}$) and has, consequently, the advantage of its simplicity. Stem arrangement within crystals could explain the fast decay in drawability upon melting and re-crystallization. Moreover, stem arrangement and their mutual interaction do play a dominant role for polar polymers as discussed below.

5.3.3.4 Solid-state processing

From the discussion in section 5.3.3.3. one might conclude that all attempts to prepare disentangled UHMW-PE structures by specific polymerisations are in vain since no advantage could be obtained in subsequent melt-processing. This conclusion is, in fact, not true. The only lesson to be learned is that the disentangled UHMW-PE precursors should never be heated above the melting temperature. Below the melting temperature, the disentangled UHMW-PE reactor powders, the so-called nascent or virgin powders, are remarkable ductile and can be processed via calandering or hot-compacting and subsequently drawn into tapes or fibrillated structures. The draw ratio of well-prepared nascent UHMW-PE powders is similar to solution-crystallized UHMW-PE samples. Processing of UHMW-PE reactor powders has been partly successful for making oriented tapes by sintering/compacting between rollers and subsequent drawing.

Kanamoto and Porter [76] developed a two-stage drawing process for reactor powders and they obtained Young's moduli over 100 GPa. Nippon Oil Company (77,78) developed and patented several solid state processing routes for making strong UHMW-PE tapes. A process consisting of three stages: compaction, roll drawing and tensile drawing, has been developed to pilot plant stage. The products obtained by this process are characterised by a high tensile (Young's) modulus (up to 120 GPa), but a moderate tensile strength (up to 1.9GPa).

All these operations have to be performed, as discussed above, below the melting temperature. If premature melting occurs, the drawability of the (partly) re-crystallized samples is lost completely. The melting point of UHMW-PE reactor powders has been a subject of controversy. It is well-known that the melting point of UHMW-PE reactor powders, as recorded in a DSC at a fixed heating rate, can be very high, close to the equilibrium melting temperature of polyethylene, 141.5°C, to be compared with 133–135°C for melt- and solution-crystallized samples respectively. This high melting temperature has been explained in literature as the result of the formation of extended-chain crystals during polymerisation [68] on the catalyst surface, similar to the growth of whiskers in polyoxymethylene [79] and in Nature, cellulose [80]. A fibrillar morphology is often observed in the case of UHMW-PE reactor powders and substantiates the view of extended-chain crystal formation during polymerisation.

In our view, this model of extended-chain crystals in nascent UHMW-PE could be applicable in certain cases, but the high melting point can also be explained in a rather straightforward manner, based on meta stable folded-chain crystals. During polymerisation, the chains grow rather independently and form metastable folded-chain crystals as discussed in section 3.3. In the limit of low catalyst activity (low temperatures), one could envisage that ultimately one chain will form a monomolecular crystal. This implies for UHMW-PE, possessing a molar mass $M_w > 10^6$ D, that metastable crystals are formed possessing not only small dimensions, in the order of 10 nm, in the chain direction (the fold length) but equally well in the lateral dimensions. Taking as an example a molar mass of 10^6 D, the lateral dimensions are typically in the order of 10 nm. Consequently, the melting point depression is large with respect to the equilibrium melting temperature.

The melting temperature T_m of folded-chain crystals is given by the well-known Gibbs-Thomson equation, modified for lamellar crystals:

$$T_m = T_m^{\,o}\,[1 - 2\sigma_e / L_f.\rho.\Delta H_m - 2\,\sigma/A.\rho.\Delta H_m - 2\,\sigma/B.\rho.\Delta H_m] \qquad (4)$$

In equation (4), T_m is the experimental melting point, $T_m^{\,o}$ is the equilibrium melting point for infinite perfect crystals (141.5°C in the case of polyethylene), σ_e is the surface free energy of the fold planes, σ the surface free energy of the lateral planes, L_f the crystal thickness in the chain direction (fold length), ΔH_m the heat of fusion, A and B the lateral crystal dimensions, and ρ the crystal density. The value of $T_m^{\,o}$ for PE is still a matter of debate: the value of 141.5°C is taken from literature and was also obtained recently in our laboratory by Dr Hoehne et al, by careful analysis of the melting behaviour of PE fibres.

L_f is usually in the order of 10 - 30 nanometers and A and B in the order of a few microns. Consequently, the last two terms in equation (4) are usually ignored. However, when A and B are small with dimensions in the order of the crystal thickness L_f, the two last terms have to be taken into account and, consequently, contribute to the melting point depression.

In our view, nascent UHMW-PE reactor powders consist of highly metastable crystals which melt at a low temperatures. During heating in a DSC these crystals melt and reorganise continuously [69] during heating, resulting in a high end-melting temperature (final melting). This high melting point has no relationship at all with the original morphology but is due to fast reorganisations upon heating in the DSC apparatus.

a

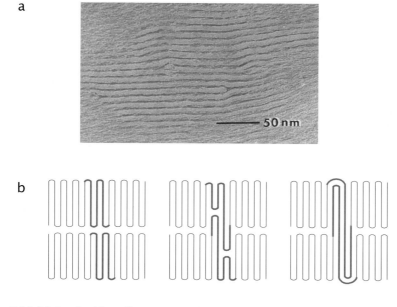

50 nm

b

Figure 5.38 (a) Stacked lamellar PE crystals and (b) model for chain (sliding) diffusion.

If fast molecular (stem) reorganisations occur at rather low temperatures, as envisaged in the case of nascent UHMW-PE reactor powders, the question arises how the drawability is preserved during compacting/sintering at temperatures below but close to the melting temperature. In section 5.3.3. it was stated that only small stem displacements are sufficient to destroy ultra-drawability.

Chain mobility below the melting temperature, in relation with drawability, has been studied in our laboratory based on model systems of well-stacked UHMW-PE single crystals. Figure 5.38(a) shows stacked UHMW-PE lamellar crystals obtained by sedimentation from dilute solutions. Upon heating these (dried) solution-cast films above ~110°C, it is observed that the lamellar thickness increases to twice its initial value, from 12.5 to 25 nm. Detailed in-situ synchrotron X-ray measurements combined with laser-Raman (longitudinal acoustic mode) demonstrate that this quantum jump in thickness is related to stem diffusion [81] across the crystal interfaces as shown schematically in Figure 5.38(b). Chain diffusion across crystal interfaces provides adhesion between crystals, a prerequisite for ultra-drawing an ensemble of individual single crystals of UHMW-PE, but the drawability is preserved since stem diffusion does not take place perpendicular to the chain direction, viz. across the crystal planes.

In conclusion, solid-state processing, viz. strictly below the melting temperature, of disentangled UHMW-PE structures is a possible route to produce high-modulus (split) fibres and tapes.

5.3.3.5 Processing via the hexagonal phase

As discussed before, UHMW-PE is an intractable polymer due to its excessive high melt-viscosity related to the high molar mass, typically $> 10^6$ D (according to ASTM definitions $M > 3. \ 10^6$ D). If one would attempt to process (extrude) UHMW-PE one would choose intuitively a processing temperature as high as possible within the limits of thermal decomposition. The result is that the extruded UHMW-PE strands show extensive melt-fracture. To one of the authors surprise (P.J.L.), it was observed in the early 80s that upon lowering the processing temperature, the extruded strands became rather homogeneous around temperatures a low as 150°C! Figure 5.39 shows the extrusion characteristics of UHMW-PE in the three characteristic temperature domains:

$\qquad$ 1) $T_{extr.} < 135°C;$
$\qquad$ 2) $135 °C < T_{extr.} < 155°C$
$\qquad$ 3) $T_{extr} > 155°C.$

At temperatures $< 135°C$, region-1, extrusion is, of course, impossible and only some sintering of individual powder particles occurs.

At high temperatures, region-3, extrusion is also not feasible due to extensive melt-fracture.

In a narrow temperature range, region-2, $135°C < T_{extr.} < 155°C$, strands could be extruded which look rather homogeneous upon visual inspection. This extrusion behaviour was independent of the initial crystallization or polymerisation history.

Since the extruded strands in temperature region 2 showed no enhanced drawability, starting from nascent reactor powder or solution-crystallized flakes, the topic of melt-extrusion in this specific temperature region was not pursued. Recently, Keller and Kolnaar [82,83] revisited this topic and they were able to show that the hexagonal phase plays a role in this region-2 extrusion process. During extrusion, the UHMW-PE powder is in contact with the cylinder and die walls and orientation is induced, in particular at the interface polymer/metal. At this interface, the 'mobile' hexagonal phase could occur at temperatures around 155°C, and this 'mobile' hexagonal interface lubricates the extrusion process of UHMW-PE strands. The core of the strands consists of compacted UHMW-PE powder particles which are just melted and poorly sintered/fused. The extruded strands, consequently, demonstrate a drawability in the solid state which is at most similar to standard melt-crystallized UHMW-PE samples, but usually the maximum drawability is lower due to poor fusion/welding of the individual UHMW-PE particles, see below. Nevertheless, the occurrence of the hexagonal phase in polyethylene is a subject which deserves more future attention in relation to processing, both for 3-D products and for fibres (1-D), as will be discussed below.

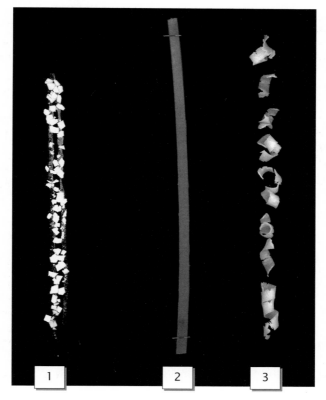

Figure 5.39   Extrusion characteristics of UHMW-PE: region 1 extrusion at temperature $T_{extr.} < 135°C$; region 2 $135°C < T_{extr.} < 155°C$; region 3 $T_{extr} > 155°C$, i.e. extrusion of UHMW-PE melts

For *isotropic* polyethylene, the hexagonal phase is usually observed at elevated pressure and temperature, in fact above the triple point Q located at 3.4 Kbar and 220°C according to the pioneering work by Bassett et al. and Wunderlich [84,85,86]. More detailed studies [87] showed later that the equilibrium triple point Q_0 is located even at higher P and T values, 5.2 Kbar and 250°C respectively, see Figure 5.40.   The equilibrium triple point was obtained by in-situ light microscopy and X-ray studies [87,88] and is higher than obtained by differential thermal analysis previously [84, 85,86]. The reason for different literature data for the triple point Q is the effect of the (initial) crystal size, to some extent similar to the melting-point depression in folded-chain crystals possessing limited crystal dimensions, notably the small fold length, as discussed before.

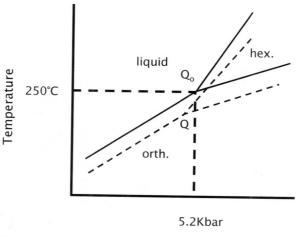

5.2Kbar

Pressure

Figure 5.40   Pressure v. temperature diagram for polyethylene (hex = hexagonal phase; orth = orthorhombic phase).

It was proven by detailed in-situ optical microscopy studies [87,88] that even in the stable orthorhombic phase, even far below the reported (experimental) triple point of Bassett and Wunderlich (84,86), that crystals grow initially in the hexagonal phase. After a certain time, c.q. crystal size, these hexagonal crystals are transformed into the thermodynamically stable orthorhombic crystal structure and crystal growth is arrested, both laterally and in the chain direction. Only in the hexagonal phase, due to enhanced chain mobility, refolding to longer fold lengths (chain-extension) is feasible and ultimately full chain extension can be obtained, depending on P, T and molar mass.

The observation that the thermodynamic stable crystal structure is reached via a metastable state of matter, in which crystals grow faster, is not unique for polyethylene but has been invoked as early as in 1897 by Ostwald, commonly expressed as Ostwald's stage rule [90,91].

The effect of the (initial) crystal dimensions on the phase stability might be utilised for processing UHMW-PE. It was shown by us [66] that solution-grown PE crystals or specific nascent reactor powders show a metastable (transient) hexagonal phase far below the equilibrium triple point, for example at pressures as low as approximately 1 Kbar! The occurrence of this metastable hexagonal phase was related to the small crystal dimensions. Annealing these crystals shows that the hexagonal phase is really metastable since the crystals transform with time into the orthorhombic structure. Increasing the P and T will result again in the hexagonal phase, and a transformation into the orthorhombic crystal structure upon annealing. This opens a novel route for sintering UHMW-PE reactor powders. It is well-known that sintering of UHMW-PE powders takes many hours up to days and the result is still a compression-moulded product containing grain boundaries due to poor sinterability, related to the high melt-viscosity and poor flow characteristics, see above. In fact, UHMW-PE is a highly intractable polymer. Following the 'hexagonal crystal route', grain boundary free products can be obtained as follows. Nascent UHMW-PE reactor powder consisting of small crystallites is pressurized and heated into the (metastable) hexagonal phase. In the hexagonal phase (partly) chain-extension will occur. Chain-extension implies disentangling and ultimately, upon further heating and increasing pressure, the degree of chain-extension and, consequently, disentangling can be controlled. Upon heating in the liquid phase, above the stable thermodynamic regime for the hexagonal phase, will result in melting. As discussed before, in section 5.3.3, upon melting the chains will adopt immediately ('chain-explosion') their random coil conformation and will interpenetrate each other, resulting in a grain boundary free material, viz. the original morphology on nano (crystal size) and micro-scale (powder particle size) is lost completely.

Apart from processing/sintering UHMW-PE for various applications, notably hip-joints and other demanding applications, the route via the metastable hexagonal phase might also be used in the future for drawing polyethylene in general into 2-D products. Ward et al. used annealing in the hexagonal phase to change the entanglement density to obtain enhanced drawability [92]. Upon annealing in the hexagonal phase, disentangling will occur as a consequence of extended-chain crystal formation as discussed above.

For *oriented* UHMW-PE, the hexagonal phase can be even observed at much lower temperatures and pressures, compared with isotropic polyethylenes. When a gel-spun/drawn UHMW-PE fibre such as Dyneema is heated under stress, for example embedded in a composite matrix or in a tensile tester under a constant load, the ultimate temperature at which the fibre fails is 155°C. This seems rather surprising at first sight since the equilibrium melting temperature of a perfect polyethylene crystal is 145.5°C. However, there is no apparent reason for an oriented 1-D polymeric structure to melt at all when heated under constrained conditions. For example, oriented gel-spun/drawn UHMW-Polypropylene fibres can be heated for a prolonged time at 230°C without any loss in mechanical properties [93]. In the case of UHMW-PE fibres, the ultimate temperature is 155°C and this temperature is related to the onset of the hexagonal phase. The orthorhombic structure in the UHMW-PE fibres transforms in the hexagonal crystal structure at $T > 155$°C and this so-called 'mobile phase' can not bear any load and, consequently, the fibre fails. It is rather unfortunate that this solid-state transition from orthorhombic into the hexagonal crystal structure occurs since it sets an upper temperature limit to the use of UHMW-PE fibres in composite applications which would have been otherwise much higher, see also section 5.5.5.

Summarising the results concerning the drawing behaviour of ultra-high molecular weight polyethylene, one can make the following conclusions for *isothermal drawing* experiments, see Figure 5.41

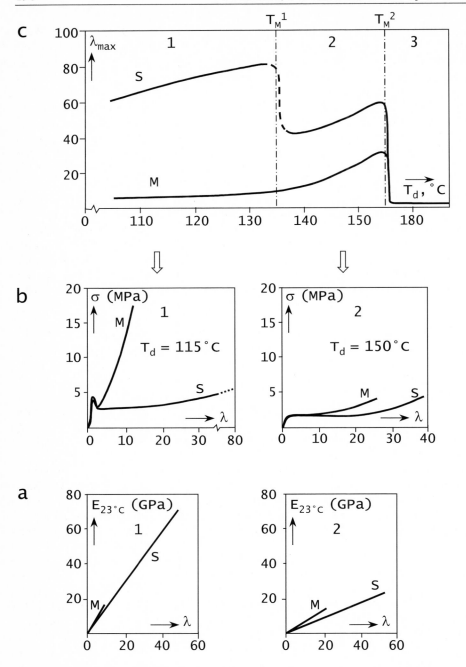

Figure 5.41   Isothermal drawing characteristics of UHMW-PE (a) 23°C Young's modulus-
draw ratio; (b) stress-draw ratio; (c) maximum draw ratio – drawing
temperature.   M = melt crystallized; S = solution crystallized.

Figure 5.41(c) summarises the drawing characteristics of UHMW-PE in the three temperature domains discussed above, for standard melt-crystallized (M) and solution-crystallized (S) samples, respectively. The drawing characteristics of nascent reactor powders are not depicted since they are similar to the solution-crystallized samples.

In region-1, below the melting temperature of folded-chain crystals T_m^1, a large difference is observed between melt-crystallized (M) vs. solution-crystallized (S) samples, see 5.41(c). Solution-crystallized samples become ultra-drawable and the drawability is dependent on the molar mass and initial polymer concentration. Figure 5.41(b) and 5.41(a) show the corresponding stress strain behaviour and properties, respectively.

In region-2, $T_m^1 < T_{draw} < T_m^2$ (155°C) there is often a noticeable difference in drawing behaviour between melt-crystallized samples, obtained via compression molding, and solution-crystallized samples. These differences are not related to a difference in entanglement network structures, but related to macroscopic effects like poor sintering in the case of melt-crystallized samples. Solution-crystallized samples are somewhat better drawable (no grain boundaries) but in both cases the drawing efficiency, in terms of the development of the tensile modulus as a function of the draw ratio, is low. In region-2 chain slippage and relaxation processes occur and effective drawing is not feasible.

In region-3, $T_{draw} > 155°C$, drawing is not possible due to the onset of the hexagonal phase.

5.3.3.6 Drawing behaviour of other flexible polymer systems

The success of melt-spinning/drawing in general and of solution(gel)-spinning in the particular case of ultra-high molecular weight polyethylene, to produce oriented polyethylene structures possessing high tensile moduli, stimulated the research activities concerning the drawing behaviour of other linear polymers, notably polypropylene, polyoxymethylene and the aliphatic polyamides, nylon 6 and nylon 66. The prime motivation for using these polymers to obtain high modulus and high strength fibres is their higher melting temperature in comparison with linear polyethylenes. In this respect, it has to be noted that a fundamental difference exists between the drawability of apolar polymers such as polyethylene and polypropylene on the one hand and polar polymers such as the polyamides, on the other hand.

In Table 5.3.3, the tensile (Young's) moduli of melt- and solution-spun, drawn polypropylene [94,95,96] and polyoxymethylene [97,98] are compared with the estimates of the maximum moduli based on X-ray measurements. The experimental values for the tensile moduli approach the maximum moduli, derived from X-ray studies, which illustrates that the concepts derived for linear polyethylenes can be used for other apolar polymers. The ultimate properties of solution(gel)-spun polypropylene fibres are, of course, limited intrinsically due to the fact that the polypropylene chain possesses a 3_1 helix conformation in the solid state, see 5.3.2.1., and consequently the upper limit of the tensile modulus is below 50 GPa. Nevertheless, also in the case of i-polypropylene the theoretical limits are approached.

Table 5.3.3: The experimental and X-ray Young's modulus of melt- and solution-spun polypropylene and polyoxymethylene fibres

Polymer	process	modulus [GPa] (experiment)	modulus [GPa] (X-ray)	Reference
i-PP	melt-spinning	20	46	[94,95]
	solution-casting	36	46	[96]
POM	melt-spinning	40	73	[97,98]

Much more interesting would be to produce fibres from polar polymers such as the polyamides, see Table 5.3.2. The high melting temperatures, 225 (nylon 6) to 310^0C (nylon 46) compared with polyethylene, $\sim 150^0$C, and the presence of hydrogen bonds, which could reduce the creep, see below, make the polyamides attractive candidates. An extensive research effort has been performed to produce high modulus and strength fibres based on aliphatic polyamides. These attempts have failed however, despite major efforts in industry to perform solution(gel)-spinning based on high molar mass polyamides. The reason for this failure is rather obvious. It was demonstrated by Postma et al. [54] that the hydrogen bonds in lamellar, solution(gel)-crystallized polyamides, are essentially static up to the melting temperature and act as barriers prohibiting draw. Solution(gel)-spinning promotes the formation of rather well-developed polyamide (single) crystals and, consequently, unfolding into extended-chain structures becomes more difficult.

Another polymer of interest is poly(vinyl alcohol), PVAL, which is commercially available in its atactic form. The small -OH side group does not prevent atactic PVAL to crystallize and the combination of a small side group and an orthorhombic crystal structure, like polyethylene, renders a high theoretical stiffness value, see Table 5.3.2. The intermediate character in terms of polarity of PVAL, more polar than polyethylene but less directed hydrogen bonds (atactic) compared with polyamides, results in an drawability in between the both extremes, respectively polyethylene and polyamides. The major difference with drawing of polyethylene is that in the case of PVAL the alpha-relaxation temperature increases with the draw ratio. In the case of polyethylene, the alpha relaxation temperature at $\sim 70 - 80^0$C, hardly increases upon drawing, in other words the crystals remain ductile, even in a highly oriented/extended structure. This property is favourable for ultra-drawing but also is responsible for creep upon static loadings, see below. In the case of PVAL, the alpha-relaxation temperature increases upon draw and fibre fracture occurs as soon as the alpha-relaxation temperature approaches the melting and/or drawing temperature [99,100]. This limits, to a certain extent, the maximum attainable draw ratio. Poly(vinyl alcohol) fibres with a tensile (Young) modulus and tensile strength of respectively ~70 GPa and ~2.5 GPa have been produced [101,102].

Similar observations were reported recently concerning another polymer with intermediate polarity, the polyketone fibres (PECO). Fibres, based on alternating copolymers of ethylene and carbon monoxide possessing a tensile modulus and tensile strength of respectively ~50 GPa and 3.5 GPa were produced by Lommerts [103].

Lommerts proposed that the maximum attainable draw ratio of semi-crystalline polymers is related to their cohesive energy density which, in principle, represents the total energy of all intermolecular interactions in a polymer. The experimentally observed relationship between maximum attainable draw ratio of semi-crystalline polymers and cohesive energy density further illustrates that enhanced intermolecular interactions in 'polar' polymers dominate their solid state drawing behaviour.

Research concerning drawing of polymers possessing an intermediate polarity is still going on. For example, the drawability of high molecular weight polyesters (PETP) has been studied extensively by Ito and Kanamoto. Moduli up to ~ 35 GPa and tensile strengths up to about 2 GPa could be obtained [104].

In conclusion, the success of producing oriented polyethylene structures possessing high moduli, 70 GPa via meltspinning/drawing and > 100 GPa via solution(gel)-spinning, could not be repeated for other flexible polymers up to now. Polyethylene is unique since the crystals are ductile and can be unfolded rather easily upon draw, provided that the PE chains are organised properly within the crystals (disentangled and regularly folded). However, one has to pay a penalty for this 'easy draw'. PE fibres show pronounced creep, as discussed below, since there is no fundamental difference, albeit at a different time-temperature scale, between drawing, to produce the fibre, and creep upon static loading.

5.3.4   Properties and applications of polyethylene fibres

5.3.4.1 Tensile strength (1-D)

Until the late seventies, the maximum tensile strength of textile and technical yarns based on flexible macromolecules was limited to approximately 1 GPa. This situation was changed with the discovery of solution(gel)-spinning of UHMW-PE fibres. Presently, UHMW-PE fibres possessing tensile strength's of 3-4 GPa are produced commercially, for example Dyneema^{TM} by DSM and Spectra^{TM} by its licensee Allied Signal. Figure 5.42 shows the properties of these high-performance polyethylene fibres in comparison with other advanced and classical (steel, glass) yarns. Due to its low density, the specific values for the stiffness and strength of polyethylene fibres are currently superior, at least at ambient temperatures. On a laboratory scale, fibres with a strength level up to 6-8 GPa can be made by optimized drawing procedures.

It is obvious that the experimental values for the maximum tensile strength of solution-spun, ultra-drawn UHMW-PE fibres (6-8 GPa) are still low in comparison with the theoretical values, viz. > 20 GPa, for an extended polyethylene chain with an infinite molecular weight. In the past, different approaches were used to describe and to interpret the origin(s) of this discrepancy between experimental and theoretical values. One aspect of this discrepancy has already been addressed in section 5.2.3, the difference between finite and infinite chains. In the case of finite chains, the overlap of and the secondary forces between the chains are of utmost importance. The molar mass distribution and in particular the number average molecular weight (chain ends) are important parameters.

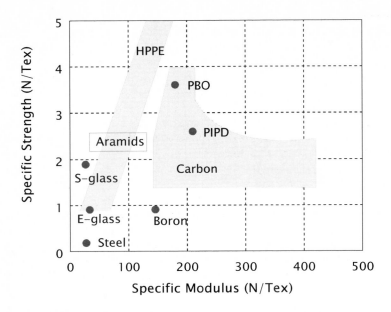

Figure 5.42   Specific strength versus specific modulus of various fibres

The influence of the weight average molar mass on the tensile strength of melt- and solution-spun, ultra-drawn polyethylene fibres was systematically investigated by Smith et al. [105]. In these studies, the tensile strength of fibres was compared at a fixed E- moduli to eliminate the influence of degree of orientation and chain extension on the tensile strength. It was shown that the tensile strength of drawn fibres increases with increasing molecular weight and an empirical relationship between the tensile strength, tensile modulus and the molecular weight was derived: $\sigma \propto E^{0.7} M_w^{0.4}$

Smith and Termonia [106] have addressed the issue of finite chains theoretically and they have developed a kinetic model. The influence of the molar mass and the effect of chain-end segregation on the theoretical tensile strength of polyethylene and aromatic polyamides was investigated by Smith and Termonia using their kinetic model. For a molar mass in the order of 10^6 D, the theoretical tensile strength is estimated to be approximately 10 GPa.

Wang and Smith proposed that the strength of polyethylene fibres is limited by stress-induced local and ephemeral melting, and infer a limiting tensile strength of 7.5 GPa, independent of the molar mass [107,108].

Summarizing, the factors determining the ultimate tensile strength of polyethylene fibres is still a matter of debate. The tensile strength values of UHMW-PE fibres, obtained via optimized laboratory experiments, 6-8 GPa, are however already close to the theoretically predicted maximum values of approximately 7.5 - 10 GPa. The question remains whether chain rupture is the limiting factor or stress-induced melting, or alternatively a solid-solid transition into the hexagonal phase.

A parameter which will not be discussed in this section is the influence of the fibre diameter on the tensile strength of polyethylene fibres. It was demonstrated convincingly by Bastiaansen that, at least in the case of solution(gel)-spun polyethylene fibres, there is no effect of the fibre diameter on tensile strength [99, 109].

All experimental studies concerning the tensile strength of polyethylene fibres were, of course, focused on achieving the maximum values for strength and stiffness in the fibre direction, a typically 1-D(imensional) problem. Unfortunately, the world is 3-D(imensional) and this fact was noticed immediately when UHMW-PE fibres were used in composite applications as will be discussed below.

5.3.4.2 Properties (3-D)

Polyethylene fibres are highly anisotropic structures. The favourable characteristics in the fibre (chain) direction in terms of stiffness and strength are not matched at all in off-axis properties. This behaviour can be understood from the so-called stiffness matrix of polyethylene. The complete stiffness matrix Cij and the compliance matrix Sij of perfect polyethylene (single) crystals, were calculated by Tashito et al [110] and are presented below.

$$
\mathrm{Cij} = \begin{vmatrix}
7.99 & 3.28 & 1.13 & 0 & 0 & 0 \\
3.28 & 9.92 & 2.14 & 0 & 0 & 0 \\
1.13 & 2.14 & 316 & 0 & 0 & 0 \\
0 & 0 & 0 & 3.19 & 0 & 0 \\
0 & 0 & 0 & 0 & 1.62 & 0 \\
0 & 0 & 0 & 0 & 0 & 3.62
\end{vmatrix} \ \mathrm{GPa}
$$

$$
\mathrm{Sij} = \begin{vmatrix}
14.5 & -4.78 & -0.019 & 0 & 0 & 0 \\
-4.78 & 11.7 & -0.062 & 0 & 0 & 0 \\
-0.019 & -0.062 & 0.032 & 0 & 0 & 0 \\
0 & 0 & 0 & 31.4 & 0 & 0 \\
0 & 0 & 0 & 0 & 61.7 & 0 \\
0 & 0 & 0 & 0 & 0 & 27.6
\end{vmatrix} \ (10^{-2}) \, \mathrm{GPa}^{-1}
$$

The compliance and stiffness in any direction can be calculated from these tensors. For example, when the b-axis of the crystal is oriented perpendicular to the uniaxial drawing direction, the tensile modulus $E(\theta)$ as a function of the test angle (between the testing direction and the chain direction) is given by :

$$E(\theta) = S_{33}^{-1}(\theta) \tag{5}$$

with:

$$S_{33}(\theta) = S_{11}\sin^4\theta + S_{33}\cos^4\theta + 2(S_{13} + S_{55})\cos^2\theta\sin^2\theta - 2S_{35}\cos^3\theta\sin\theta - S_{15}\cos\theta\sin^3\theta \tag{6}$$

From equation (5), the Young 's Modulus in the chain direction, $E(0) = S_{33}^{-1}(0)$, can be calculated to be 312 GPa.

Figure 5.43 shows the E-Modulus as a function of θ and demonstrates the highly anisotropic characacter of the orthorhombic polyethylene crystal. The dramatic drop in the E-modulus, even at small test angles with the chain direction, is mainly caused by the low shear moduli of polyethylene, an advantage for ultra-drawing, see section 5.4, but detrimental for off-axis properties of oriented polyethylene fibres. For comparison, the orientation dependence of graphite and PPTA is also plotted and last but not least of glass (isotropic).

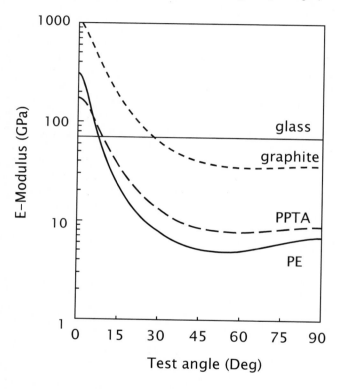

Figure 5.43   Dependence of E-modulus on test angle for various fibres

The low transverse and shear moduli of polyethylene are due to the absence of specific interactions perpendicular to the chain (only weak van der Waals bonding). The absence of strong interactions between adjacent chains causes also the strength of the fibres to be highly anisotropic. a major drawback in structural composite applications.   It was shown by Peijs [111] that the low interlaminar shear strength, ILLS, obtained from a three point bending test of unidirectional composites containing UHMW-PE fibres, see Table 5.3.4, is inherently due to the poor shear and compressive properties of polyethylene fibres. Consequently, UHMW-PE fibres can not be used in structural composite applications.

Table 5.3.4

Fibre	ILSS (MPa)
UHMW-PE (untreated)	13
UHMW-PE (corona/plasma treated)	20-30
Aramid	45-70
E-glass	75-95
Carbon	80-120

5.3.4.3 Properties and major applications

In the previous sections the properties of solution-spun, ultra-drawn UHMW-PE fibres were discussed in terms of their tensile (Young) modulus and tensile strength which show impressive values at short loading times but rather extensive creep at prolonged loading times, see below. These properties are, of course, relevant in a large number of applications such as fishing nets and/or lines, ropes and composites. Solution-spun, ultra-drawn UHMW-PE fibres also possess a number of other favourable properties such as a high impact resistance, high cutting- resistance, low dielectric constant, low dielectric loss, low stretch, high heat conductivity in the fibre direction, and a high sonic modulus. Especially these 'secondary' properties are an important advantage in applications such as ballistic applications (bulletproof vests), helmets (impact), hybrid composites (impact), gloves (cutting-resistance), fishing lines (low stretch), loudspeaker cones (sonic modulus) and radomes/sonar domes (dielectric properties).

Figure 5.44 presents the most important commercial applications of gel-spun UHMW-PE fibres.

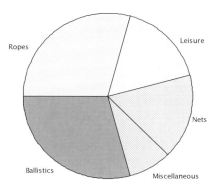

Figure 5.44   Applications of UHMW-PE fibres

It shows that most applications are either in ropes/nets or in products for ballistic protection. Long-term properties related to deformation and creep are relevant to or limiting for all rope and cable applications and for nearly all applications listed under the heading miscellaneous. Tables 5.3.5 presents more details concerning properties and applications of gel-spun PE fibres.

Table 5.3.5: Physical properties of solution(gel)-spun fibres

Density	970 – 980 kg/m^3
Crystallinity	80 - 90 %
Speed of sound	10 –12.10^3 m/s
Relative dielectric constant	2.3 –2.4
Dielectric loss factor	10^{-4}
Melting point (maximum)	155°C
Thermal conductivity	20-40 W/mK

Properties which are limiting the use of the fibres in specific applications are: creep (ropes, composites), low compression strength (composites), low melting temperature (composites), low adhesion (composites), and low transverse strength (ropes, composites). A low transverse strength is a common property of all highly drawn polymeric fibres, the other properties are specific to polyethylene fibres. Improvement of one ore more of those limiting properties, can be expected to increase the range of possible applications.

5.3.5   Limiting properties of polyethylene fibres

5.3.5.1   Melting temperature

The melting temperature of polymeric materials has been a topic of extensive study in the past and still, up to now, almost 50 years after the discovery of linear polyethylene, the equilibrium melting temperature T_m^0 of polyethylene is still a matter of debate and literature data range from 141 – 146 °C: we took a value of 141.5°C. The equilibrium melting temperature refers to the melting under atmospheric pressure of an infinite perfect PE crystal, in equilibrium with its surrounding melt. Solution(gel)-spun and ultra-drawn polyethylene fibres approach these perfect PE crystals and, consequently, one would expect these fibres to melt close to the equilibrium melting temperature, viz 145-146°C.

When oriented solution(gel)-spun PE fibres are heated in a differential scanning calorimeter (DSC) , multiple melting peaks are recorded in the temperature range from 142 – 155 °C, or at even higher temperatures when high heating rates are employed. This complex melting behaviour has been analysed by us in the past [1] and can be explained straightforwardly.

When crystalline polymeric fibres are constrained, for example in a DSC pan or embedded in a composite (matrix), the melting of the fibre depends on the type and amount of constraints. The melting temperature, in a first approximation, is given by: $Tm = \Delta H/\Delta S$ and the chains in the fibre can not melt when the fibre is constrained, because the chains can not adopt a random coil conformation. Consequently, ΔS is low and the melting temperature increases. However, when the PE fibres are chopped into small pieces and suspended into silicon oil in the DSC pan, to avoid any external constraints, a melting temperature of 141.5^{0}C will be recorded, as recently measured by Dr Hoehne in our laboratory.

An increased melting temperature has been has been observed for solution(gel)-spun polypropylene [93] and these PP fibres can be annealed for many hours at temperatures as high as 210 ^{0}C, so 60 degrees above the standard melting temperature, provided that they are constrained (loaded).

In the case of UHMW-PE fibres, this enhanced melting by external constraints is not possible due to the onset of the hexagonal phase at 155 ^{0}C. Above this temperature, the orthorhombic crystals in the fibre transform into the hexagonal structure. In the hexagonal phase the chain mobility is high and the fibre can not sustain any load and fails [1].

5.3.5.2 Long term mechanical properties: creep

Soon after the gel-spun fibres were produced, it became apparent that the exceptionally good short-term mechanical properties were not matched with equally good long-term properties. In tensile tests at high strain rates, solution(gel)-spun PE fibres behave like 1-dimensional diamond, possessing superior specific tensile moduli and tensile strengths. The high work-to-break (energy absorption) at relatively low strain-at-break, 3-6 %, results in UHMW-PE fibres which are ideal candidates for ballistic protection applications such as bullet proof vests and protection shields for armoured vehicles.

The mechanical properties of solution(gel)-spun PE fibres at prolonged time scales were studied extensively in our laboratory by Govaert [112] and more recently by Jacobs [115]. They found that the mechanical properties of solution (gel)-spun PE fibres at prolonged time-scales are rather similar to melt-spun PE fibres especially from a phenomenological point of view. The creep behaviour of these melt-spun PE fibres has been studied extensively by Wilding and Ward [116-119] and their results are summarised in Chapter 5.1

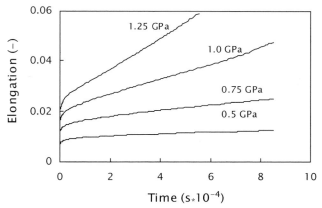

Figure 5.45   Ambient temperature Elongation (creep) of Dyneema® SK66 UHMW-PE fibre possessing a tensile (E-) modulus of ~100 GPa

In tensile tests at low strain rates, the UHMW-PE fibres exhibit a drop in Young's modulus and they exhibit a yield stress which is highly dependent on the strain rate. In fact, it is shown that the tensile strength (yield stress) drops by at least an order of magnitude at prolonged time scales.

Some representative creep data of UHMW-PE fibres, in this case of a Dyneema® fibre (SK66) possessing a tensile modulus of ~ 100 GPa are shown in Figure 5.45: initially the elongation increases proportionally to the logarithm of time. The creep compliance of the fibre is plotted as a function of stress, temperature and loading time in figure 5.46. In this figure the compliance ($E^{-1} = \varepsilon/\sigma$) is plotted vs. the logarithm of time. The slope of the initial part of the graphs appears to be independent of both the temperature and the stress (Figure 5.46) which indicates that the process is linear viscoelastic. For long loading times the creep process is non-linear, as is demonstrated by the diverging graphs in Figure 5.46.

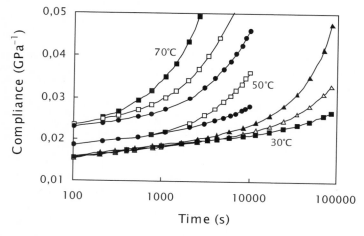

Figure 5.46   Creep compliance of Dyneema® SK66 UHMW-PE fibre at various temperatures and stresses • 0.25GPa; □ 0.4GPa; ■ 0.5GPa; Δ 0.75GPa; ▲ 1GPa

Govaert [112] quantitatively investigated the separation of the total elongation of a gel-spun polyethylene in reversible and irreversible creep contributions. It was shown that the viscoelastic contribution to the total deformation is reversible and this in contrast to the non-viscoelastic contribution i.e. the irreversible creep therefore behaves as a flow process. It should be noted here that both the reversible, visco-elastic and the irreversible, non-visoelastic deformation occur simultaneously. At short time scales, the visco-elastic, reversible response dominates the deformation and this is in contrast with prolonged time scales where the irreversible deformation is predominantly observed.

The visco-elastic, reversible mechanical response of UHMW-PE fibres was related to the structure and morphology of the fibres by Govaert and Lemstra. In practice, UHMW-PE fibres are not perfect and the crystallinity is ~ 80 - 90%. The amorphous component can be due to residual entanglements, chain ends, remaining folds and, according to Govaert and

Lemstra [114], to chains which are out of register, as shown schematically in Figure 5.47. Since chains can not cross during the drawing operation, some chain segments will not be in a parallel register in the fibre and they will contribute to recoverable creep, see below. Govaert and Lemstra [114] attribute the reversible, linear viso-elastic creep to the tensioning and retraction of (non-crystalline) chain segments that are out of register.

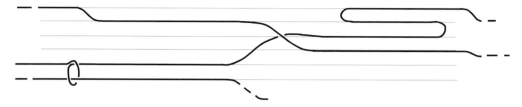

Figure 5.47   Chain alignment in a polyethylene fibre

As discussed previously, the flow creep becomes important at longer loading times. This pronounced, irreversible creep is not too surprising when we take into account that UHMW-PE fibres are made by ultra-drawing and in the drawing process folded-chain crystals are transformed into chain-extended structures. In order to avoid fibre breakage in the drawing operation, the fibres are never drawn to their maximum draw ratio λ_{max} and , since there is no 'lock-in' mechanism, the fibres can de drawn further in any subsequent tensile loading experiment, such as a creep test.   Consequently, there is no basic difference between the drawing operation and a creep test, especially at elevated temperatures. Moreover, a pronounced, irreversible creep is expected even in the hypothetical case of a perfectly oriented (i.e. infinitely drawn) fibre with a perfectly elastic mechanical response at low deformations (which corresponds to a total elimination of visco-elastic, reversible contributions to the mechanical behaviour). Termonia and Smith modelled such a hypothetical fibre and it was shown that the stress-strain behaviour of these polyethylene fibres exhibits a strong dependence of the breaking stress on the strain rate [106]. In their model, however, they visualise a PE fiber as a perfect array of extended chain molecules which is, however, not the reality. Their modelling has been very useful to understand the time dependence of the tensile properties, the importance of chain overlap and intermolecular interactions, see Figures 5.34 and 5.35, but can not describe the actual creep behaviour of PE fibres since notably the reversible, visco-elastic creep is dominated by PE molecules which are not in a perfect crystalline array but out of register, see above.

To further demonstrate the similarity between solution (gel)-spun and melt-spun fibres, we have plotted some experimental data concerning both fibres in a similar way. A so-called Sherby-Dorn plot is used (Figure 5.48) which was used previously by Ward-et al in the analysis of the creep of melt-spun polyethylene fibres. In such a Sherby-Dorn plot, the creep strain rate is given as a function of the strain [120]. Initially, the creep rate is high, it decreases with increasing elongation and levels off at an elongation of a few percent. The creep rate for the fibre attains a constant value, which is referred to as the plateau creep rate. Of course, the initial part of the curve, in which the strain rate drops, corresponds to the reversible, visco-elastic regime while the plateau regime corresponds to the irreversible, non-linear domain.

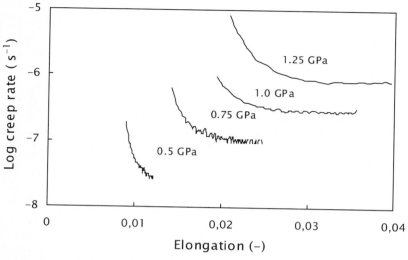

Figure 5.48   Sherby-Dorn plot for the creep data of Figure 5.46

The irreversible (flow) creep is a stress-assisted, thermally activated process, involving chain slip through the crystals in the fibre which is enabled by diffusion of defects, such as Reneker type defects or, as suggested by Boyd [121,122], short twist comprising 12 CH_2 units. It was shown by Ward for melt-spun PE fibres that the plateau creep rate (on a logarithmic scale) as a function of the applied stress is not linear, as is expected from a single thermally activated process. These results have been confirmed by Jacobs for solution(gel)-spun PE fibres. Ward suggested that at least two thermally activated processes act in parallel. He differentiates between a process involving chain slip through the crystals (process 2), as discussed above, and a network process (process 1) which becomes more pronounced at low stress levels, with increasing molar mass and with increasing copolymer content. One has to realise that these two molecular processes are based on models and to some extent intuition. It is impossible to discriminate between the two models using advanced characterisation methods. Figure 5.47 gives, to some extent, an idea between chains in crystalline register and chains which are out of register and contribute to the network.

Analysing creep data and pointing out similarities between melt-spun and solution(gel)-spun fibres is a valuable approach but, of course, the ultimate aim is to find sophisticated routes to improve the long-term behaviour, viz. to decrease creep. In order to explore possible routes to decrease creep one should have a clear picture of the molecular processes contributing to creep.

Assuming that creep involves two separate molecular mechanisms as discussed above, the next question is how to decrease the creep rate. Basically, there are two possibilities:

a)   modification of the polymer (before drawing);
b)   modification of the fibre (after drawing).

a)   *modification of the polymer (before drawing)*
Following the approach by Ward, Bastiaansen [123] and Ohta [124,125] studied the possibilities to reduce the flow creep, using PE with short chain branches, up to 12.5 methyl groups/1000 C atoms. The basic idea is that (small) side groups will reduce molecular slip within the crystals (hence will have an effect on process 2). The results confirmed this concept and at a *comparable* E-Modulus, the plateau creep rate decreases with increasing comonomer content. However, the maximum drawability also decreases with increasing comonomer content and, moreover, the drawing stress increases. This observation is in accordance with the statement, made above, that there is no fundamental difference between the drawing process and a creep experiment, "easy draw implies easy creep" [126.]. Consequently, PE-copolymers is not a real solution to improve the intrinsic creep behaviour of PE fibres .

b)   *modification of the fibre (after drawing)*
It is well-known that polyethylene can be crosslinked by electrons or gamma irradiation. Charlesby et al. performed pioneering experiments concerning crosslinking of polyethylene and other polymers and electron-beam radiation is nowadays a common technology to improve the properties of polymer products, notably PE films (Cryovac) and crosslinked PE shrinkable tubes (Raychem). High energy radiation such as Gamma-rays and (accelerated) electrons (0.5 – 3 MeV) has two major effects on polymer systems. On the one hand, grafting and crosslinking occurs whereas on the other hand chain scission will take place. The ratio between crosslinking and scission in the case of isotropic polyethylene is about unity. Crosslinking takes place in the amorphous zones whereas in the crystalline domains crosslinking can not occur since the chains are separated too far apart (0.4-0.6 nm) to be able to form a covalent C-C bond (0.154 nm). Ward et al. could successfully crosslink their melt-spun fibres using acetylene as a 'prorad' [127] and reduce the creep. In the case of solution(gel)-spun PE fibres radiation treatment has been less successful which is easy to understand. The solution(gel)-spun fibres are highly crystalline and, consequently, chain scission dominates.

Jacobs [115] recently used UV to crosslink solution(gel)-spun PE fibres in combination with benzophenone as a sensitiser/initiator. Benzophenone was diffused into the highly drawn PE fibres using supercritical CO_2 as a carrier. It was shown that the solution(gel)-spun PE fibres were obtained with a high gel content which illustrates that the fibres were indeed chemically crosslinked. A drastic decrease in plateau creep rate was observed and, simultaneously, the short term Young's modulus and strength of the fibres were preserved. However, also in this case the impregnation route is less applicable in the case of super-drawn PE fibres.

All known methods for improving the creep resistance, including these described above, require the presence of a certain degree of imperfection in the fibre. Creep improvement therefore implies a compromise between short term and long term mechanical properties.

5.3.6   Conclusions

It has been shown in this chapter that UHMW-PE fibers can be made possessing impressive strength and stiffness values, see Figure 5.42, especially when their specific values are taken into account. These short-term values for strength and stiffness have not been challenged and matched by any other fibre based on flexible polymer molecules. Polyethylene is the 'primus inter pares' thanks to the availability of high molar mass base material, the enhanced drawability after removing the constraints limiting drawability, the absence of specific interactions such as hydrogen bonds and the small cross-sectional area of the PE chain. The penalty one has to pay for all these beneficial characteristics is that the oriented polyethylene fibres are prone to creep. Basically there is no difference between drawing and creep experiments ("easy draw-easy creep"), i.e. there is no lock-in mechanism after draw. Consequently, UHMW-PE fibres are less suitable for applications, such as reinforcing structural composites. The high work-at-break of these fibres provides, however, unique possibilities for applications where impact resistance is important.

References

1. Lemstra, P.J., Kirschbaum, R., Ohta, T., Yasuda, H. Developments in Oriented Polymers-2, Ward, I.M. (Ed.), (1987) Elseviers Applied Science, London, pp. 39- 79
2. Lammers, M., Klop, E.A., Nordholt, M.G., Sikkema, D.J., Polymer, 1998, 33, 5999; Lammerts M., Ph.D. thesis, ETH Zürich, 1998
3. W-Fang Hwang, Proc. Int. symp. Fibre Sci. Technol. (ISF), Hakone, 1985, 39
4. Carothers, W., Hill, J.W., J.Am.Chem.Soc. 1932, 54, 1586
5. Abitz, W., Gerngross, O., Hermann, K. Naturwissenschaften, 1930, 18, 754
6. Bunn, C.W., Trans. Faraday Soc., 1939, 35, 483
7. Storks, K.H , J. Am. Chem. Soc., 1938, 60, 1753
8. Keller, A., J. Polym. Sci., 1955, 17, 351
9. Jaccodine, R., Nature, 1955, 176, 302
10. Fischer, E.W. , Z. Naturforsch., 1957, 12a, 753
11. Till, P.H., J. Polym. Sci., 1957, 24, 301
12. Keller, A., Phil. Mag., 1957, 2, 1171
13. Meijer, K.H. and Lotmar, W., Helv. Chim. Acta, 1936, 19, 68
14. Treloar, L.R.G. , Polymer , 1960, 1, 95
15. van der Vegt, A.K., Smith, P.P.A. , Adv. Polym. Sci. Monograph 1967, 26, 313
16. Southern, J.H., Porter, R.S., J., Macrom. Sci. 1970, B4, 541; Porter, R.S., Southern, J.H., Weeks, N.E., Polym. Eng. Sci., 1975, 15, 213
17. Zachariades, A.E., Porter, R. S. "The Strength and Stiffness of Polymers", Zachariades, A.E. and Porter, R.S. (Eds.), (1983) Marcel Dekker, New York, pp. 1 -51
18. Andrews, J.M. and Ward, I.M, J. Mater. Sci. , 1970, 5, 411
19. Capaccio, G., Ward, I.M. , Polymer 1974, 15, 233
20. Capaccio, G., Ward, I.M. , Polymer 1975, 16, 243
21. Capaccio, G. , Gibson, A.G., Ward, I.M. "Ultra-High Modulus Polymers", Ciferri, A. and Ward, I.M. (Eds.), (1979) Elseviers Applied Science, London,Ch. 1

22. Juergenleit, W., US patent 3,048,465 (1956)
23. Zwick, M., patent application NL 6501248 (1965)
24. Blades, H., White, J.R. , US patent 3,081,519 (1963)
25. Mitsuhashi, S., Bull. Text. Res. Inst. (J), 1963, 66 , 1
26. Pennings, A.J., Kiel, A.M. , Kolloid Z.Z. Polymere, 1965, 205, 160
27. Frank, F.C., Proc. Royal Soc., London Ser. A, 1970, 319 , 127
28. Peterlin, A., J. Polym. Sci. , 1966, B4, 287
29. De Gennes, P.G. , J. Chem. Phys. 1974, 60, 15
30. Keller, A., Odell, J.A., Colloid Polym. Sci., 1985, 263, 181
31. Pennings, A.J , J. Polym. Sci. Polym. Symp. , 1977, 59 , 55
32. Zwijnenburg, A. Ph.D. Thesis, (1978) University of Groningen
33. DSM/Stamicarbon, U.S.Patent 4,137,394
34. Mackley, M. NRDC Eur. Patent 22681
35. Smith,P., Lemstra, P.J. UK Patent 2,051,661 (1979)
36. Smith, P., Lemstra,P.J., Makrom. Chem., 1979, 180, 2983
37. Smith,P., Lemstra, P.J.and Booij,H.C., J. Polym. Sci., Phys. Ed., 1982, 20, 2229
38. Smith, P., Lemstra, P.J. J. Mater. Sci., 1980, 15, 505
39. Tashiro, K.,Kobayashi, M., Tadakoro, H., Macromolecules, 1978, 10, 914
40. Odajima, A., Madea, T. J. Polym. Sci. part C, 1966, 34 , 55
41. Nakamae, E.K., Nishino, T., Advances in X-Ray analysis, 1992, 35 , 545
42. Nishino, T., Ohkubo, H., Nakamae, K., J., Macromol. Sci.-Phys , 1992, B31 191
43. Nakamae, K., Nishino, T., Ohkubo, H., J. Macromol. Phys., 1991, B30, 1
44. Nakamae, K., Nishino, T., Ohkubo, H., Matsuzawa, S., Yamura, K., Polymer 1992, 33, 2281
45. Nakamae, K., Nishino, T., Yokoyama, F., Matsumoto, T., J. Macrom. Sci.-Phys., 1988, B27(4), 404
46. Sakurada, I., Ito, T., Nakamae, K., Bull. Inst. Chem. Res. Kyoto Univ., 1966, 44, 77
47. Nakamae, K., Nishino, T., Hata,K., Kobunshi Ronbonshu, 1985, 42, 241
48. Nakamae, K., Nishino, T., Shimiizu, Y., Hat, K., Polymer, 1990, 31, 1909
49. He, T., Polymer 1986, 27 , 253
50. Kelly, A., Macmillan, N.H., "Strong Solids", Clarendon Press, Oxford (1986)
51. Zhurkov, S.N., Int. J. of Fracture Mechanics, 1965, 1, 311
52. Termonia, Y, Smith, P., Macromolecules , 1987, 21, 835
53. Termonia, Y., Smith, P., Polymer,1986, 27, ,1845
54. Postma, A.R., Smith, P., English, A.D., Polymer Comm., 1990, 444
55. Capaccio, G., Crompton, T.A., and Ward I.M., J. Polym. Sci., 1976, 14, 1641
56. Lemstra, P.J, DSM, internal report
57. Lemstra, P.J. and Smith, P., Brit. Polym. J., 1980, 12 , 212
58. Peterlin,A., personal communication
59. Ward ,I.M., "Developments in Oriented Polymers" 2nd Ed., 1988, Elsevier, New York
60. Dirix, Y., Tervoort, T.A., Bastiaansen, C.W.M., Lemstra, P.J., Text. Inst., 1995, 86/2 , 567
61. Pinnock, P.R. and Ward, I.M., Brit. J. Appl. Phys., 1966, 17, 575
62. Pazur, R.J., Ajii, A. , Prud'homme,R.E., Polymer, 1993, 34,.4004
63. Muenstedt, H. Coll. Polym. Sci., 1981, 259, 966
64. Leblans, J.R., Bastiaansen, C.W.M., Macromolecules,1989, 22, 3312

65. Bastiaansen, C.W.M., Meijer, H.E.H., Lemstra, P.J., Polymer 1990, 31, 1435
66. Rastogi.S, Kurelec, L., Lemstra, P.J., Macromolecules, 1998, 31, 5022
67. Rastogi, S., Koets, P. ,Lemstra, P.J., patent appl. NL 98/00093
68. Smith, P., Chanzy, H.D., Rotzinger, B.P., Polym. Comm. 1985, 26 , 258
69. Engelen, Y.M.T., Lemstra, P.J., Polym. Comm., 1991, 32, 343
70. De Gennes, P.G., J. Chem. Phys., 1971, 55, 572
71. Klein, J., Briscoe, B.J., Proc. Royal Soc. Lond., 1979, B356, 53
72. Sadler, D.M., Barham, P.J., Polymer, 1990, 31, 36
73. De Gennes, P.G., C.R. Acad. Sci. Paris, t. 312/II, 1995, 363
74. Lemstra, P.J., van Aerle, N.A.M.J., Bastiaansen, C.W.M. Polymer Journal, 1987, 19, 97
75. Keller, A., Faraday Discussions of the Chemical Society , 1979, 68 145
76. Kanamoto, T., Ohama, T., Tanaka, K. Takeda, M. Porter,R.S., Polymer, 1987 28 1517
77. Otsu, O., Yoshida, S., Kanamoto, T. and Porter, R.S., Proceedings PPS, Yokohama (1998)
78. Eur. Patent Appl. EP 376 423 and EP 425 947
79. Iguchi,M., Polymer, 1993, 24 , 915
80. Iguchi, M. in "Integration of Polymers Science & Technology", Eds. P.J. Lemstra and L.A. Kleintjens, Elseviers Appl. Sci., 1991, 371
81. Rastogi, S., Spoelstra, A.B., Goossens, J.G.P. and Lemstra, P.J., Macromolecules, 1998, 30, 7880
82. Kolnaar, J.W.H., Ph. D. thesis Bristol (1993)
83. Kolnaar, J.W.H., Keller,A., Polymer 36 (1995) 821
84. Bassett, D.C., Khalifa,A., Turner,B., Nature (London), 106 (1972), ibid 240 (1972)
85. Bassett, D.C., Polymer 17 (1976) 460
86. Wunderlich,B., Grebowicz, J., Adv. Polym. Sci. 60 (1984) 1
87. Hikosaka, M., Rastogi, S., Keller, A. and Kawabata, H., J. Macrom. Sci. Phys. B31 (1992) 87
88. Rastogi, S., Hikosaka, M. , Kawabata, H. and Keller, A., Macromolecules 24 (1991) 6384
89. Hikosaka, M., Rastogi, S., Keller, A. and Kawabata, H., J. Macromol. Sci. Phys. B., 31 (1992) 87
90. Ostwald, W., Z. Physik. Chem. 22 (1862) 286
91. Keller, A., Hikosaka, M., Rastogi, S., Toda, A., Barham, P.J. and Goldbeck-Wood, G., J. Materials Sci. 29 (1994) 2579
92. Maxwell, A.S., Unwin, A.P., Ward, I.M., Polymer 37 (1996)3293
93. Bastiaansen,C.W.M., Lemstra, P.J. , Makrom. Chem., Macromol. Symp., 28 (1989) 73
94. Cansfield, D.I.M., Capaccio, G., Ward, I.M., Polym. Eng. Sci., 16 (1976) 721
95. Taylor, W.N., Clark, E.S., Polym. Eng. Sci., 18 (1978) 518
96. Peguy, A., Manley, R. st J., Polym. Commun., 25 (1984)39
97. Capaccio, G., Ward, I.M., Brit. Pat Apl. 52644/74 (1973)
98. Clark, E.S., Scott, L.S., Polym. Eng. Sci. 14 (1974) 682
99. Bastiaansen, C.W.M., "Materials Science & Technology", Eds. Cahn, R.W., Haasen, P. and Kramer, E.J., 18 , VCH, Chpt. 11, 551 (1997)
100.Garrett,P.D., Grubb,D.T., J. Polym. Sci., Polym. Phys. Ed. B26 (1988) 2509

101.Kwon, Y.D., Kavesh, S., Prevorsek, D.C., U.S. Patent 4 440 771 (1984)
102.Schellekens, R., Rutten, H., Lemstra, P.J., EP 212 757 (1986)
103.Lommerts, B.J., Ph.D thesis, University of Groningen (1994)
104.Ito, M., Takahashi, K., Kanamoto, T., J of Appl. Polym Sci., 40 (1990) 1257
105.Smith, P., Pijpers, J. and Lemstra, P.J. , J.of Polym. Sci. , Polym. Phys. Ed, 20, (1983)2229
106.Termonia, Y.,Smith, P., in "High Modulus Polymers", Eds. Zachariades A.E., Porter R.S., Marcel Dekker, N.Y., 326   (1996)
107.Smith, K.J. jr, Wang, J. Polymer 40 (1999) 7251
108.Wang, J., Smith, K.J. jr , Polymer 40 (1999) 7261
109.Bastiaansen, C.W.M., Ph. D thesis, Eindhoven University of Technology (1991)
110.Tashiro K., Kobayashi M. and Tadokoro H., Macromolecules, 11 (1978), 914
111.Peijs, A.A.J.M., Ph. D. thesis, Eindhoven University of Technology (1993)
112.Govaert L.E., Ph.D Thesis, Eindhoven University of Technology (19920
113.Govaert L.E., Bastiaansen C.W.M., Leblans P.J.R., Polymer 34 (1993) 534
114.Govaert. L.E., Lemstra P.J., Coll. Polym. Sci, 270 (1992) 455
115.Jacobs, M. J.N., Ph.D. thesis, Eindhoven University of Technology (1999)
116.Wilding, M.A., Ward, I.M., Polymer 19 (1978) 969
117.Wilding, M.A., Ward, I.M., Polymer 22 (1981) 870
118.Wilding, M.A., Ward, I.M., Plastics & Rubber Proc. Appl., 1 (1981) 167
119.Wilding, M.A., Ward, I.M., J. Mater. Sci. 19 (1984) 629
120.Ward, I.M., Wilding, A.M., J. Polym. Sci., Polym. Phys. Edn. 22 (1984) 561
121.Boyd, R.H., Polymer 26 (1985) 323
122.Boyd, R.H., Polymer 26 (1985) 1123
123.Bastiaansen C.W.M., EP 269 141 ((1987)
124.Ohta Y., J. Polym Sci., B, Polym. Phys., 32 (1994) 261
125.Ohta Y., Yasuda H., Kaji A, Polym. Preprints Japan 43 (1994) 3143
126.Conclusion of a literature study by our Ph.D. student Mrs. Xue
127.Rasburn, J., Klein P.G., Ward I.M, J. Polym. Sci., B Polym. Phys., 32 (1994) 1329

6 Development of Molecular Orientation during Biaxial Film Tentering of PET[*]

J-F Tassin
Chimie et Physique des Matériaux Polymères, UMR CNRS 6515
Université du Maine
Avenue Olivier Messiaen
72085 Le Mans Cedex

6.1 Introduction

Polyethylene terephthalate (PET) films find wide ranges of applications especially as substrates for magnetic storage and in the packaging market. It is clearly established that mechanical as well as end-use properties of the final films are mainly controlled by the structure and orientation at a molecular level, which originate from the various thermal and deformation conditions experienced by the film throughout its process. Therefore, numerous studies have been devoted to the characterisation of molecular orientation, mechanical properties and end-use properties such as gas permeability or hydrolysis, showing the importance of being able to afford an accurate characterisation of molecular orientation and structure in oriented samples.

The determination of the chain deformation processes throughout the film processing conditions requires both the production of well defined samples, meaning that the state of the initial sample and the drawing conditions are well known and properly characterised, and an extensive characterisation of the samples which can only be achieved by using a combination of several techniques. The former criterion is not so easy to fulfil as it will appear later, and therefore relatively few studies focussing on the molecular processes occurring during the biaxial stretching process are available. Furthermore, the loss of uniaxial symmetry adds several practical as well as theoretical difficulties in characterising the state of molecular orientation in the samples.

During the last ten years, collaborative studies between Rhône-Poulenc, Prof. Ward's group in Leeds University, Prof. Monnerie's group at the ESPCI in Paris and us have dealt with accurate and extensive characterisations of model biaxially oriented PET films that closely mimic the industrial processing conditions. For this purpose, a general strategy has been chosen, which consists in obtaining model samples on laboratory scale devices so that the initial state of the film as well as the stretching conditions are well defined and controlled. Starting samples usually originate from production or pilot lines and are also thoroughly characterized. This method has been used to follow the development of molecular

---

[*] *This chapter is dedicated to Prof. Lucien Monnerie, on the occasion of his retirement.*

orientation and structure in the normal and the so-called inverse stretching processes of PET films.

In the normal process, shown schematically in Figure 6.1, the molten PET film, extruded through a flat die, is quenched on a cold roll, leading to a rather thick amorphous film. It is first stretched longitudinally between a series of rolls at a temperature in the range 85-110 °C which is slightly above the glass transition temperature of PET. The industrial process usually involves temperature control of the film through infrared heaters located between the slow and the fast rolls. This stretching process is not uniaxial since it is conducted at constant width (pure shear deformation). In this case, although the draw ratio is imposed by the ratio of the tangential speed of the fast rolls over that of the slow rolls, the stretching process is conducted under a constant drawing force, instead of a constant stretching rate as for commonly working laboratory stretching machines. This implies that the film will choose its kinetics of deformation so that the stretching will be carried out under minimum force conditions [1-2]. The film is then drawn transversely to the machine direction, thanks to a series of clamps, located on each side of the film. In this stage, the deformation kinetics is imposed by the rate and the profile on which the clamps are moving. The stretching temperature is a few degrees above that of the first stretching process (in the range 100-140 °C). Finally, the biaxially stretched film undergoes a heat setting treatment at temperatures slightly below the melting temperature of PET crystals. This annealing step may, in practice, involve several temperature zones on the industrial line as well as additional transverse or longitudinal stretchings but limited to very small draw ratios, which can be used by the film producer to adjust the final properties of the film.

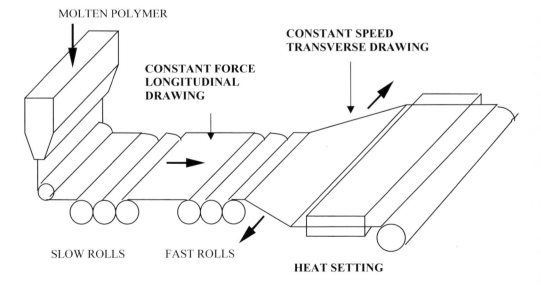

Figure 6.1 Schematic representation of the normal or balanced process of biaxial drawing of PET films

In the inverse sequence (Figure 6.2), the first stretching, applied to an amorphous film is transverse with respect to the direction of the line, whereas the second one is carried out along the longitudinal or machine direction and acts on a partially crystallised film. However, the transverse and longitudinal stretching steps keep their individual characteristics as being strain and stress controlled respectively. As in the normal sequence, the process ends with an heat-setting treatment.

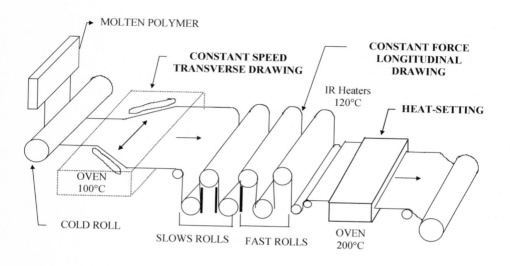

Figure 6.2 Schematic representation of the inverse process of biaxial drawing of PET films

The three main steps of both kinds of stretching processes have been studied using model samples [3-15]. In the present chapter, we will review the main results concerning the development of structure and morphology throughout these processes, on the basis of our own studies as well as those from other groups.

In a first part, the minimum information dealing with the description of the materials, and the most frequently used experimental techniques for the characterisation of oriented films as well as important definitions will be given.

In a second part, the behaviour of the polymeric material in the first stretching of both industrial processes will be described and compared, whereas the second stretching step will be discussed in the third part. Finally, the molecular mechanisms involved during the high temperature annealing step will be presented.

6.2   Definitions, Materials and Experimental Characterisation Techniques

6.2.1   Quantitative characterisation of orientation

In order to follow the development of structure and molecular orientation during stretching, it is necessary to quantify the degree of molecular orientation in a polymer film. This can be done in a general way by using the Euler angles (θ, ϕ, ψ) and a normalised orientation distribution function $N(\theta,\phi,\psi)$ where $N(\theta,\phi,\psi)d\theta d\phi d\psi$ represents the fraction of molecular units (assuming orthorhombic symmetry) having their directions located in the elementary solid angle centred around the direction defined by the Euler angles. Since this function can rarely be measured, it is more common to expand it in a series of generalised spherical harmonic functions, and to consider the projections along the various vectors [16]. In practice, second order averages afford useful and physically tractable information about the molecular orientation [17-18]. They allow to quantify the orientation of molecular direction u with respect to any macroscopic direction X_i associated to the polymer film. Therefore, hereafter, the orientation will be quantified by will be quantified by the average value of the second order Legendre Polynomial:

$$P_2^{u/X_i} = P_{200}^{u/X_i} = \frac{1}{2}\langle 3\cos^2\theta_{u,X_i} -1\rangle$$

where θ_{u,X_i} is the angle between the molecular direction u and the macroscopic axis X_i and the brackets denote an average over all the molecular units.

As far as macroscopic directions of the film are concerned, X_1 will refer to the first drawing direction, X_2 to the second drawing direction and X_3 to the direction normal to the plane of the film. Obviously, the directions X_1 and X_2 are exchanged with respect to the machine and transverse directions of the film, when the normal sequence or the inverse sequence will be considered. The angle in the plane of the film between the X_1 and X_2 directions will be noted ϕ_p and arbitrarily $\phi_p = 0$ corresponds to the X_1 direction.

The PET molecular unit involves many relevant molecular directions the orientation of which can be characterised using various techniques :
-    the chain axis direction, which is clearly defined inside the crystalline phase or for trans conformers in the amorphous one [19] (abbreviated as the c direction in the following)
-    the C_1-C_4 direction of the aromatic ring, which is slightly tilted with respect to the chain axis [20-22] (abbreviated as C_1-C_4)
-    the phenyl ring normal (abbreviated as n or ring)

Examination of X-ray diffraction patterns reveals in oriented films various intense crystallographic reflections, the indexation of which is based on the unit-cell determined by Daubeny et al. [19]:

- ($\bar{1}05$) for which plane normal is close to the chain axis direction
- (100) for which plane normal is close to the normal to the phenyl ring
- (010) for which plane normal determines a third independent direction

6.2.2 Experimental techniques: Characterization of the crystalline phase

The volume fraction crystallinity is usually measured from a density gradient column [25-26], or from the mean refractive index [23], or sometimes from DSC measurements of the melting endotherm [24]. All methods agree with the evolution of crystallinity during the polymer process, although the absolute values might differ from one method to another.

6.2.2.1 X-Ray diffraction

As far as orientation of the molecular units in the crystalline phase are concerned, the ($\bar{1}05$) and the (100) diffracting planes are usually considered and the orientation of the respective plane normals can be calculated with respect to any macroscopic direction of the film.

The quantitative use of X-Ray data involves rather long experiments using a conventional laboratory set-up and requires that several pieces of stretched film are stacked together in order to afford sufficient intensity. This problem can be overcome by using synchrotron facilities and on-line measurements are becoming available [27-28]. The ($\bar{1}05$) reflection (associated to a Bragg angle of $2\theta_{\bar{1}05} = 43.2°$) is usually studied in transmission whereas the (100) reflection (Bragg angle of $2\theta_{100} = 25.5°$) requires studies in the reflection mode. In these studies the sample is rotated around two axes, in order to sweep the relevant directions of the crystallographic axes with respect to the main directions of the film [29]. The calculation of P_{200} values requires accurate determination of the baseline intensity before numerical integration of the corrected diffracted intensity. The procedure used to collect the diffracted intensity and to compute the order parameter P_{200} of these two directions with respect to the principal directions of the sample is accurately described elsewhere and we refer the interested reader there for further details [10].

X-ray diffraction is also commonly used to estimate the dimensions of the crystals along the directions perpendicular to the ($\bar{1}05$), (100) and (010) planes, through the angular broadening of the diffraction peaks, using the Sherrer formula [29]. Taking into account the orientation of the crystalline blocks inside an oriented polymer film, hereafter, we will often refer to these dimensions as the *length*, the *thickness* and the *width* of the crystallites respectively. It is important to note, that depending on the orientation of the sample with respect to the observed diffraction vector, the dimensions of crystals located at a particular direction, with respect to the film plane, can be measured affording useful information about the deformation mechanisms.

6.2.2.2 Birefringence

The three principal refractive indices of the films are usually measured using an Abbe refractometer under polarized light, a monochromatic polarized light retardation technique or more recently a multiwavelength polarized light retardation measurement using various incident angles [30].

In the case of PET, within a good approximation, the polarizability tensor of the monomeric unit can be considered as uniaxial with the symmetry axis oriented along the normal to the phenyl ring [9,22,31]. Under this condition, the orientation (P_{200}) of the normal to the phenyl ring with respect to 3 principal axes of the film can be calculated either from the macroscopic polarizabilities which can be obtained from the Lorentz-Lorenz equation using the measured refractive indices [22], or directly from them [9].

The orientation refers to an average over the crystalline and non-crystalline material.

6.2.2.3 Infrared dichroïsm

Infrared spectroscopy has been mainly used to characterize molecular orientation in oriented PET films as well as conformational changes occuring during stretching. The most commonly used infrared bands together with the information they can bring to the characterization of PET film are listed in Table 6.1.

In biaxially oriented samples, orientation measurements require the 3 principal absorbances to be known, so that tilted films experiments are necessary for transmission measurements [22, 32-33], although direct measurement of the absorbance along the thickness direction can be achieved using IR microscopy after microtoming of the films [34]. Attenuated total reflection (ATR) spectra can also be collected to overcome the problem of saturation of strong bands [35-38] and photoacoustic detection has also been used [39]. Recently a front surface specular reflection technique has been developed in order to access to the orientational state of thick samples or to allow the use of strongly absorbing bands [40-41].

6.2.2.4 Raman spectroscopy

The Raman band appearing at 1616 cm^{-1} is suitable for the measurement of the orientation of the C1-C4 axis of the phenyl ring. The orientation represents an average over the crystalline and the non-crystalline phases. The accurate determination of the orientation in biaxially oriented films is quite painful and few studies have been reported [8,20,22]. It can be noted that Fourier Transform Raman spectroscopy can be used during deformation to study conformational changes [42].

Table 6.1 Most commonly used bands in IR spectroscopy of PET

Wavenumber (cm^{-1}) and intensity	729 (s)	875 (w)	972 (w)	1018 (s)
Attribution	Aromatic CH out of plane bending	Aromatic CH out of plane bending	Trans glycol C-O stretching	In-plane benzene ring CH deformation
Information	Orientation of the normal to the phenyl ring	Orientation of the normal to the phenyl ring	Content of trans conformers	Orientation of the C_1-C_4 axis of the benzene ring
Technique	Front surface specular reflection	Transmission	Transmission	Transmission if thin enough or front surface specular reflection
Wavenumber (cm^{-1}) and intensity	1340 (s)	1370 (w)	1410 (s)	1725 (vs)
Attribution	CH$_2$ wagging of trans conformers	CH$_2$ wagging of gauche conformers	Ring vibration	Carbonyl stretching vibration
Information	Trans conformer content, orientation of trans conformers	Gauche conformers content	Reference band	
Technique	Transmission	Transmission	Transmission, ATR	Front surface specular reflection

6.2.2.5 Fluorescence polarisation

Fluorescence polarisation has been used in two ways. The first one involves introduction of rigid fluorescent probes before film processing. A frequently used probe is VPBO, for which it has been shown that the orientation of the probes reflects that of a few (2-3) consecutive segments in trans conformation in the amorphous phase [43]. The second one is based on the intrinsic fluorescence of PET [44-46]. Its origin is attributed to the groundstate-stable dimers of PET which are exclusively formed in the non crystalline regions.

Under biaxial orientation conditions, 6 polarised fluorescence intensities need to be measured to compute the second and the fourth moment of the orientation distribution function. However, the procedure is quite tedious [47] and despite the non-uniaxial character of the polymer films, a reduced number of intensities is usually measured [12] and the data are treated as if the sample would possess a uniaxial symmetry [48, 49].

6.3   First Stretching Process

6.3.1 Normal sequence

As quoted in the first part of this chapter, the longitudinal drawing of the amorphous film in the normal sequence involves a constant-force (CF) deformation of the film. In order to understand the stretching process under this condition a laboratory experiment has been set-up. The kinetics of the process [1] have been studied, and an extensive structural analysis of the samples has been carried out [2,3,6-9].

6.3.1.1 Kinematics of constant force stretching [1]

Amorphous PET samples, with a length of 10 mm, a width of 120 mm and a thickness of 200 μm, were stretched, in a specially designed stretching machine, by applying to the sample a constant dead weight corresponding to a nominal stress ranging between 1 to 10 MPa in a temperature range between 80 and 110°C. The deformation kinetics which can be quite rapid have been followed using a high speed video camera through ink grids placed on the sample. In the centre of the film, the deformation is effectively carried out at a constant width, corresponding to pure shear. Figure 6.3 shows examples of deformation kinetics, and Figure 6.4 depicts the extensional rate defined as $\dot{\varepsilon} = \dfrac{1}{l}\dfrac{dl}{dt}$, where l is the length of the sample at time t.

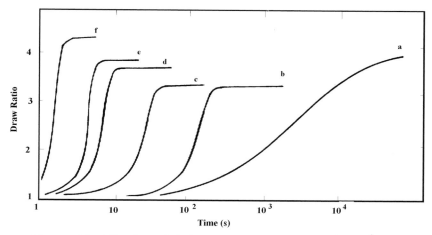

Figure 6.3   Kinetics of deformation at 80°C of PET films under various loads expressed in MPa : (a) 1.47, (b) 3.72, (c) 4.9, (d) 6.17, (e) 8.0, (f) 10.29

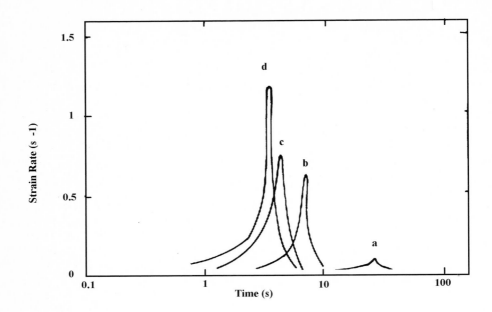

Figure 6.4 Evolution of the strain rate during stretching at 800C
under various loads in MPa : (a) 4.9, (b) 6.17, (c) 7.35, (d) 8.62

The deformation starts slowly, then the rate of deformation increases, goes through a
maximum and decreases again until the deformation reaches a plateau value, called λ_p
hereafter. The evolution of the plateau draw ratio with the applied stress and the stretching
temperature is given in Figure 6.5.

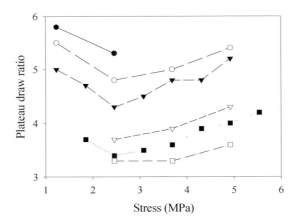

Figure 6.5 Evolution of the plateau draw ratio with the applied stress
for 80°C (□), 85°C (■), 90°C (∇), 95°C (▼), 100°C (○), 105°C (●)

Two regimes are apparent. At low stresses, rather large draw ratios are obtained in relatively long times. The samples produced under these conditions will be called "stretched with flow". As the load increases, the plateau draw ratio decreases until it reaches a minimum value. It might be seen that the corresponding stress is rather temperature independent. It corresponds to the equivalent of a natural draw ratio. In a second regime, at higher stresses, the plateau draw ratio increases with the applied load, as one would expect from a crosslinked elastomer. These conditions will be called "crystallizing conditions". The higher the temperature, the higher the plateau draw ratio in both regimes. This behaviour reveals the existence of chain relaxation processes during stretching [50]. The existence of plateau draw ratio can be explained by an orientation induced crystallisation of the samples which obviously happens during the stretching process. It generates, thanks to the connection of the chains through the crystalline blocks, a molecular network which can be stretched until it becomes in equilibrium with the applied stress. It is important to realize that under constant load stretching, high strain rates can be reached if the stretching temperature and the applied stress are high enough. In this case, high orientation level are expected inside the sample. On the other hand, for low stresses ("stretching with flow"), the stretching is dominated by relaxation phenomena, leading to a relatively high deformation of the samples with almost no molecular orientation, until the applied true stress becomes high enough so that extensional rates overcome the reciprocal of the chain relaxation time. Under these conditions, chain orientation is generated inside the film and induced crystallisation may occur. Finally, large draw ratios are obtained and it is anticipated that the orientation will be weaker at a total draw ratio than in the case of "crystallizing conditions".

The Doi-Edwards model of chain relaxation in polymer melts [51] has been used to give a molecular description of the relaxation processes occurring during constant force stretching [50], which possesses the unusual feature of beginning with low deformation rates. Therefore, the very first stages of the deformation involve the reptation of the chains as long as $\dot{\varepsilon}\tau_c < 1$, where τ_c is the reptation time. In practice, at temperatures close to Tg, the reptation times are very long (on the order of 10^4 s at 80°C), so that this process occurs only at the very beginning of stretching. After this, retraction of the chain inside its deformed tube is able to occur, and the amplitude of this highly non-linear step is enhanced by an increasing stretching of the sample. The retraction process leads to an increase in the average number of monomers between entanglements. This decrease in the entanglement network density alllows higher strain rates to be reached, so that the retraction process becomes less and less efficient and the trapped subchains orientate strongly. When the orientation becomes higher than a critical value, induced crystallisation appears, preventing any additional large scale relaxation motions of the chains. Therefore, the subchains permanently trapped by crystallites will continue to deform (and eventually to crystallise) until the entropic reaction equals the applied stress.

The description explains the existence of a loose entanglement network prior crystallisation for samples stretched with low stresses. At large stresses, retraction appears restricted to small deformations, and does not significantly alter the network. A quantitative treatment, based on the rubber-like elasticity, substantiates this statement [50]. Furthermore, recent experiments following during stretching the development of crystallization using synchrotron radiation effectively show that crystallization occurs as the draw rates become comparable with the time scale of the retraction process [52].

6.3.1.2 Development of structure and orientation: Samples drawn at the plateau draw ratio

The samples obtained under crystallising stretching conditions present some common and general tendencies recalled hereafter [3, 6-9]. The measurement of crystallinity of samples that have reached their plateau draw ratio, under crystallizing conditions, but with various loads and temperatures, shows that crystallinity is only controlled by the draw ratio, as recently confirmed by other constant force experiments [49]. In the crystalline phase, the chain axis orientation (($\overline{1}05$) planes normals) shows a high degree of orientation towards the machine direction which increases with temperature and stress. The chain axis orientation with respect to X_1 as well as that of normal to the phenyl ring ((100) planes normals) with respect to X_3 are only function of the plateau draw ratio. Obviously, the phenyl ring normal in the crystalline phase tends to become more and more perpendicular to the stretching direction, but more interesting is the behavior of the (100) planes normals with respect to X_3, illustrated in Figure 6.6. An increase in the draw ratio leads to an higher orientation of the phenyl rings in the plane of the film. The draw ratio apparently controls this orientation although high draw ratios are attained with high temperatures and large loads. This planar orientation of the phenyl rings in the plane of the film is observed as soon as the uniaxial symmetry of the stretching process is lost, and has been observed by many authors [2,25,53,54]. Such behaviour has also been observed on other aromatic ring-containing polyesters like polyethylene naphthalate (PEN) [55,56].

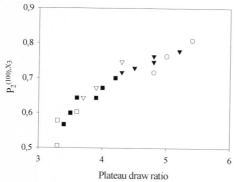

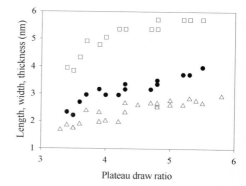

Figure 6.6 : Orientation of the (100) planes normals with respect to X_3 plotted versus the draw ratio. The symbols refer to the same temperatures as in Figure 6.5.

Figure 6.7 : Length (□), width (●) and thickness (Δ) of the crystals versus the plateau draw ratio

The draw ratio is also controlling the size of the crystalline blocks along their main directions, although the thickness is essentially increasing with the stretching temperature, and can therefore be used as a fingerprint of this experimental parameter (Figure 6.7). The number of crystals, calculated from the ratio of the crystallinity to the volume of crystalline blocks, decreases as the temperature increases. At temperatures close to Tg (85°C), the number of crystals increases slightly with the applied load. At intermediate temperatures (90-95°C), it remains constant whereas it tends to decrease at the most elevated temperatures (105°C). The evolution of the number of crystals during the stretching procedure has been followed by stopping the deformation at various draw ratio for different samples. Two temperatures (90°C and 100°C) were studied. It clearly appears that the number of crystals is not an increasing function of the draw ratio. Therefore, we can deduce that the increase of crystallinity, observed during stretching, originates from the growth of the crystals and that no additionnal nucleation appears as soon as the orientation induced crystallisation has occurred.

The edge-through observation of the crystalline organization using small angle x-ray scattering shows, in agreement with other studies [53,57], that the crystalline blocks are gathered into layers whose normals are inclined at 70° with respect to the draw direction. The comparison with other works [53] shows that the processing conditions, especially the extensional draw rate, might affect the angular position of this large scale organization.

Fluorescence polarization has been used to study the orientation of rigid fluorescent probes during stretching. Figure 6.8 shows the evolution of the order parameter versus the draw ratio for samples stretched at 90 and 100°C. Although differences between the two temperatures are rather small, it can be seen that, before crystallisation, higher orientation is noted at 90°C, reflecting more efficient relaxation processes as the temperature is increased. After the onset of crystallisation, a rather regular increase of the orientation is noted at 90°C whereas a more pronounced jump is observed at 100°C followed by a smoother increase at the highest deformations. The lowest rate of orientation at high extensions has been recently confirmed by Salem [49].

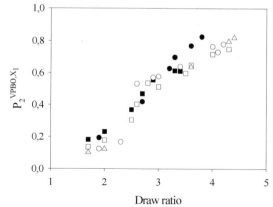

Figure 6.8 Orientation of the long axis of VPBO molecules with respect to the draw direction versus the draw ratio.
90°C (■) 2.42 MPa and (●) 4.92 MPa;
100°C (▢) 2.42 MPa, (○)3.69 MPa, (△) 4.92 MPa.

Infrared and Raman spectroscopy have been used to study conformational changes occurring during stretching as well as the evolution of the orientation, both informations being averaged over the crystalline and non-crystalline phase [7]. The fraction of trans conformers has been evaluated and is presented as a function of the plateau draw ratio in Figure 6.9.

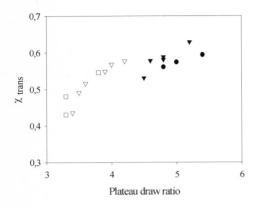 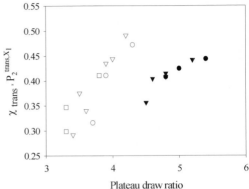

Figure 6.9 Evolution of the fraction of trans conformation during stretching (□) 85°C, (▽) 90°C, (▼) 100°C, (●) 105°C

Figure 6.10 Evolution of the average orientation of the chain axes during stretching(□) 85°C, (▽) 90°C, (○) 95°C, (▼) 100°C, (●) 105°C

The crystallinity and the fraction of trans conformers, χ_{trans}, are increasing with the plateau draw ratio, a behavior noted by various authors although the stretching conditions are somewhat different [58,59]. However, two domains of temperature can be distinguished. Indeed, a small relaxation seems to appear between 95 and 100°C, although χ_{trans} has a large contribution from the crystalline phase. This is even more clear, if we compute the orientation of the chain axes averaged over the crystalline and non crystalline phases as being the product $\chi_{trans} \times P_2^{trans}$ where P_2^{trans} represents the orientation of the trans conformers with respect to the draw direction, since it has been shown that the orientation of the gauche conformers is totally negligible [59]. Figure 6.10 depicts a clear relaxation between 95 and 100°C and also a slight relaxation within each temperature domain.

The existence of a molecular network governing the deformation behaviour of PET has been recognized for a long time by Ward and coworkers [60-61]. Following these ideas, and taking into account the pure shear deformation tensor, the average orientation of a statistical segment should be related to the draw ratio λ by the following expression if the assumption of a network type deformation is fulfilled :

$$P_{200} = \frac{1}{10N}(2\lambda^2 - \lambda^{-2} - 1)$$

where N is the number of statistical segments between junctions.

Figure 6.10 confirms the validity of the network assumption and shows that two types of network, each one linked to a different temperature domain, might exist. For temperatures below 95°C, a number of 6 monomers between junctions can be calculated whereas a much looser network seems to be formed above 95°C, containing 20 monomers between crosslinks.

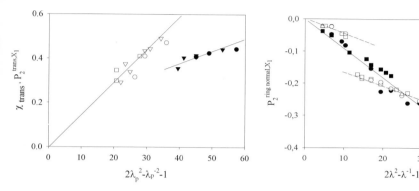

Figure 6.10   Network type deformation for the average orientation of the chain axes (same symbols as Figure 6.9)

Figure 6.11   Network type deformation for the average orientation of the ring normal (same symbols as Figure 6.8)

Since it can be argued, that the deformation path depends on temperature and applied load in this kind of stretching, "quenched" samples have been studied. The orientation of the normal to the phenyl ring, calculated from refractive indices measurements, plotted in Figure 6.11, confirms the existence of the two types of network which control the molecular orientation, that obtained at elevated temperatures being almost 3 times looser than closer to Tg. The value of 6 monomers between junctions is very close to the average distance between entanglements in PET. It can thus be deduced that the entanglement network is controlling the orientation at temperatures not too high with respect to Tg. On the other hand, at higher temperatures, the crystalline blocks act as junctions. In this case, the crystallinity, as well as the length of the crystallites, allow us to estimate a number of 15 monomeric units between crystallites, in reasonable agreement with the order of magnitude obtained from orientation measurements.

From this analysis, it can be deduced that essentially the amorphous phase is sensitive to these changes in the parameters controlling the network. Indeed, no relaxation is observed in the crystalline phase, whereas a clear evidence of relaxation is seen as far as the average orientation is concerned.

6.3.2 Inverse sequence

The constant speed drawing of amorphous PET films has been the subject of many studies, several of them being devoted to constant width rather than uniaxial stretching [25,26,63]. Only the main characteristics of the resulting films will be described here. The major event in this stretching of amorphous film slightly above the glass transition temperature is the appearance of orientation induced crystallization during stretching. As an illustration, Figure 6.12 shows the evolution of the crystallinity with respect to the draw ratio at various temperatures. At the highest temperature, where samples could be obtained in a sufficiently large range of draw ratios, the crystallinity shows the usual sigmoidal shape. It can be seen that crystallization appears at lower draw ratios as temperature decreases, in agreement with previous studies [24-26,36,63-67]. This behavior can be explained by the existence of a critical orientation for induced crystallization which decreases as temperature increases [63,64].

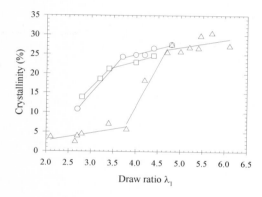

Figure 6.12 Crystallinity versus draw ratio at different temperatures (□) 95°C, (○) 105°C, (Δ)115°C

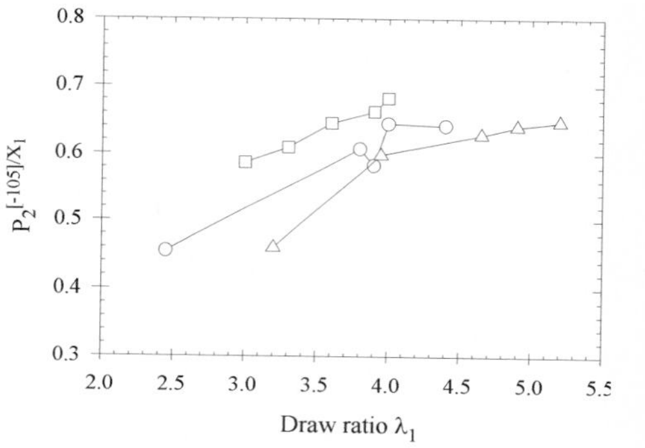

Figure 6.13 Chain axis orientation with respect to the X_1 direction versus draw ratio at different temperatures (□) 95°C, (○) 105°C, (Δ)115°C

The orientation of the chain axes (more precisely $(\overline{1}05)$ planes normals) with respect to the stretching direction is given in Figure 6.13. The chain axes appear to align more and more towards the draw direction as the draw ratio increases or the temperature decreases. Figure 13 shows the importance of relaxation phenomena in this type of stretching, even on crystalline orientation, since the decrease of orientation with temperature cannot be solely explained by the shift in the appearance of orientation induced crystallization.

It is interesting to note that as the molecular weight of the chain decreases, a lower orientation even in the crystalline phase is observed under comparable conditions of temperature and draw ratio. This behavior emphasizes the role of chain relaxation processes on the orientation of the crystalline phase [13,61,66-70].

The length of the crystalline blocks is also affected by relaxation phenomena. At a given draw ratio, shorter crystals are formed as the temperature increases and the final length of the crystals is lower, although the kinetics of thermal crystallization are more rapid and higher extensions have been obtained.

The evolution of the number of crystalline blocks as a function of the draw ratio shows two different behaviors depending on the temperature. At low temperatures, the number of crystals is essentially constant, meaning that after a nucleation step, further stretching favors their growth. At the highest temperature (115°C), the number of crystals undergoes an almost twofold increase between $\lambda = 4$ and $\lambda = 5.2$. Therefore, it can be deduced that, after some crystals are formed as soon as the critical orientation has been reached locally, inhibition of relaxation processes in the amorphous phase takes place, so that the critical orientation for crystallization can be reached at other locations upon further stretching, inducing new crystalline blocks.

More information on the deformation can be obtained by studying the orientation of the normal to the phenyl ring averaged over all the material and by plotting it versus the crystallinity of the corresponding sample. Figure 6.14 shows two regimes, whatever the draw temperature. In the first one, corresponding to relatively low draw ratios, crystallites are formed, and the normal to the phenyl ring shows a more and more pronounced perpendicular orientation with respect to the draw direction, which reflects the high orientation observed in the crystalline phase and its increasing volume fraction. In a second regime, the crystallinity does not increase significantly, but the average orientation is still more and more pronounced, pointing towards the significant development of orientation in the non crystalline phase. This retarded orientation of the amorphous phase has also been pointed out by Yoshihara [65].

Interestingly, in this type of stretching, the concept of a molecular network, controlling the development of orientation seems also to apply. Indeed, we first pointed out above that, after orientation induced crystallisation, the amorphous phase undergoes an important increase of its orientation. Moreover, stress-strain curves registered during stretching and depicted in Figure 6.15 show an important increase of stress after a pseudo-plateau and an inflection point where crystallisation occurs. This large increase of the stress is attributed to the extension of chain segments between junction points. Essentially, two types of network are observed depending on the temperature conditions, which are already clearly apparent in the large deformation part of the stress-strain curves. At temperatures below 105°C, quantitative examination of stress-strain curves [13] yields a network modulus of 8 MPa, instead of only 3.4 MPa at 115°C. A looser network is thus obtained at more elevated temperatures. It can be thought that both chain entanglements and crystallites act as junction points at low temperatures and essentially crystallites at higher ones.

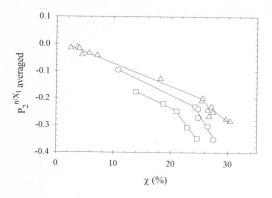

Figure 6.14 : Average orientation of the phenyl ring normal versus the crystallinity (□) 95°C, (○) 105°C, (Δ)115°C

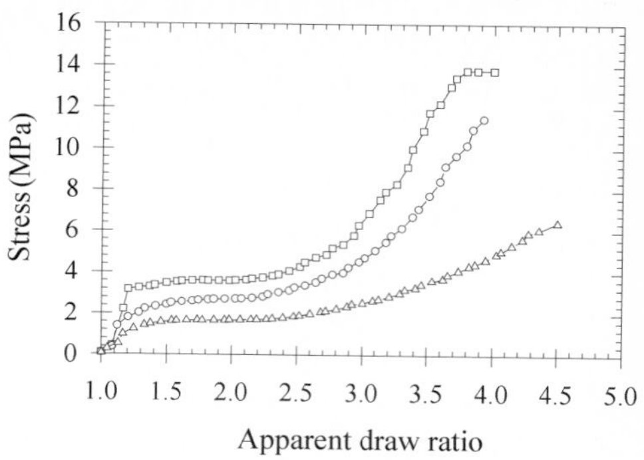

Figure 6.15 : Engineering stress versus apparent draw ratio (□) 95°C, (○) 105°C, (Δ)115°C

6.3.3 Comparison between constant rate and constant force drawing of amorphous samples

In this section, we would like to emphasize the main differences between constant force stretching and imposed strain rate deformation. The influence of draw ratio and stretching temperature on the orientation and morphology of the crystalline phase seems to be quite different in the two types of stretchings, as confirmed recently by Salem [49].

For samples stretched under constant load which have reached the equilibrium deformation λ_p, the orientation of the chain axes (($\overline{1}05$) planes normals) with respect to the stretching direction as well as that of the normal to the benzene rings (((100) planes normals) with respect to the X_3 direction, was an increasing function of the equilibrium draw ratio λ_p, whatever the stretching temperature or the applied load. In the case of samples for which deformation is stopped before λ_p (quenched samples), the same behavior is observed : i.e. the orientation is independent on the time (selected through the temperature and applied load) required to reach the given deformation.

This is obviously not the case with samples stretched at a given strain rate. At a given draw ratio, the orientation decreases as temperature increases, demonstrating the major role of relaxation phenomena. Accordingly, the chain axis orientation in the crystalline phase is much lower in this latter type of stretching. The fact that, in constant force stretching, the orientation is governed by the equilibrium draw ratio might be due to a rough compensation as temperature increases between the decrease in relaxation time and an increase of the deformation kinetics.

A more subtle differential behavior concerning the (100) planes normals can also be noted. In constant force stretching, the chain axes in the crystalline phase show such a high orientation with respect to the draw direction that the [100] direction is essentially perpendicular to X_1. An increase of the draw ratio leads to an increase of orientation of the benzene rings in the plane of the film, so that P_{200}^{100/X_2} has to decrease with draw ratio.

On the contrary, in constant strain rate deformation P_{200}^{100/X_2} is almost constant due to a relaxation of the chain axis with respect to the draw direction (which induces a less perpendicular orientation of the [100] direction with respect to X_1) and a relaxation of preferential orientation of the benzene rings in the plane of the film.

The difference in the deformation kinetics between the two types of stretching may also generate differences in the nucleation and growth processes of the crystals. Whereas under constant force the number of crystals decreases as the temperature increases, it seems to be independent of this parameter in strain rate controlled deformations. Furthermore, in this latter case, at high temperatures, the number of crystals per unit volume increases during stretching. Such behavior has never been encountered in constant force stretching.

The crystalline orientation is essentially governed by the draw ratio in constant force stretching. This is not to say that relaxation phenomena do not occur in this type of stretching, since a relaxation has been detected on the amorphous orientation especially between 95° and 100°C, i.e. over a limited temperature range. The same type of relaxation seems to occur also in constant strain rate stretching. For instance, a strong change in behavior is noted between 105° and 115°C. The network modulus differs by a factor around 2.5 in the two domains of temperatures.

6.4   Transverse Stretching of One-Way Drawn Samples

We now turn to the evolution of structure and molecular orientation during the second drawing stage of the PET film processing. The normal as well as the inverse sequence have been investigated using model samples that we prepared from films having undergone the first stretching process on a pilot plant.

6.4.1   Transverse stretching in the normal sequence

When drawing monodrawn samples in order to mimic the second drawing of the process two main difficulties have to be overcome. The first one is that the film has to be brought rapidly to the stretching temperature in order to prevent thermal crystallization of the sample. Before stretching, a preheating treatment has thus been applied. It usually consists in a 30 s heating at 87°C followed by a 3 s heating at the stretching temperature. It was checked that these conditions allow thermal equilibration of the sample. However, a change in the structure of the sample due to the preheating step cannot be avoided. Therefore, the data corresponding to the undrawn sample refer to the state of orientation and structure at the end of the preheating rather than the initial one-way drawn sample.

The second one is related to the symmetry of the deformation which must be pure shear. However, in the second drawing step, because of the initial orientation of the film perpendicular to the direction of stretching the film has a strong tendency to shrink along this direction. Therefore, tiny clamps have been fitted to the film in order to keep the width of the film constant throughout stretching. The deformation of the film was checked through an ink grid, and parts of the films located around the centre, where deformation is homogeneous, were cut for further analysis.

The main events occurring during the transverse stretching will first be described, then the influences of the initial state of structure and orientation of the one-way drawn sample as well as the stretching temperature will be discussed.

6.4.1.1 Description of the main phenomena occurring during the transverse stretching [4,10-12]

The starting one-way drawn film has a draw ratio of $\lambda_1 = 3.4$ and has been produced at 127°C. It is typically a so-called plateau drawn sample. Rectangular pieces were taken from this film and further stretched transversally to the first drawing direction in a specially designed drawing machine at the Saint Fons Research Center of Rhône Poulenc. The stretching temperature was 120°C and draw ratios of 2.2, 3.6 and 4.5 were obtained. The film draw speed was set at 12mm/s, corresponding to a strain rate decreasing from 0.58 to 0.14 s^{-1}.

The crystallinity appears to be constant during stretching with an average value of 30±2%. This observation means that all the structural modifications are conducted under a constant crystalline volume fraction, implying that if new crystals are formed, old ones necessarily have to melt at least partially.

The evolution of the X-ray diffracted intensity of the $(\overline{1}05)$ reflection is plotted in Figure 6.16. Initially ($\lambda_2 = 1$) only crystals with the chain axis oriented along the X_1 direction of the film are present (X_1 crystals). As the stretching proceeds, the fraction of crystals lying along the X_1 direction decreases rapidly and diffracted intensity along the X_2 direction increases, reflecting the appearance of a new population of crystals (X_2 crystals). Although the diffracted intensity at intermediate angles between the principal directions of the plane of film contains a contribution from the $(0\overline{2}4)$, there are several evidences for the existence of an other population of crystals oriented at "intermediate" angles (I crystals).

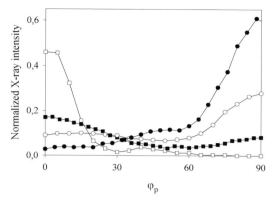

Figure 6.16 : X-ray diffracted intensity by the ($\overline{1}05$) planes, versus the direction in the plane of the film (0° corresponds to X_1 direction)

Quantitatively, the orientation of the ($\overline{1}05$) planes normals with respect to the main directions of the film is plotted in Figure 6.17a. The chain axes in the crystalline phase appear to lie strongly in the plane of the film ($P_{200}^{\overline{1}05/X_3} = -0.47$). As stretching proceeds, the orientation is progressively balanced from the X_1 towards the X_2 direction. The reorientation phenomena are rather efficient, since an equilibrated state of orientation (in terms of P_{200}) is reached for a second draw ratio lower than the first one, as observed by others [71,72] but, although the sample has been stretched up to higher draw ratios than the one-way drawn film, the final orientation is less pronounced along the X_2 direction than it was originally along X_1.

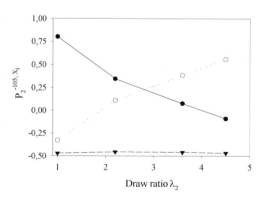

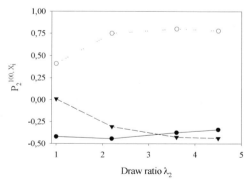

Figure 6.17a : Orientation of the chain axis in the crystalline phase with respect to the three principal direction versus the draw ratio :X_1 (●); X_2 (○); X_3(▼)

Figure 6.17b : Orientation of the normal to the phenyl ring in the crystalline phase with respect to the three principal direction versus the draw ratio :X_1 (●); X_2 (▼); X_3(○)

The normal to the phenyl ring direction in the crystalline phase also undergoes important changes. The orientation of the (100) planes normals is plotted versus the draw ratio in Figure 6.17b. It can be seen that the second stretching induces a further alignment of the phenyl ring in the plane of the film, in agreement with other reports [25,53,72]. The distribution of the normal to the rings which did not show a strong anisotropy in the X_2X_3 planes, shows after stretching a strong perpendicularity with respect to X_2. This is a consequence of both the orientation of the chain axes along X_2 and the highly planar orientation of the rings.

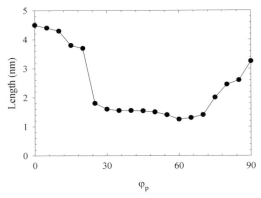

Figure 6.18: Length of the crystals versus the direction in the plane of the film

Further study of the angular broadening of the ($\overline{1}05$) diffraction peak allows us the characterise the length of the crystals as a function of their position in the plane of the film. Figure 6.18 shows (for a film initially slightly different film stretched at a draw ratio of 2.7 along X_2) the existence of the three populations of crystals that were described above. Long crystals are still present oriented along $X_1\pm20°$ (X_1 crystals), quite small crystals (I crystals) exist at intermediate angles (between 20 and 70°) whereas the new crystals preferentially oriented along X_2 (X_2 crystals) have a length which increases as the become more oriented along X_2.

The evolution of the length of the X_1 and X_2 oriented crystals is given in Figure 6.19. The length of the X_1 crystals decreases during stretching and becomes hardly measurable at the highest draw ratios, since the diffracted intensity is quite weak. At the same time, X_2 crystals are formed. Under these conditions, the first appearing X_2 crystals are only slightly smaller that the original X_1 crystals.

Different behaviour is noted concerning the width of these crystals (size along the (010) planes normals), as shown in Figure 6.20. Large draw ratios are required before a strong decrease of the width of the X_1 crystals is observed whereas that of the newly formed X_2 crystals is small and increases strongly with the draw ratio.

The thickness of the crystals (size along the (100) planes normals) increases (in average) from almost 4 to 6 units during stretching.

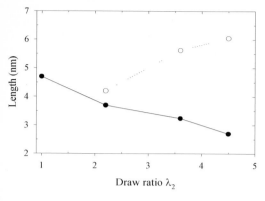

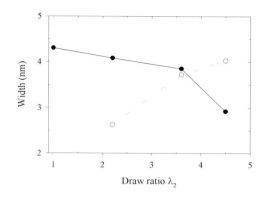

Figure 6.19 : Length of the crystals oriented with the chain axis along the (●) X₁ or (○) X₂ direction versus the draw ratio

Figure 6.20 : Width of the crystals oriented with the chain axis along the (●) X₁ or (○) X₂ direction versus the draw ratio

The behaviour of the amorphous phase can be obtained by combining orientation measurements from refractive indices (average orientation) and from X-ray diffraction (crystalline phase) or directly using fluorescence polarisation.

The orientation of the normal to the phenyl ring is plotted versus the draw ratio in Figure 6.21. It must be pointed out that, since the (100) direction does not fully correspond to the phenyl ring normal, only a general tendency can be discussed. Figure 6.21 shows that, starting with an essentially uniaxial orientation of the normal to the phenyl ring in the amorphous phase, the main effect of the second drawing is to bring a perpendicular orientation of the normal to the phenyl ring with respect to the X_2 direction, as could be expected from a preferential orientation of the chain axes along this direction. In the same time, the normal to the ring tends to orient parallel to the other directions (X_1 and X_3), which can be seen as a direct consequence of the previous statement.

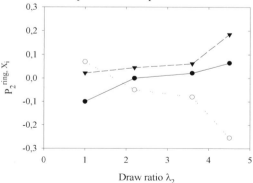

Figure 6.21 Orientation of the normal to the phenyl ring in the amorphous phase with respect to (●) X_1, (○) X_2 and (▼) X_3 versus the draw ratio

Fluorescence polarisation measurements were carried out on samples containing VPBO molecules. Figure 6.22 shows the most probable distribution of the probes in the plane of the film, obtained from the measurements of $\cos^2$ (VPBO/X_1) and $\cos^4$ (VPBO/X_1) using a procedure detailed in ref. 12. As deformation proceeds, the probe axes distributions become larger and a rotation of the large part of the probes is observed. Unlike in the crystalline phase, there is no evidence for a bidistribution of the axes of the probes along the X_1 and X_2 directions. We can also deduce from Figure 6.22, that an almost isotropic distribution (in the plane of the film) is obtained for a draw ratio intermediate between 1.6 and 3.1. These results, linked to the orientation of rigid probes, differ significantly from those coming from the intrinsic fluorescence of PET, where a biaxial distribution of the chains in the amorphous phase is noted [48].

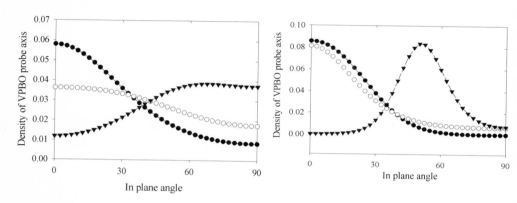

Figure 6.22 Density of VPBO molecules oriented along a given angle in the plane of the film : draw ratio $\lambda_2 = 1(\bullet)$; $= 1.6$ (O); $= 3.1$ ($\blacktriangledown$)
Film with small crystals

Figure 6.23 Density of VPBO molecules oriented along a given angle in the plane of the film : draw ratio $\lambda_2 = 1(\bullet)$; $= 1.9$ (O); $= 3.9$ ($\blacktriangledown$)
Film with large crystals

6.4.1.2 Influence of the structure of the one-way drawn film

Depending on the processing conditions, one-way drawn films with quite different structures and levels of orientation can be obtained. More precisely, films with rather small crystals and moderate levels of orientation are obtained at low temperature and moderate draw ratios if the time allowed for stretching is short, so that the plateau deformation is not reached. In the same line, films drawn at the plateau draw ratio can have large crystals and a higher molecular orientation.

We now emphasise the main consequences of these differences. Only the key figures will be reported here. We refer the interested reader to a more complete report [11]. The decrease in the length of the X_1 oriented crystals is almost independent of their initial size, but the width of the crystals only decreases in the case of large ones. However, the existence of initially large crystals leads, at a given λ_2, to smaller (both in length and width) new, X_2 oriented crystals. The decrease of orientation of the chain axes in the crystalline phase with respect to X_1 (or equivalently the increase of that with respect to X_2) is more pronounced in the case of the films having initially short crystals (and also a lower initial orientation).

These main features indicate that apparently the reorientation phenomena along the draw direction are favoured by a less constrained environment coming from a lower crystallinity, the presence of smaller crystalline blocks and a weaker molecular orientation. Except the decrease of width of X_1 oriented crystals during the transverse drawing, the reorientation mechanisms do not seem much different as far as the crystalline phase is concerned. Only, reorientation becomes more difficult and requires higher draw ratios to be achieved in the case of films having initially a high level of orientation.

This tendency also appears in the amorphous phase. As an illustration, we have plotted in Figure 6.23, the distribution of the long axes of the VPBO molecules as deduced from fluorescence polarisation measurements for films having long crystals which can be compared with the data of Figure 6.22 corresponding to a film with short crystals. It is first interesting to note that the reorientation of VPBO molecules appears difficult at low draw ratios. At larger draw ratios, a rather narrow distribution centred at intermediate angles in the plane of the film is observed in the case of films having initially long crystals in contrast with less structured films which show a much broader and more along X_2 oriented distribution. It is from these data quite clear that a pseudo-affine deformation mechanism may apply for the reorientation of rigid probes in the amorphous phase in the case of films with small crystals but not in the other situation.

6.4.1.3   Influence of the draw temperature

Films with initially small crystals were stretched at 100 and 120°C whereas films with large crystals were stretched at 120 and 140°C. Here again, we limit ourselves to the description of the main effects of temperature.

The influence of the draw temperature can be understood as the result of the competition between two processes, molecular relaxation and crystallization. Low stretching temperatures imply slow relaxation processes and low mobility, leading to slow crystallization kinetics; reorientation and breakdown of the crystals is difficult owing to low molecular mobility. Conversely, at elevated temperatures, the molecular mobility is higher and the tendency of the chains to crystallize during orientation is greater. Larger crystals are therefore created at such temperatures. However, reorientation of the one way drawn films is difficult because it is restricted by the strong tendency of the chains to crystallize. It appears that there is an optimum temperature, located around 120°C, where the mobility is high enough to allow rotation of small crystalline units, but low enough to avoid rapid crystallization.

6.4.2   Transverse stretching in the inverse sequence

The same strategy as above has been used to study the molecular behaviour of one-way drawn films, transversely stretched in the conditions of the inverse sequence, i.e. under a constant load. Model samples were prepared under various conditions of temperature and applied load, which should be set above the corresponding yield stress of the one-way drawn sample. As in the case of amorphous samples, this type of stretching is characterized by an unusual kinetics, a typical example of which is given in Figure 6.24. The stretching process starts with low strain rates. A maximum is observed and finally, at longer times, the stretching process stops itself showing that a maximum (and equilibrium) value of the draw ratio is obtained for a given condition of temperature and applied load. It can be noted that, as compared to the previously described constant rate drawing, high strain rates (a few s^{-1}) are obtained.

Films with a draw ratio lower than the plateau value can be simply obtained by stopping mechanically the deformation before it reaches the plateau value. We refer to such samples as "quenched samples". Their study affords interesting information about the deformation mechanisms.

Both temperature and applied load increase the deformation kinetics and the equilibrium deformation when they increase. However, the influence of the applied load is the most pronounced.

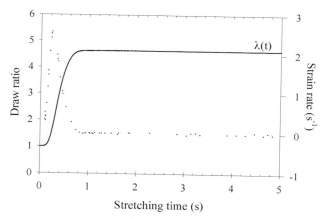

Figure 6.24.   Stretching kinetics under constant force:
temperature 115°C, engineering stress 20 MPa

Also in this type of stretching the reorganization of the structure is carried out at a constant crystallinity. The evolution of the orientation of the chain axes with respect to the 3 principal directions of the film is given in Figure 6.25.

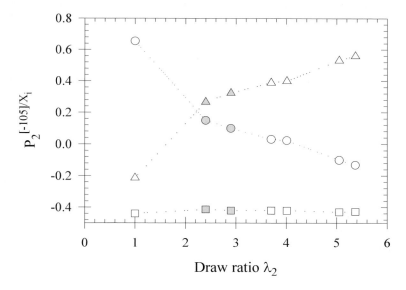

Figure 6.25. Orientation of the chain axes in the crystalline phase versus the draw ratio : temperature : 125°C ; grey symbols : "quenched samples"

The chain axes remain oriented in the plane of the film as the deformation proceeds. Their orientation, which was originally along X_1, gradually moves along X_2. Quantitatively, the orientation along X_2 is less than the previous one along X_1, although the draw ratio is higher. Nevertheless, the reorganization mechanisms appear very efficient since moderate draw ratios (on the order of 2.2) are sufficient to create a biaxially equilibrated orientation (in terms of P_{200}).

The calculation of P_{200} from X-Ray data is however somewhat obscuring the details of the molecular processes occurring in the crystalline phase. Figure 6.26 shows the diffracted intensity (in the plane of film) versus the considered direction in this plane, for various draw ratios. As the draw ratio increases, the measured intensity towards X_1 decreases rapidly. Even at moderate draw ratios, the crystals initially oriented along X_1 have almost disappeared. At the same time, an increase of intensity is observed along X_2 and a significant part of the crystalline chain axes is lying at intermediate directions between X_1 and X_2. A relative maximum, which moves slightly towards the X_2 direction, is detected at intermediate angles.

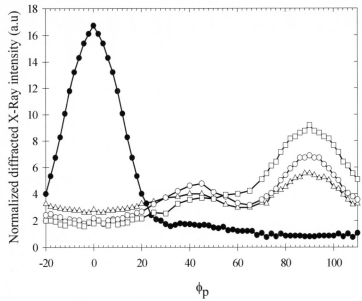

Figure 6.26. Distribution of diffracted intensity in the plane of the film
(draw ratios : $\lambda_2 = 1$ (●) ; 2.4 (○); 2.9 (△); 3.7 (▢))

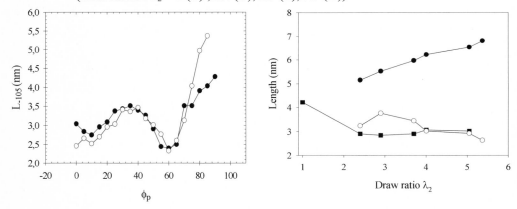

Figure 6.27 Length of the crystals versus
their position in the plane of the film
(draw ratios : (●) 2.4 and (○) 2.7)

Figure 6.28 Length of the various types
of crystals (■) X_1, (●) X_2,(○) I versus
draw ratio

Here again, various populations of crystals can be discerned. The length of the crystals as a
function of their position with respect to the main directions of the film is plotted in Figure
6.27. Even at small draw ratios, the length of the crystals lying around the intermediate
maximum is important (but slightly less than the initial one) and X_2 oriented crystals show
a size which becomes rapidly higher than that of those initially oriented along X_1. The

evolution during drawing of the length of various types of crystals is plotted in Figure 6.28. The length of the crystals oriented along X_2 increases with draw ratio, meaning that drawing favors the growth of these crystals at the expense of those lying at intermediate angles (since the transformation is carried out at an almost constant crystallinity). The length of the X_1 oriented crystals decreases during the first moments of the drawing process and remains constant afterwards. However, the number becomes lower and lower as the stretching proceeds.

6.4.2.1 Influence of the stretching temperature

The influence of temperature on this transverse stretching process possesses a strong analogy with that previously described. An increase in temperature leads to somewhat longer X_2 oriented crystals, but also to crystals remaining along X_1. This behaviour can be explained by the higher tendency for crystallization at elevated temperatures, and probably also by the release of tension acting on the X_1 oriented crystals because of the thermal relaxation of stress acting on amorphous segments. The orientation in the crystalline phase appears to be independent on the stretching temperature, except for the fact that higher temperatures will allow higher draw ratios to be reached.

Infrared dichroïsm measurements have been used to characterize the orientation of the normal to the phenyl ring averaged over both the crystalline and non-crystalline phases, and to calculate the contribution of the amorphous phase after weighted subtraction of the crystalline orientation. Figure 6.29 shows an increasing orientation of the normal to the phenyl ring perpendicular to the draw direction X_2. It can be seen that no relaxation with temperature is obvious between 105 and 115°C, but that some relaxation is present at 125°C, especially at the highest draw ratios. It is interesting to note that a similar observation has been made on the constant force stretching of an amorphous film.

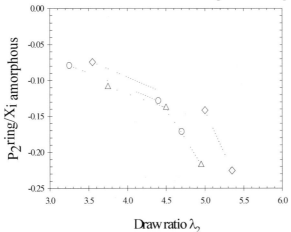

Figure 6.29   Orientation of the normal to the phenyl ring in the amorphous phase with respect to the X_2 direction versus the draw ratio ; stretching temperature 105°C (O), 115°C (◇), 125°C (Δ)

6.4.3 Comparison between constant rate and constant force transverse drawing

In this section, we would like to emphasize the main differences between these two types of drawing, corresponding to the second stretching of the industrial processes. Since the starting one-way drawn films are different, a direct and quantitative comparison is difficult if not impossible. We will therefore focus on the mechanisms that can be thought to happen through the detailed characterization of the samples.

A first difference, which can be of practical importance, concerns the achievable draw ratios at least observed on model samples. Under constant speed drawing, it has been experimentally noted that only limited extensions can be achieved before film breakage (up to $\lambda_2 = 4$ under an initial nominal strain rate equal to $0.25s^{-1}$). Constant force stretching may lead to higher draw ratios by increasing temperature or applied load.

The state of orientation of the chain axes and its evolution with the draw ratio also appear to be slightly different. The X-ray diffracted intensity of X_1 crystals (($\overline{1}05$) planes normals aligned along X_1) decreases smoothly with draw ratio in the constant rate drawing leading to bimodal distributions of diffracted intensity even if the draw ratio is as large as 4. Under these conditions, the number of chain axes (in the crystalline phase) oriented along the X_1 or X_2 direction is of the same order of magnitude. For a given draw ratio, more chain axes are still aligned towards the first stretching direction in constant speed drawing as compared to constant stress extension, where we have seen that even at rather low extensions ($\lambda_2 = 2$) the X-ray diffracted intensity of the X_1 crystals is very weak.

As an illustration, we have plotted in Figure 6.30 a comparison of the X-ray intensity diffracted by the ($\overline{1}05$) planes between two films, stretched at constant stress or a constant rate, up to an equivalent draw ratio ($\lambda_2 = 2.9$ for constant stress and $\lambda_2 = 3.1$ for constant rate), originating from two one-way drawn films which show after the preheating step the same value of $P_{200}^{\overline{1}05/X_1} = 0.68$. It is obvious that the reorienting mechanisms acting on the crystalline phase towards the draw direction are much more efficient in the case of constant force drawing. Moreover, no evidence for a local maximum of intensity at intermediate angles exists in the case of constant rate deformation whereas this appears very clearly in the constant force stretching.

Differences between the two processes also exist as far as the crystal length is concerned. For imposed strain rate drawing, the length of crystals oriented along X_1 decreases and the length of new crystals aligned along X_2 increases slowly. Crystals, lying either along X_2 or at intermediate directions, are smaller than those initially present (aligned along X_1) as soon as they can be experimentally detected. In contrast, under constant stress stretching, we have shown that rather long crystals oriented along X_2 can be observed even at moderate draw ratios.

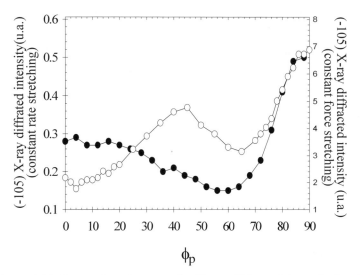

Figure 6.30 Comparison between the X-ray diffracted intensity by the
(-105) planes for a constant force stretched film (◯) and a constant
rate stretched one (●)

In both types of stretching, the changes of molecular orientation, the formation of a new population of crystals oriented along the stretching direction as well as the disappearance of that oriented along the first stretching direction occur at a constant crystallinity. This observation means that total or partial "melting" of existing crystals occurs simultaneously with the formation of the new crystalline blocks. The increase of the size with the draw ratio implies that they are formed rapidly, during the first stages of the drawing, and that they grow further.

The existence of small crystalline blocks lying at intermediate angles with respect to the main directions of the film in the case of films stretched under constant rate suggests that they originate from a partial erosion (or melting) along the length, and eventually the width of the X_1 oriented crystals. The rotation of these tiny crystalline blocks along the X_2 direction is obviously easier, requiring much less co-operativity than that of the parent blocks. As soon as a fraction of these small crystals is almost aligned along the new draw direction, their growth is observed along all their principal directions by a mechanism which can be a stacking of oriented parts of chains on these nuclei.

This rotation phenomenon is partially revealed by the behaviour of the rigid fluorescent probes. Furthermore, a more constrained environment makes the rotation more difficult, in such a way that the peeling mechanisms acting on the width of the crystalline blocks are efficient and lead to a decrease of the size along this direction in the case of sample bearing initially large crystals.

The results obtained for constant force drawing indicate different mechanisms for the formation of the new crystalline network. The new crystals can be thought to be formed mainly by the rotation of rather large crystalline blocks previously oriented along X_1. These crystals can grow further and align along X_2 as the draw ratio increases. Since the stretching is carried out under constant crystallinity, some initial crystals (aligned along X_1) have to disappear, that is to say that the partial melting of the X_1 aligned crystals still exists. Examination of the evolution of their length as a function of the draw ratio indicates that it is rather efficient, since only rather small crystals remain along this direction. A sketch of our current view on the deformation processes in both types of stretching is given in Figure 6.31.

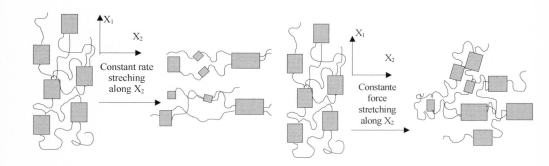

Figure 6.31 Schematic representation of the reorientation mechanisms in both processes. Note the small size of intermediate crystals as well as the larger number of crystals remaining along the X_1 direction in constant rate stretching as opposed to constant force stretching.

The main explanation of these two different types of behaviour has to be found in the very different deformation kinetics in constant rate or constant stress drawing. Large strain rates are reached with constant force drawing, meaning that only rapid relaxation phenomena can occur. Therefore, the chains in the non-crystalline part of the material, which are the only ones able to show any mobility, appear quite stiff. They are therefore efficient in transmitting the stress inside the material, allowing the crystalline blocks to break or to rotate. Indeed, in this kind of stretching, any relaxation phenomenon is partly compensated by an increase in the deformation kinetics and the consequently higher draw ratios.

6.5   High Temperature Annealing

The last step in the industrial film process consists in a high temperature annealing treatment. Typical temperatures are in the range 180-220 °C, and various temperature zones are often used. The duration of the annealing process is on the order of a few tens of seconds.

The annealing process is known to improve the mechanical properties of the film (modulus, shrinkage), especially if it is carried out at fixed dimensions. The influence of the heat-setting treatment has been the subject of various studies on one-way drawn films [31,73,74], or on biaxially oriented films [48,75-81].

All authors agree with an increase of the crystallinity of the sample and an improvement of the overall state of orientation in the material. The increase of crystallinity is explained by the growth of the crystals originating from the previous stretchings. Although the orientation of amorphous parts of chains is partially relaxed by the annealing treatment, the decrease of amorphous orientation is rather weak. This is usually explained by the crystallisation of amorphous segments and the restricted mobility in the surroundings of the crystals. The orientation of the chain axis in the crystalline phase does not appear much modified, but the tendency of the phenyl rings in this phase to lie in the plane of the film is increased. The increase of average orientation is thus explained by the further crystalline growth and possible fusion of adjacent crystalline blocks having a similar orientation.

In a detailed study, Gohil and Salem [48] consider the evolution of the amorphous and crystalline phases during a 5 min annealing at 200°C. Annealing appears to almost double the crystallinity and wipe out the differences that were existing at the end of the drawing process. They observe a redistribution of the amorphous orientation as determined by intrinsic fluorescence polarization measurements. In the case of samples with a low second draw ratio ($\lambda_2 = 2.5$), the relaxation of the amorphous orientation preferentially occurs along the X_2 direction, because of a remaining crystalline structure oriented along X_1 which inhibits relaxation of amorphous chains along this direction. This relaxation along X_2 generates ipso-facto an increase of the amorphous orientation along X_1. Samples drawn with a higher second draw ratio, show that the relaxation of amorphous orientation occurs in both directions (along X_1 and X_2), leading to a more isotropic distribution of the amorphous chain segments.

In another study [77], Gohil points out the existence of an optimal heat-setting temperature in terms of mechanical properties of the film. This temperature corresponds to roughly 180°C. Below this temperature, for which the rate of relaxation is lower than the rate of crystallization, heat treatment increases the size of the crystals leading to the development of a highly physically crosslinked structure, with a constrained amorphous phase through which the load is easily transferred to the crystallites. Above this temperature, melting of metastable crystals may occur, allowing the development of more relaxed molecules, as well as crystallization with a lamellar morphology, producing a decrease of the modulus.

In our own studies [5,15], we have been interested in studying the structural changes occurring during the annealing treatment, looking in detail at the influence of the heat setting temperature as well as that of the annealing time. The same strategy as above has been used. Model samples were prepared from a biaxially oriented film, obtained on an inverse processing plant with the following characteristics (first stretching temperature = 107°C, $\lambda_1 = 4.0$, second stretching temperature = 110°C, $\lambda_2 = 4.0$). The inverse sequence allows an efficient quenching after the longitudinal drawing and no heat-setting was applied to the film. Although this sequentially drawn film possess the same draw ratios along the X_1 and X_2 directions, a preferential orientation exists along the direction of the longitudinal drawing (X_1).

Square pieces were cut from this film, glued inside a thick cardboard frame in order to prevent any retraction during the heat setting process. A preheating at a temperature of 100°C during 5s was applied to the film (which will consist of a reference state), prior being submitted to an annealing treatment at a fixed temperature for a given duration. The film was then thermally quenched using room temperature blown air, at the end of the prescribed annealing time.

6.5.1 Influence of annealing time

The influence of the annealing time was studied at 200°C for durations varying between 3 and 20 s.

The main event of this high temperature annealing process (200°C) is the important growth of the crystals along the 3 investigated directions, as quoted by a number of authors. The volume of the crystals (roughly estimated as the product of their size along 3 directions) is plotted versus the annealing time in Figure 6.32. The largest increase is observed during the first 3 seconds of the thermal treatment. Since the sizes do not increase significantly after 5 seconds, it can be argued that the equilibrium thermodynamic sizes are almost reached in that time.

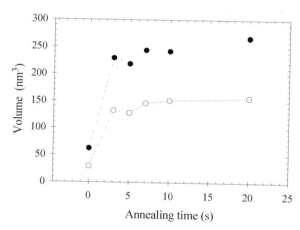

Figure 6.32 Volume of the crystals versus annealing time at 200°C for (●) X_2 oriented crystals, (○) X_1 oriented crystals

Differences in the growth rate between the various types of the crystals can be detected. For instance, the width of the X_2 oriented crystals shows a larger increase than that of the X_1 oriented ones, whereas the length of the X_1 and intermediate crystals increases more than that of the X_2 crystals. This apparent growth of the crystals reflects an increase in crystallinity inside the sample, observed in many studies. The diffracted intensity also increases accordingly.

Valuable information can be deduced directly from the diffracted intensity of the $(\bar{1}05)$ planes. Its evolution (corresponding to $(\bar{1}05)$ plane normals lying in the plane of the film) is plotted versus the azimuthal direction for various annealing times in Figure 6.33.

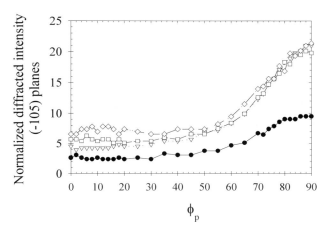

Figure 6.33. Diffracted intensity in the plane of the film for various annealing times (●) t = 0; (∇) t = 3s; (□) t = 7s; (◊) t = 20s.

At short times a strong increase in the intensity along X_2 is noted which produces a larger number of crystalline chain axes pointing towards this direction.

At longer times, the changes are essentially observed along the other directions of the film plane, especially along X_1. The observation of the out of plane diffracted intensity shows a very small tendency for chain axes to lie even more in the plane of the film for X_2 crystals. It cannot however solely account for the large increase of the in-plane intensity at 90° shown in Figure 6.33. The evolution of the 3 order parameters with respect to the 3 principal directions of the film versus the annealing time is plotted in Figure 6.34. It confirms of course the observations quoted above, although the differences are rather weak.

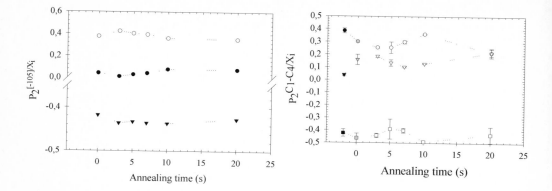

Figure 6.34 Orientation of chain axes in the crystalline phase versus annealing time at 200°C, with respect to($\bullet$) X_1; ($\circ$) X_2; ($\blacktriangledown$) X_3

Figure 6.35 Average orientation of C1-C4 axes versus annealing time at 200°C, with respect to (∇) X_1; ($\circ$) X_2; ($\square$) X_3

A slightly more perpendicular orientation with respect to the X_3 direction is observed, together with a somewhat more complicated behaviour along X_2. An increase of orientation along X_2 is observed during the first stages, which can be correlated to the rapid increase of the intensity along this direction. It does not reflect the fact that order is improved along this direction, but that at short times a stronger increase in the number of oriented crystals is seen along X_2 with respect to the other directions of the film.

In the same way, the decrease of orientation with respect to X_2 at longer times is arising from the growth of the crystals that are not oriented specifically along the principal directions of the film, but instead isotropically distributed in between.

The orientation of the C1-C4 axes of the phenyl rings, averaged over all the molecular units inside the material, as measured by infrared dichroïsm using tilted film, is plotted versus the annealing time in Figure 6.35. Even with the expected less oriented contribution of the amorphous segments, C1-C4 axes are strongly perpendicular to X_3. It can thus be concluded that even in the amorphous phase the phenyl rings are lying in the plane of the film. As far as the orientation with respect to X_2 is concerned a lower averaged orientation is observed, which can be accounted for by a lower orientation in the amorphous phase along this direction as is usually observed. The behaviour of the orientation with respect to X_2 is more complex and shows, as compared to that of the crystalline chain axis, the opposite trend at short times. Whereas crystalline chain axes become more oriented along X_2, the average orientation shows a decrease in the same time. For longer annealing times, whereas as a relaxation of $P_{200}^{\overline{1}05/X_2}$ is observed, a stronger orientation, averaged over the crystalline and non-crystalline phases is measured before a relaxation is finally observed at even longer times.

This new and unexpected behaviour affords information about the mechanisms occurring in the amorphous phase. Comparison between Figure 6.34 and 6.35 shows that short annealing times tend to relax amorphous segments (that are preferentially oriented along the last draw direction X_2) whereas longer ones tend to stretch them. The molecular processes at the origin of these tendencies are not straightforward. Indeed the annealing step does not involve dimensional changes along any direction which could generate a driving force for orientation or relaxation. Moreover, the structure of the biaxially oriented samples is commonly described as a network where amorphous segments are trapped between crystalline blocks, in such a way that free dangling amorphous segments only represent a small volume fraction of the material. This network structure induces a high degree of connectivity inside the stretched material, so that the explanation of the previous observations must be connected to the clear and strong changes observed in the crystalline phase.

The strong relaxation of the amorphous phase orientation with respect to X_2 observed at short time can be connected with the increase in the number of crystals oriented along X_2. During the growth process, trans conformers present in the amorphous phase are transferred into the crystalline blocks. This generates an enrichment in gauche conformations of the amorphous phase, thereby generating a loss in orientation along X_2. This idea is sketched in Figure 6.36, where it is shown that the increase of the length of the crystals along the chain axis direction necessarily releases tension on amorphous segments since a relatively longer chain contour is available to link them. It could be argued that the same phenomena also hold for X_1 oriented crystals, as well as for intermediate crystals. However, the volume fraction of X_2 crystals is much higher than that of X_1 crystals, generating therefore the above described behaviour.

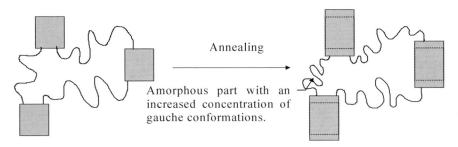

Annealing

Amorphous part with an increased concentration of gauche conformations.

Figure 6.36 Schematic drawing showing how the increase in length of the crystalline blocks might induce relaxation in the amorphous phase

At longer times, as it has been mentioned by many authors and further verified in our studies, an increase of the width of the crystals is noted (along (010) planes normals). The formation of folded crystals (sometimes called lamellae) has been argued [81] although being not the unique possible mechanism for explaining this type of crystal growth. In this case also, the connectivity of the structure is partly responsible for the increase of orientation in the amorphous phase. As it is shown in Figure 6.37, the lateral growth of the crystals by formation of folds at their surfaces leads to stretching of amorphous chain segments along X_2, and can therefore explain the observed tendency of the average orientation. At even longer times, no strong changes in the dimensions of X_2 crystals can be seen. However, in this time scale X_1 crystals are still appearing as it is apparent from Figure 6.33 and probably others are also growing since the average size does not decrease. In this direction, the formation of folded crystals is even more likely than along X_2. We can thus expect from the crystal growth mechanism the formation of stretched chain segments along X_1, which leads to a relaxation along X_2.

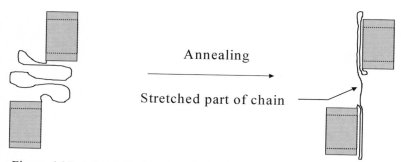

Figure 6.37 Schematic drawing showing how the increase in width of the crystalline blocks might stretch a chain in the amorphous phase

6.5.2   Influence of the annealing temperature

After preheating the sample at 100°C for 5s, it was suddenly brought to the annealing temperature. The annealing time was set at 5s and the influence of temperature was studied between 180 to 230°C. It is important to notice that hereafter our reference is the preheated film. We can only compare final states with each other. The graphs where properties will be plotted versus temperature have not to be interpreted as the thermal "history" of the sample. As in the previous part, the crystalline structure and the molecular orientation of the samples will be described, and attempts will be made to correlate the observed tendencies with matrices on a molecular scale.

The dimensions of the crystalline blocks along the previously defined directions are plotted versus annealing temperature in Figure 6.38a, b, c. Whatever the considered direction, the heat setting step increases the size of the crystals as compared to the preheated samples. Despite some unavoidable scatter in the data two regimes of temperature seem to appear which where also apparent in a previous study on uniaxially stretched films [80]. Indeed, along all the directions, the size of the crystals does not seem to depend on temperature in the range $180 \leq T \leq 210°C$. But for higher temperatures a strong increase in size is observed especially for the width and the thickness. The length is obviously less affected. The two domains of temperature are clear as well as the strong growth observed above 210°C, essentially with respect to the width and thickness. It is likely that at the highest temperature the rapid increase of the size originates from the merging or welding of smaller crystals.

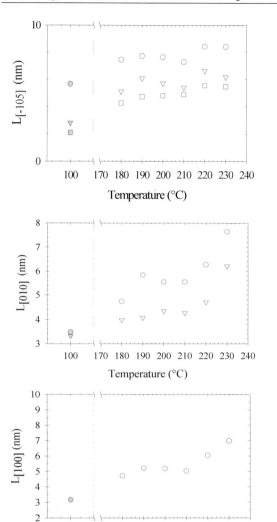

Figure 6.38a   Length of the various types of crystals versus annealing temperature (▢) I crystals, (○) X_2 crystals, (∇) X_1 crystals

Figure 6.38b Width of the various types of crystals versus annealing temperature (○) X_2 crystals, (∇) X_1 crystals

Figure 6.38c   Thickness of the crystals versus annealing temperature

The diffracted intensity from the ($\bar{1}05$) planes is plotted versus azimuthal angle for different temperatures in Figure 6.39. It clearly shows that even at the lowest temperature an increase in the number of diffracting planes is observed whatever the angle. At 230°C the increase seems essentially to originate from the X_2 crystals.

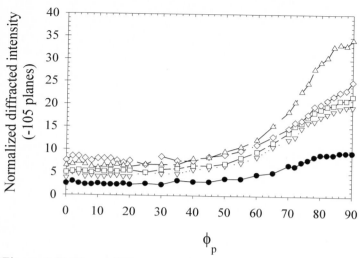

Figure 6.39 X-ray diffracted intensity in the plane of the film for (-105) planes for preheated sample at 100°C (●) and annealed samples at 180°C (∇), 200°C ($\square$), 220°C ($\diamond$) and 230°C ($\triangle$).

The growth rate of the crystals does not seem to depend strongly on their orientation with respect to the film of the plane. Indeed, out of plane intensity distributions are superimposable when normalized by their maximum value. In terms of second order moments, the temperature effects are quite weak (cf. Figure 6.40). Orientation is essentially constant whatever the temperature, although slightly higher along X_2 than in the preheated sample. In particular, there is no strong sign of changes even in the high temperature regime. However, a careful examination of the diffracted intensity reveals that the number of crystals oriented along X_2 increases significantly. In the same time, X_1 oriented crystals appear. These two effects might be cancelling each other in terms of the order parameter P_{200}, which does not mean that nothing happens.

Second order moments of the C1-C4 axes distribution function with respect to the principal directions of the film are given in Figure 6.41. In order to illustrate the tendencies occurring during the preheating step, the data corresponding to the biaxially oriented not heat-set sample have been added. A slight disorientation of the C1-C4 axes with respect to the plane of the film occurs as the temperature is increased. As far as the orientation with respect to X_2 is concerned, a relaxation is first observed for the lowest temperatures. As temperature increases less relaxation is observed. This unexpected behaviour might be understood by the combination of a relaxation which is showed down at elevated temperatures and a reorientation mechanism which becomes more and more intense as the temperature increases. In order to substantiate this assumption, the fraction of trans conformers along each principal direction of the film (extracted from the 973 cm^{-1} absorption band) has been plotted in Figure 6.42. Although the data originates from another infrared band that that used to calculate the orientation of the C1-C4 axes, the same tendencies as in Figure 6.41 appear. Whatever the annealing temperature, an increase in the concentration of trans

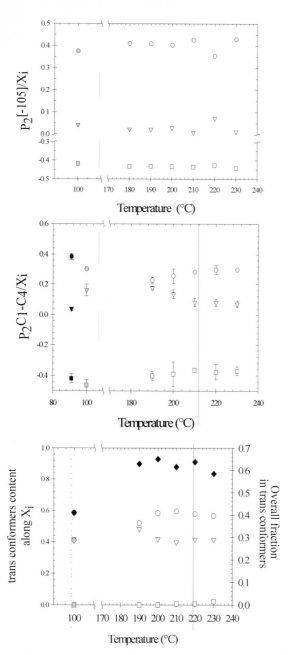

Figure 6.40 Orientation of the (-105) planes normals with respect to the principal directions of the film (X_1 ($\triangledown$), X_2 ($\bigcirc$) et X_3 ($\square$)) versus annealing temperature. Grey symbols refer to the preheated film at 100°C.

Figure 6.41 Average orientation of the C1-C4 axis with respect to the principal directions of the film (X_1 ($\triangledown$), X_2 ($\bigcirc$) et X_3 ($\square$)) versus annealing temperature. Black symbols refer to the initial biaxially stretched film, grey symbols to the preheated film at 100°C.

Figure 6.42 Concentration in trans conformers ($\blacklozenge$) and their fraction along X_1 ($\triangledown$), X_2 ($\bigcirc$) et X_3 ($\square$) versus the annealing temperature. Grey symbols refer to the preheated film at 100°C.

conformers is noted with respect to the reference film. It can be attributed to the growth of the crystals which obviously generates trans conformers as well as thermally induced conformational changes. The slightly less perpendicular orientation of the C1-C4 axes is corroborated by the increased fraction of trans conformations oriented along the X_3

direction. Taking into account the constant value of $P_{200}^{\overline{1}05/X_3}$, it can be deduced that C1-C4 axes in the amorphous phase lose part of their in-plane orientation, revealing some relaxation.

Despite the unbalanced $(\overline{1}05)$ planes diffracted x-ray intensity between the X_1 and X_2 directions, the trans conformers seem rather well spread along these 2 directions. It can be deduced that the amorphous phase contains more trans conformers aligned along X_1 rather than along X_2 (the latter having a strong tendency to be incorporated in the crystalline blocks).

As compared to the reference film, a relaxation of average orientation with respect to X_2 is observed at the lowest temperatures. It can be interpreted as a consequence of the increase of the length of the crystals and of their width by a mechanism which does not imply chain folding. As temperature increases, this relaxation appears less and less although the growth of the crystals is still present, and eventually disappears completely. This might be interpreted by the occurrence of the mechanisms sketched in Figures 6.36 and 6.37, only the weight of each mechanism depends on temperature. More precisely, at the highest temperatures, the relaxation of the amorphous phase orientation originating from an increase of the gauche content between crystalline blocks is cancelled by the stretching induced by folds located at the surface of the crystals. As temperature increases, it has been shown by several authors that the formation of lamellae becomes increasingly probable. This explains, at least qualitatively, the experimental observations.

6.6   Conclusions

In this chapter, we have given a description of the development of molecular orientation and crystalline organization in PET films during two different film processing conditions, which both of which involve sequential drawings followed by an heat-setting treatment. It turns out that in both cases, there is a strong memory in terms of preferential directions of orientation of the second drawing step, especially in the case of the inverse sequence where very efficient reorientation mechanisms have been observed, whereas more balanced properties are observed in the normal sequence. This appears as an illustration of the consequence of the very different deformation kinetics involved in either constant speed or constant force stretching.

The particular kinetics of constant force stretching opens new possibilities in industrial processing through the introduction of two consecutive longitudinal drawing steps instead of one. Under convenient temperatures and draw ratios conditions, quite large draw ratios can be obtained at the end of these longitudinal steps [82], leading of course to higher production rates but also to a modification of the properties of the films.

Finally, we focussed here on orientation behavior and almost discarded the use of stress-strain curves, as a signature of the state of the material. Valuable information can be obtained from the knowledge of the stress, and moreover, the development of constitutive models, able to predict the mechanical response of the polymer film during biaxial drawing appears as an new stimulating challenge [83,84,85].

Acknowledgments:

It is a pleasure to thank Rhône-Poulenc, Rhône-Poulenc Films and Toray Plastics Europe for financial support, kind access to specific equipment as well as fruitful and stimulating discussions with Drs J. Beautemps, G. Lebourvellec, G. Lorentz, F. Bouquerel and D. Veyrat. This work owes much to the efforts of P. Lapersonne, J. B. Faisant de Champchesnel[†], M. Vigny throughout their Ph. D. thesis. It is a pleasure to thank them warmly, as well as Prof. L. Monnerie who initiated and contributed to much of this research. Part of the work has also benefited from collaborations with Prof. I. M. Ward and Dr. D. I. Bower, who are acknowledged with gratitude.

References

1. G. Lebourvellec, J. Beautemp, J. P. Jarry, *J. Appl. Polym. Sci.*, **39**, 319 (1990)
2. G. Lebourvellec et J. Beautemps, *J. Appl. Polym. Sci.*, **39**, 329 (1990)
3. P. Lapersonne, *Ph D. Thesis, University Paris VI* (1991)
4. J.B. Faisant de Champchesnel, *Ph D. Thesis, University Paris VI* (1994)
5. M. Vigny, *Ph D Thesis, University du Maine, Le Mans*, (1997)
6. P. Lapersonne, J.F. Tassin, L. Monnerie, J. Beautemps, *Polymer*, **32**, 3331 (1991)
7. P. Lapersonne, D.I. Bower, I.M. Ward, *Polymer*, **33**, 1266 (1992)
8. P. Lapersonne, D.I. Bower, I.M. Ward, *Polymer*, **33**, 1277 (1992)
9. P. Lapersonne, J.F. Tassin, L. Monnerie, *Polymer*, **35**, 2192 (1994)
10. J.B. Faisant de Champchesnel, D.I. Bower, I.M. Ward, J.F. Tassin, G. Lorentz, *Polymer*, **34**, 3763 (1993)
11. J.B. Faisant de Champchesnel, J.F. Tassin, D.I. Bower, I.M. Ward, G. Lorentz, *Polymer*, **35**, 4092 (1994)
12. J.B. Faisant de Champchesnel, J.F. Tassin, L. Monnerie, P. Sergot, G. Lorentz, *Polymer*, **38**, 4165 (1997)
13. M. Vigny, J. F. Tassin, A. Gibaud, G. Lorentz, *Polym. Eng. Sci.*, **37**, 1785, (1997)
14. M. Vigny, J. F. Tassin, G. Lorentz, *Polymer*, **40**, 397,(1999)
15. M. Vigny, J. F. Tassin, D. Veyrat, *Macromol. Symp.*, in press
16. I.M. Ward, *Advances in Polymer Sciences*, **66**, 81 (1985)
17. D. I. Bower, *J. Polym. Sci., Polym. Phys. Ed.*, **19**, 93, (1981)
18. D. I. Bower, *Polymer*, **23**, 1251, (1982)
19. R. de Daubeny, C.W. Bunn, C.J. Brown, *Proc. Roy. Soc. London (A)*, **226**, 531 (1954)
20. D. I. Bower, D. A. Jarvis, I. M. Ward, *Polymer*, **24**, 1459, (1986)
21. D. I. Bower, D. A. Jarvis, E. L. V. Lewis, I. M. Ward, *Polymer*, **24**, 1481, (1986)
22. D.A. Jarvis, I.J. Hutchinson, D.I. Bower, I.M. Ward, *Polymer*, **21**, 41 (1980)
23. A.J. De Vries, C. Bonnebat, J. Beautemps, *J. Polym. Sci., Polym. Sym.*, **58**, 109 (1977)

[†] Deceased on November 27, 1996

24. A. Ajji, J. Guèvremont, K.C. Cole, M.M Dumoulin, *Polymer*, **37**, 3707 (1996)
25. P. Chandran, S. Jabarin, *Adv. Polym. Techn.*, **12**, 119, 133, 153 (1993)
26. D.R. Salem, *Polymer*, **33**, 3182, 3189 (1992)
27. H. Hirahata, S. Seifert, H. G. Zachmann, K. Yabuki, *Polymer*, **37**, 5131, (1996)
28. D. J. Blundell, D. H. MacKerron, W. Fuller, A. Mahendrasingam, C. Martin, R. J. Oldman, R. J. Rule, C. Riekel, *Polymer*, **37**, 3303, (1996)
29. L.E. Alexander, *X-Ray Diffraction Methods in Polymer Science,* John Wiley & Sons (1969)
30. A. Ajji, J. Guèvremont, R. G. Matthews, *Int. Symp. on Orientation of Polymers*, Montreal, 1998
31. A. Kaito, K. Nakayama, H. Kenetsuna *J. Polym. Sci., Part B, Polym. Phys. Ed.*, **26**, 1439 (1988)
32. A. Cunningham, G.R. Davies, I.M. Ward, *Polymer*, **15**, 743 (1974)
33. A. Cunningham, I.M. Ward, H.A. Willis, V. Zichy, *Polymer*, **15**, 749 (1974)
34. N. J. Clayden, J. G. Eaves, L. Croot, *Polymer*, **38**, 159, (1997)
35. D. J. Walls, *Appl. Spectrosc.*, **45**, 1193, (1991)
36. D.J. Walls, J.C. Coburn, *J. Polym. Sci., Part B, Polym. Phys.*, **30**, 887 (1992)
37. P. Yuan, C. S. P. Sung, *Macromolecules*, **24**, 6095, (1991)
38. K. H. Lee, C. S. P. Sung, *Macromolecules*, **26**, 3289, (1993)
39. L. Quintanilla, J.C. Rodriguez-Cabello, T. Jawhari, J.M. Pastor, *Polymer*, **34**, 3787 (1993)
40. K.C. Cole, J. Guèvremont, A. Ajji, M.M. Dumoulin, *Appl. Spectrosc.*, **48**, 1513 (1994)
41. K.C. Cole, H. Ben Daly, B. Sanschagrin, K. T. Nguyen, A. Ajji, *Polymer.*, **40**, 3505, (1999)
42. J.C. Rodriguez-Cabello, J.C. Merino, M.R. Fernandez et J.M. Pastor, *J. Raman Spectrosc.*, **27**, 23 (1996)
43. J. H. Nobbs, D. I. Bower, I. M. Ward, *J. Polym. Sci., Polym. Phys. Ed.*, **17**, 259, (1979)
44. M. Hennecke, A. Kud, K. Kurz, J. Fuhrmann, *J. Colloid Polym. Sci.*, **265**, 674, (1987)
45. B. Clauss, D. R. Salem, *Polymer*, **15**, 3193, (1992)
46. B. Clauss, D. R. Salem, *Macromolecules*, **28**, 8328, (1995)
47. P. Lapersonne, J. F. Tassin, P. Sergot, L. Monnerie, G. Lebourvellec, *Polymer*, **30**, 1558, (1989)
48. R.M. Gohil, D.R. Salem, *J. Appl. Polym. Sci.*, **47**, 1989 (1993)
49. D. R. Salem, *Polymer*, **39**, 7067, (1998)
50. G. Lorentz, J.F. Tassin, *Polymer*, **35**, 3200 (1994)
51. M. Doi, S. F. Edwards, *The Theory of Polymer Dynamics*, Clarendon Press, Oxford, (1986)
52. D. J. Blundell, R. J. Oldman, W. Fuller, A. Mahendrasingam, C. Martin, D. H. MacKerron, J. L. Harvie, C. Riekel, *Polymer Bulletin*, **42**, 357, (1999)
53. M. Cakmak, J. Spruiell, L. White, J.S. Lin, *Polym. Eng. Sci.*, **27**, 893 (1987)
54. M. Cakmak, J.L. White, J.E. Spruiell, *Polym. Eng. Sci.*, **29**, 1534 (1989)
55. M. Cakmak, Y. D. Wang, M. Simhambhatla, *Polym. Eng. Sci.*, **30**, 721, (1990)
56. S. Murakami, Y. Nishikawa, M. Tsuji, A. Kawaguchi, S. Kohjiya, M. Cakmak, *Polymer*, **36**, 2291, (1995)
57. B.J. Jungnickel, M. Teichgraber, C. Ruscher, *Faserforschung Textiltechnik*, **24**, 423, (1976)

58. J. Guèvremont, A. Ajji, K.C. Cole, M.M. Dumoulin, *Polymer*, **36**, 3385 (1995)
59. A. Ajji, K.C. Cole, M.M. Dumoulin, J. Brisson, *Polymer*, **36**, 4023 (1995)
60. P.R. Pinnock, I.M. Ward, *Bristh Journal of Applied Physics*, **15**, 1559 (1964)
61. P.R. Pinnock, I.M. Ward, *Transactions of the Faraday Society*, **62**, 1308 (1966)
62. D.R. Salem, *Polymer*, **35**, 771 (1994)
63. G. Lebourvellec, L. Monnerie, J.P. Jarry, *Polymer*, **27**, 856 (1986)
64. G. Lebourvellec, L. Monnerie, J.P. Jarry, *Polymer*, **28**, 1712 (1987)
65. N. Yoshihara, A. Fukushima, Y. Watanabe, A. Nakai, S. Nomura, H. Kawai, *J. Soc.
      Fiber Sci. & Tech. (Japan)*, **37**, T-387 (1981)
66. F. Rietsch, B. Jasse, *Polym. Bull.*, **11**, 287 (1984)
67. F. Rietsch, *Eur. Polym. J.*, **26**, 1077 (1990)
68. J.C. Rodriguez-Cabello, J.C. Merino, L. Quintanilla, J.M. Pastor, *J. Appl. Polym. Sci.*,
      62, 1953 (1996)
69. B. Gunther, H.G. Zachmann, *Polymer*, **24**, 1008 (1983)
70. A. Mahendrasingam, C. Martin, W. Fuller, R. J. Oldman, D. J. Blundell, J. L. Harvie,
      D. H. MacKerron, C. Riekel, P. Engström, *Polymer*, **40**, 5553, (1999)
71. R.M. Gohil, *J. Appl. Polym. Sci.*, **48**, 1635 (1993)
72. R.M. Gohil, *J. Appl. Polym. Sci.*, **48**, 1649 (1993)
73. S.K. Sharma, A. Misra, *J. Appl. Polym. Sci.*, **34**, 2231 (1987)
74. T. Thistlethwaite, R. Jakeways, I.M. Ward, *Polymer*, **29**, 61 (1988)
75. T.C. Ma, C.D. Han, *J. Appl. Polym. Sci.*, **35**, 1725 (1988)
76. H. Chang, J. M. Schultz, R. M. Gohil, *J. Macromol. Sci., Phys.*, **B32**, 99, (1993)
77. R. M. Gohil, *J. Appl. Polym. Sci.*, **52**, 925, (1994)
78. P. Varma, E. A. Lofgren, S. A. Jabarin, *Polym. Eng. Sci.*, **38**, 237, 245, (1998)
79. Y. Maruhashi, T. Asada, *Polym. Eng. Sci.*, **32**, 481, (1992)
80. Y. Maruhashi, T. Asada, *Polym. Eng. Sci.*, **36**, 483, (1996)
81. C. Ramesh, V. B. Gupta, J. Radhakrishnan, *J. Macromol. Sci., Phys.*, **B36**, 281, (1997)
82. D. R. Salem, *Int. Symp. on Orientation of Polymers*, Montreal, 1998, to appear in
      Polym. Eng. Sci.
83. C. P. Buckley, D. C. Jones, D. P. Jones, *Polymer*, **37**, 2403, (1997)
84. A. M. Adams, C. P. Buckley, D. P. Jones, *Polymer*, **39**, 5761, (1998)
85. R. G. Matthews, R. A. Duckett, I. M. Ward, *Polymer*, **38**, 4795, (1997)

7. Rolling and Roll-Drawing of Semi-Crystalline Thermoplastics

A. Ajji and M. M. Dumoulin
Industrial Materials Institute, National Research Council Canada
75, Boul. De Mortagne, Boucherville, Québec
Canada J4B 6Y4

7.1   Introduction

This chapter discusses the rolling and roll-drawing processes, in particular for semi-crystalline thermoplastics, with special emphasis on polyethyleneterephthalate (PET). This process has the advantages of being continuous, capable of high production rates, and able to achieve high deformation ratios with some degree of biaxial orientation (double orientation).   The roll-drawing process allows the extent of biaxial orientation to be controlled by adjusting the tension applied by drawing.

These processes are discussed in details and some experimental results obtained with a four-station roll-drawing set-up are presented.   The effect of process parameters such as temperature, speed, gap and tension are presented systematically for a number of polymers: polyesters (PET), polyolefins (PE and PP) as well as some engineering and speciality resins such as polyamides (PA-6 and PA-11), and a polyketone (PEEK). Aspects reviewed include relaxation, structure development in terms of orientation and crystallinity as a function of draw ratio, draw ratio as a function of process parameters and finally mechanical properties as a function of draw ratio.

7.1.1 Why orient polymers?

Orientation can be induced in polymers either in the melt, rubbery or solid states, depending on the material and extent of property enhancement desired. When orientation takes place in the melt or rubbery states and depending on time, an isotropic morphology is often preferred.   In order to impart significant performance improvements, the molecules should be oriented in the solid state. Solid state deformation can take place in a number of processes, most of which are treated elsewhere in this book. The current chapter focuses on the solid state rolling and roll-drawing processes, and in particular their use with semi-crystalline thermoplastics.

The replacement of some conventional structural materials has led material scientists and engineers to look at new materials and processes to enhance the properties of some existing materials. Conventional structural high performance materials (metals, composites,...) have some shortcomings including weight, recyclability, adhesion of the reinforcement, etc. Hence, the development of ultra-high modulus products is very attractive in view of their significantly lower density. For example, steel is about 8 times more dense than polyethylene, PE. However, the specific modulus (modulus to density ratio) is significantly higher for highly oriented polymers than for metals in general, as shown in Table 7.1.

The carbon-carbon bond is the strongest bond known to date. If one could somehow process a material containing this bond so that these bonds are all perfectly aligned in one direction, this would lead to a material with an extremely high modulus and strength. The theoretical tensile modulus of PE - calculated as 240-300 GPa [1] - should approach or even surpass that of steel (208 GPa). However, the theoretical values were considered out of reach because all known polymers had moduli two orders of magnitude lower. The reason for such a low modulus is that the polymer assumed a random entangled and twisted configuration which has a low load bearing capability. It has been realized that the highest modulus and strength would result from an anisotropic structure of highly oriented, extended and densely packed chains. Indeed, PE has been processed into fibers exhibiting moduli of 100-200 GPa, thereby indicating that the above mentioned theoretical values can be approached.

Table 7.1: Physical properties of selected materials.

Material	Density (gm/cm^3)	T_g (°C)	T_m (°C)	Isotropic modulus (GPa)	Crystal or maximum modulus (GPa)	Maximum specific modulus (10^6 m)
PE	0.95	-120	130	0.4-2	300	31.6
PP	0.9	-18	165	1-2	60	6.7
PET	1.45	70	255	3.2	108	7.5
PA-6	1.13	50	220	2.8	160	14.2
PA-11	1.05	50	180	1.8	160	15.2
PEEK	1.32	150	335	3.6	60	4.5
Glass fibre	2.4	1713	---	75	75	3.1
Aluminium	2.8	---	660	70	70	2.5
Steel	7.8	---	1100-1450	210	210	2.7

7.1.2   Orientation processes and solid state deformation of semi-crystalline polymers

Much effort has been put in the past two decades on the development of orientation in semi-crystalline polymers [2-4]. Particular achievements have been made in the development of ultra high modulus PE fibers, with moduli of the order of 240 GPa from gel spinning techniques, which induce very high degrees of orientation to the macromolecules. Other techniques have been developed for solution and melt processes for fiber spinning and oriented films [5]. The highly oriented PE fibers are already used in some applications such as marine ropes, protective and antiballistic clothing [4]. However, their use in structural applications is rather limited due to their low melting temperature and poor creep properties. Some improvements in the temperature stabilization of the fibers and enhancement of their creep properties through crosslinking are possible but with limited benefit. In fact, the structural applications market is still dominated by composites despite many disadvantages (cost, high density, fabrication, environment, etc,). Hence, development of oriented polymers showing the performance in the range of those of composites or better is promising. Table 7.1 gives some information on the maximum modulus and strength of some semi-crystalline polymers, as well as their melting point.

Orienting semi-crystalline polymers below their melting point produces extended chain crystals having fewer chain folds and defects. In contrast with melt processing of polymers, the rearrangement of the spherulitic crystal structure below the melting point demands high levels of energy. Thus solid state forming requires very high power. The production rates remain low, to avoid polymer fracture and stay within reasonable power requirements. In addition, the product shapes are relatively simple. There are several solid state processes currently investigated, such as tensile and die drawing, hydrostatic and ram extrusion, rolling, roll-drawing and compression [5,6].

Polyolefins have been the subject of much attention so far [7-10]. Their main drawback, already mentioned above, is their low melting temperature, limiting the range of applications. To extend the number of potential uses, one could turn to polymers with a higher melting point, such as polyoxymethylene (POM), polyvinylidene fluoride (PVDF) [11] and polymethylpentene (PMP). These polymers show an α-relaxation temperature above which the intracrystalline interactions are considerably reduced. This allows to achieve very high draw ratios by solid state deformation (draw ratios above 20 were often obtained). Other semi-crystalline polymers which do not show an α-relaxation can also be solid state formed. They include polyethyleneterephthalate (PET), polyamides (PA), polyethersulfone (PES) and polyetheretherketone (PEEK) [12]. Orientation of these materials can lead to very high mechanical properties [12]. PET's high melting point, T_m (250°C) and chemical resistance make it very interesting for a number of structural applications. Similarly, polyamides (PA) show high toughness, chemical resistance, high drawability, a wide range of melting temperatures, and a low friction coefficient. They are well suited for a wide range of engineering applications. On the other hand, poly(etheretherketone) (PEEK), a wholly aromatic polymer, has attracted the attention of many researchers because of its excellent mechanical and thermal properties (T_g=145°C and T_m=334°C). Despite its high melting point, PEEK can be melt processed into films and fibres by conventional methods. Many attempts have been made to orient PEEK using stretching techniques such as drawing in an oven, die drawing [13], zone-drawing [14] and solid state extrusion [15-17].

As mentioned above, production rates of solid state deformation processes are quite low. Apart from tape rolling, the processes used up to now were discontinuous and slow. An exception to this is the die drawing process developed by Ward et al. [5]. Thus, producing oriented profiles having relatively large dimensions by continuous rolling and at relatively high production rates will be an important practical improvement in solid state forming of polymers.

Roll-drawing of polymeric materials has been used to improve their mechanical and optical properties through orientation. Solid state roll-drawing has been used successfully to produce high modulus sheets with semi-crystalline polymers, such as PE, PP and POM [18-19]. However, the examination of the process and its application to engineering polymers was limited. Berg et al. [20] and Sun et al. [10] studied the 3-D structure-property relationships by measuring the three dimensional mechanical properties of several rolltruded polymers: PP, PVDF and propylene/ethylene block copolymers. They found that mechanical strength was enhanced triaxially for all these materials upon rolltrusion. The material's stiffness also exhibited 3-D improvement, depending on the polymer type and processing conditions. These results were discussed in terms of structural models. They found that mechanical property enhancement not only occurs in the principal draw direction, but also in the transverse directions. A variety of observations were made during tensile and compressive tests. Deformation bands, anisotropic yielding and ductile to brittle behaviour were observed. These results were presented along with a morphological model developed to account for the behaviour of the unique triaxially oriented polymers.

Yang et al. [21] studied the structure, transitions and mechanical properties of PP oriented by roll-drawing. Deformation of PP was conducted at 158°C on sheets 10 mm thick and 150 mm wide. Draw ratios between 3.6 and 16 were obtained. The increase in the static tensile modulus parallel to the draw direction to 12 GPa and a small increase in the perpendicular modulus were consistent with the development of a reinforcing phase of parallel fibres having the maximum achievable crystalline modulus. Polarised-light microscopy showed that the replacement of the initial spherulites by an oriented structure was complete at a draw ratio of about 4. The structure became more fibrous at higher draw ratios, distinct fibres being visible in scanning electron micrographs of fracture surfaces. The tensile strength in the draw direction reached 400 MPa at the highest draw ratio, while the perpendicular tensile strength remained roughly constant at 30 MPa. Dynamic mechanical properties were measured in flexural mode. A broad maximum in E" at 80-100°C, due to a crystalline dispersion, developed as the draw ratio increased, more prominently in the parallel cut specimens. The maxima in E" were accompanied by corresponding decreases in the storage modulus, E'. These suggest that roll-drawing causes better packing and orientation in the amorphous fraction and changes the structure of crystalline material towards perfect alignment.

Kaito et al. [22] investigated the roll-drawing and crystallisation of UHMWPE. They found that the mechanical properties of UHMWPE sheets were much improved by crystallisation during the roll-drawing process at T=140°C. The sheets roll-drawn at 135-140°C exhibited c-axis orientation in the draw direction and (100) alignment in the sheet plane. However, at T=100°C, the elastic motion of the amorphous chains induces the twinning of lattice, which enhances the transition to the (110) alignment in the sheet plane. The dynamic storage modulus below the glass transition temperature showed good correlation with crystallinity and orientation functions, while taut tie molecules and thick crystallites play an important role in the storage modulus above that temperature.

Higashida et al.[23] studied the mechanical properties of uniaxially and biaxially rolled PE and PP sheets. They found that the tensile strength of the rolled PP sheet, initially elongated by a factor of 5 in longitudinal direction (L) and then elongated by 1.5 in the transverse direction (T), reached almost 100 MPa, and the value was three times as large as that of the initial material in all directions in the plane. This combination of rolling was the condition to get substantially the same rolling elongation in both L and T directions because the uniaxially rolled sheets become thicker by shrinking when reheated for biaxial rolling and the substantial second rolling elongation is larger than 1.5. The trend observed for molecular orientation correlated with the tensile strength characteristics of uniaxially and biaxially rolled sheets.

Chen *et al.* [9] studied the properties of oriented PP laminates. They observed that the longitudinal strength and modulus of the oriented PP (OPP) increase with an increase in draw ratio. However, the transverse properties remain relatively unchanged. In that study, multidirectional OPP laminates (with (0)2, (0/90)s and (0/±45)s lay-ups) were prepared to obtain sheet materials with improved properties in more than one direction in the plane. Hot plate welding was used to produce these translucent and recyclable laminates. The in-plane properties of the laminates were successfully predicted with classical laminate theory, which is commonly used to predict the properties of fibre reinforced materials. These laminates can be quasi-isotropic and were found to have improved modulus (up to 6 GPa for (0/90)s laminates) and strength (up to 150 MPa for (0/90)s laminates) as well as exceptionally good impact toughness.

7.2   Rolling and Roll-drawing

7.2.1   Introduction

The rolling process was first developed for processing metals. It involves the passage of a slab of material between a pair of rolls and is shown schematically in Figure 7.1. The gap between the rolls is substantially smaller than the thickness of the slab, therefore imparting a reduction in thickness and increases in the other two dimensions. Hot or cold rolling is being widely used for producing steel sheets. For polymers as well as metals, the process affects not only the dimensions, but also the microstructure, affecting the final properties. In general some slippage occurs between the polymer profile and the rolls, the amount of slippage depending on polymer type, rolling conditions and geometry. Smaller roll diameter will lead to higher slippage, which obviously reduces the output and maximum achievable reduction ratio. As can be appreciated from Figure 7.1, a larger roll diameter results in smaller deformation rates in the material, smaller contact angle and longer residence time. Larger rolls can therefore be more conducive to easier operation and higher outputs.

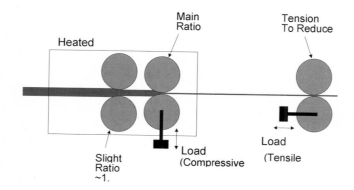

Figure 7.1- Schematic of the roll-drawing process

In the roll-drawing process, some tension is applied on the material emerging from the rolls, normally using other rolls. The process is thus significantly different from rolling as the tension applied by drawing allows control of the changes in dimensions and, most importantly, the extent of biaxial orientation. The roll-drawing process has the advantages of being continuous, capable of high production rates, and able to achieve high deformation ratios with some degree of biaxial orientation (double orientation).

Rolling has been investigated mostly with semi-crystalline polymers and to a much lesser extent with amorphous polymers because of their rapid relaxation and the limited advantages obtained. In fact, Bahadur and Henkin reported a slight increase in density for Polycarbonate (PC) arising from deformation during rolling [24]. Thakkar and Broutman used rolling to orient PC and reported an orientation effect on fracture toughness [25]. Oh and Kim studied the effect of rolling orientation on the brittle-ductile transition on the fracture of PC samples which had been subjected to a reduction of 10% achieved through several rolling passes [26]. They concluded that the brittle-ductile transition may be ascribed to microstructural changes brought about by rolling, but not necessarily to molecular orientation per se. Shin-ichi Matsuoka showed on the other hand that cold-forged products obtained from preliminarily-rolled rods of HDPE, PP, POM, ABS and PVC have extremely good dimensional stability when subjected to heat treatment in a boiling-water bath [27]. This process was applied to make small machine parts such as knobs. Kubisiak and Quesnel [28] investigated the fracture of oriented general-purpose polystyrene. Specimen microstructure was controlled using roll-drawing. Mechanical results in tension on sharp-notched samples displayed a 4X difference in longitudinal vs. transverse values. Shrinkage measurements showed that contraction in the machine direction amplifies with increasing draw ratio. Craze formation was visible only in those specimens tested in the longitudinal direction and fracture mechanisms had strong orientation dependence.

7.2.2   Roll-drawing of semi-crystalline polymers

Bigg et al. [7] examined the behaviour of several PP and HDPE resins during solid state rolling and the resulting properties. They noted that the stability of the process was highly dependent on the quality of feed materials: the process did not smooth out local imperfections such as thickness variations, voids or pinholes but would rather amplify them. They achieved reduction ratios of 10.8:1 for PE and 6.4:1 for PP. For both polymers, the maximum achievable reduction ratios and resulting modulus were higher with increasing rolling temperature.

The most important factor in solid state orientation of polymers is the maximum achievable draw ratio, λ_m. In general, the higher the draw ratio λ, the better will be the mechanical properties of the oriented material. Generally, λ values in the range of 10 for polyolefins and 4 for engineering resins have been obtained. In some cases, a larger λ can be achieved at some point in the process. However, the temperature being maintained above T_g of the material, some relaxation occurs, leading to a decrease in λ of up to 1. Figure 7.2 shows λ obtained with a minimum tension as a function of the calculated λ_c for PET and PEEK. The tension was maintained at the minimum for these tests, just enough to direct the profile and avoid fracture after compression. It can be observed from this figure that λ is lower than λ_c in all cases, indicating that relaxation takes place just after the compression between the rolls. This relaxation becomes more significant as λ increases. It should be noted that for the smaller gaps, the difference between λ and λ_c can be of the order of 8 to 10. Similar observations were made with polyolefins and PA's. It is also to be observed from this figure that λ_t (determined in the thickness direction) is higher than that measured in the length direction (λ_l). This indicates that a slight degree of biaxial orientation is obtained for low tension levels. Very high λ values (above 25) were obtained for polyolefins in various studies. Much smaller values are to be expected for engineering resins, because the hydrogen bonding and molecular interactions present cause lower drawability. λ around 6 was obtained for PET when tension was increased to its maximum and the very rapid cooling was applied after deformation in the last rolling station.

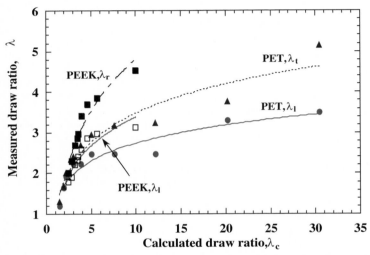

Figure 7.2- Draw ratios for rolled polymers as a function of λ_c for PET and PEEK.

The profile's initial speed was kept constant in all experiments. Therefore, for each reduction of the gap, the roll speed has to be increased to compensate for the larger deformation and maintain the throughput constant. Figure 7.2 shows for PEEK a λ_r value computed from the ratio of roll speed to that of the profile. In the case of perfect adhesion between the rolls and the profile, this ratio should be the same as that measured in the length direction. However, as seen in Figure 7.2, this λ_r is always higher than λ_l, indicating that slippage occurs between the rolls and the deforming material. Similar results were obtained with the other polymers.

7.2.3   Structure development

The performance improvements sought through rolling and roll-drawing are due to modifications in structure, specifically crystalline and molecular orientation. Crystallinity measurement results for two different roll-drawn HDPE's are presented in Figure 7.3.

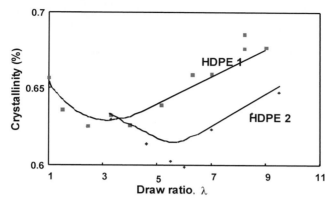

Figure 7.3- Crystallinity of roll-drawn HDPE-1 and HDPE-2 as a function of draw ratio.

First, at low draw ratio, a decrease in crystallinity is observed, up to draw ratios of 3 to 6, depending on the HDPE. Beyond this extreme, crystallinity increases with higher draw ratios. This behaviour can be explained by the deformation process. The melt-cooled specimens initially showed a spherulitic structure for the crystalline phase. The initial processing stages involve a slight deformation of the spherulites. As deformation continues, the spherulites are destroyed, thus inducing the observed crystallinity reduction, and the structure is gradually transformed into the final fibrillar structure at high draw ratios. The crystallinity results are shown in Figure 7.4 for PET, PA-6 and PEEK.

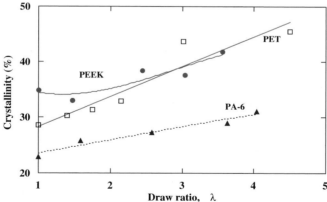

Figure 7.4- Crystallinity of roll-drawn PET, PA-6 and PEEK as a function of draw ratio.

The crystalline fraction increases upon drawing because the alignment of the polymer chains favours the formation of ordered crystalline structure, except for PEEK for which a slight decrease is observed at low λ. The maximum crystallinity is still below the maximum value that can be attained for these polymers, indicating that further orientation is possible.

7.2.4   Mechanical properties

Mechanical performance improvements are the impetus for solid state rolling and roll-drawing. Improvements sought usually revolve around stiffness or strength. Mechanical properties improvements are presented in Figures 7.5-7.8 in terms of the tensile modulus and strength respectively. For PET, moduli and tensile strength above 10 GPa and 400 MPa respectively were obtained (although moduli of up to 18 GPa were obtained). PEEK showed moduli similar to those for PET but higher tensile strength. Both polymers showed a similar slope of the modulus vs λ plot. The modulus for PAs is lower than that obtained for PET and PEEK, particularly in the case of PA-11. The slope of the modulus vs. λ plot is the same for both PAs, lower than that for PET and PEEK. However, tensile strength values for PA-6 show a sharp increase vs. λ, much sharper than for PA-11, resulting in values even higher than for PET at high draw ratios.

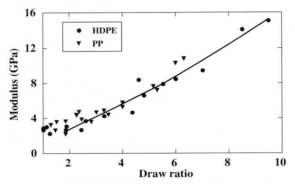

Figure 7.5   Longitudinal tensile modulus of roll-drawn HDPE and PP as a function of λ.

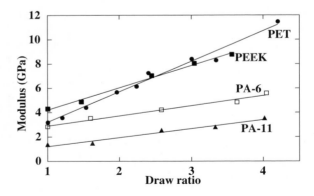

Figure 7.6   Longitudinal tensile modulus of roll-drawn PET, PA-6, PA-11 and PEEK as a function of λ.

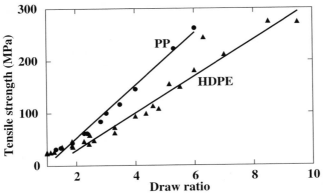

Figure 7.7   Longitudinal tensile strength of roll-drawn HDPE and PP as a function of λ.

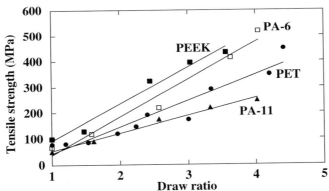

Figure 7.8   Longitudinal tensile strength of roll-drawn PET, PA-6, PA-11 and PEEK as a function of λ.

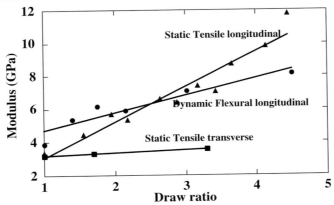

Figure 7.9   Longitudinal tensile, flexural and transverse tensile modulus obtained for PET as a function of draw ratio.

Finally, in Figure 7.9, the modulus as a function of draw ratio obtained for PET in the longitudinal direction is compared in the tensile and flexural modes. Both show a significant increase as a function of draw ratio, with a steeper slope for the tensile modulus. This is logical since the material has a much higher orientation in the longitudinal direction but the flexural modulus involves properties in all directions. On the same figure the tensile transverse modulus is also plotted as a function of draw ratio. A slight increase is observed which indicates some slight biaxial orientation and confirms the plane strain nature of the deformation.

7.3   A Case Study: PET

7.3.1   Orientation of PET

Numerous studies on PET orientation have appeared. To understand the orientation behaviour of PET, it is necessary to start from the simplest structure, i.e. from the amorphous state, which is possible for this polymer because of its relatively high glass transition temperature, Tg, (around 70°C) and slow crystallisation. It has been reported that orientation of amorphous PET above the glass transition temperature is mainly due to sliding of chains, described as flow drawing [29], and does not induce any birefringence in the films. However, other orientation measurement techniques were not used to confirm the observation. In fact, many structural modifications, such as conformational changes, strain-induced crystallisation and variation in crystal size and distribution are involved when orienting PET.

Lapersonne et al. [30] studied the uniaxial-uniplanar deformation of PET films. They found that the orientation of the crystallographic axes and the size of the crystalline blocks are well rescaled by the macroscopic draw ratio, which is a function of the drawing temperature and the applied stress. The large scale organisation of the crystalline phase is not determined by the same parameter but changes mainly with temperature. In another study, Lapersonne et al. used measurements of refractive indices to follow the uniaxial planar deformation of PET films by characterisation of their orientation [31]. They calculated the second moment of the orientation distribution function of the phenyl ring normal with respect to the three principal directions of the sample. Under their stretching conditions, the phenyl rings showed a strong tendency to orient their plane parallel to the plane of the films. The orientation of the chain axis could also be evaluated from the same data. A good correlation was observed between this indirect measurement and infrared dichroism.

The deformation of amorphous isotropic PET films at temperatures slightly above Tg under constant load has been described from a molecular point-of-view [32]. Deformation was qualitatively related to chain relaxation phenomena occurring before stress-induced crystallisation, then followed by equilibration of a rubber-like network, the junction points consisting of both trapped entanglements and crystalline units. The structure of this network is characterised by the number of segments between physical crosslinks. This parameter was calculated by comparing the predictions of the rubber elasticity theory (without Gaussian approximation) with the experimentally observed draw ratios under given conditions of temperature and loads. It was shown that small loads induce soft networks leading to high draw ratios. The predictions of the molecular orientation derived from this treatment were in good agreement with birefringence data on a wide range of samples.

The deformation mechanism of amorphous PET was studied as a function of molecular weight and entanglement density in predrawn films [33]. The deformability was shown to be affected by stress-induced crystals, which might act as network points. The tensile modulus and strength of drawn films were directly related to draw ratio and molecular weight. The improved draw efficiency with higher molecular weights was explained by the suppression of both disentanglement and relaxation of oriented amorphous molecules during deformation. Both are significant in the development of structural anisotropy and continuity along the draw direction. Thus relaxation appears to be an important factor.

Gordon et al. [34] performed a study of uniaxial and constant-width drawing of PET films. From the measurements of the changes in refractive index, drawing stress and peak shrinkage stress, it was concluded that the behaviour could in all cases be described by the deformation of a molecular network. This behaviour is similar to that observed in previous studies on PET fibres and films.

In recent works [35-38], we used different techniques (refractive index, front surface FTIR measurements and X-ray diffraction) to characterise the orientation and structure of initially amorphous drawn PET films. Contributions of the crystalline and amorphous phases were separated and the existence of a mesomorphic phase was postulated for the interpretation of the data. A good correlation between a double dichroic ratio from FTIR measurements with the tensile modulus of oriented PET was also found.

7.3.2   Relaxation and recovery during rolling

In the roll-drawing process, orientation can be increased by reducing the roll gap or increasing the tension. However, significant shape recovery due to relaxation can be observed. It is a major factor, because the deforming force is totally removed as the polymer sheet leaves the roll, which allows essentially free relaxation. The recovery process was not examined directly during roll-drawing, because of the difficulty of carrying out stress or orientation measurements in real time. While on-line orientation measurements are in principle possible, they must be carried out near the rolls, where most of the recovery occurs, which is technically difficult to achieve. Thus, the deformation and relaxation processes were examined in a simpler experimental set-up: tensile drawing. It is relatively simple to make stress and orientation measurements under tensile drawing conditions, during either drawing and relaxation. This section will therefore deal with the orientation and relaxation of samples undergoing tensile deformation. The birefringence was monitored by a multi-wavelength technique which uses the wavelength dependence of the transmitted light intensity to determine absolute birefringence [39]. The relaxation behaviour was studied for samples held at constant length, while maintained at the drawing temperature. To examine the effects of relaxation on the final oriented material, two sets of samples were produced. After drawing, some samples were quenched rapidly to minimize relaxation; these samples are called `as-drawn`. The second set of samples, on which real time measurements were made, were allowed to relax for 30 minutes; these samples are called `relaxed`. The relaxation mechanisms can be investigated by comparing the differences in the structure and orientation of the as-drawn and relaxed samples.

7.3.2.1 Orientation and constrained relaxation

True stress-strain curves, from the drawing of PET at 80°C, are shown in Figure 7.10. In these tests, amorphous PET samples were drawn to $\lambda=3.5$ at different initial strain rates. As the draw rate was increased, the drawing behaviour changed. At low rates, there was no peak in the true stress, which indicates rubber-like behaviour. At high rates, there was a peak in the true stress, which corresponds to the onset of necking. The magnitude of the peak stress also increased with the rate. Over the remainder of the true stress-strain curves, the true stress increased with strain rate. This is in agreement with previous work on PET [40-42].

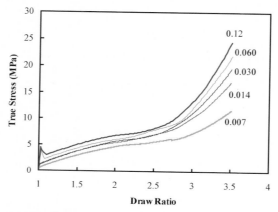

Figure 7.10 True stress-strain curves from drawing isotropic amorphous PET at different strain rates to $\lambda=3.5$ at 80°C.

When investigating the behaviour of PET, it is important to consider the effects of crystallinity. The dependence of the relaxation behaviour on crystallinity was examined by monitoring the relaxation at two different draw ratios, $\lambda=2.0$ and $\lambda=3.5$. At $\lambda=2.0$, there is significant orientation, but almost no strain-induced crystallisation; these samples are essentially amorphous. At $\lambda=3.5$ (the highest draw ratio obtainable on the birefringence rig), there is significant orientation and crystallisation.

Stress relaxation curves, obtained after drawing to $\lambda=3.5$ at 80°C, are shown in Figure 7.11. The true stress is plotted against the logarithm of the relaxation time because this allows the behaviour at low times to be observed clearly. All the samples, independent of draw rate, showed three regions to these curves. The true stress decreased continually, but the rate of decrease was greatest at times below 1 second. This is probably due to the rapid recovery of a viscous contribution to the drawing stress, which has been observed by several workers [43-44]. The magnitude of the initial fall in stress also increased with draw rate, which is indicative of viscous behaviour. At times between 1 and around 100 seconds, the true stress decreased at a significantly lower rate, whereas relaxation accelerates again at longer times. The curves were not linear, indicating that the relaxation is not caused by a single rate activated processes. The stress relaxation curves for $\lambda=2.0$ were essentially the same as for $\lambda=3.5$, so they are not shown here.

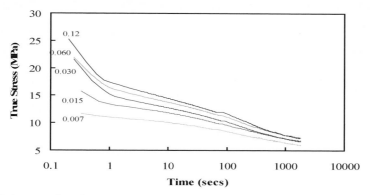

Figure 7.11 Stress relaxation curves of samples drawn at different strain rates to λ=3.5 at 80°C.

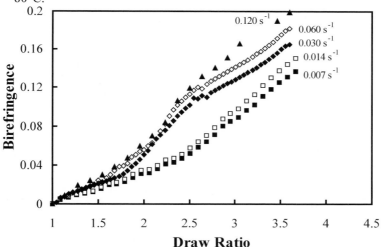

Figure 7.12 Birefringence measured during drawing of isotropic amorphous PET at 80°C and different strain rates.

Figure 7.12 shows the real time birefringence measurements made during the drawing of amorphous PET at 80°C to λ=3.5. Regardless of the initial strain rates, the birefringence increased with draw ratio. There is an increase in the slope for draw ratios above 2.5, which is probably due to the onset of strain induced crystallization. The inflexion point moved to lower draw ratios as the initial strain rate increased, suggesting that strain induced crystallization occurs at lower draw ratios. At the highest rates, the slope decreased again above λ=2.5, which corresponds to a decrease in the crystallization rate, reported in previous works [40-45].

The real time birefringence measurements made during the stress relaxation tests, at λ=3.5, are shown in Figure 7.13. The birefringence initially decreased, for times up to 30 seconds. At longer times, the birefringence was constant although the stress continued to decrease.

The birefringence decreased as the chains in the oriented amorphous material move, reducing the extension. During this process, some of the oriented amorphous or intermediate material can crystallise. At longer times, the birefringence was constant because the crystallisation, significant at λ=3.5, inhibits further relaxation. The crystallites anchor the amorphous chains in place, reducing the freedom of the chains to move, and so reducing the effects of the relaxation.

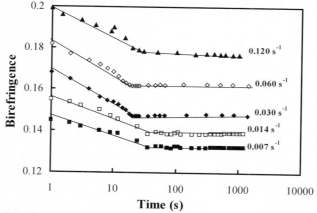

Figure 7.13 Birefringence measured on-line during stress relaxation of PET drawn at 80°C and different strain rates to λ=3.5

Real time birefringence data, from the relaxation at λ=2.0, is shown in Figure 7.14. For all the drawing rates, the birefringence decreased almost linearly with the logarithm of time. This was probably due to the low level of strain induced crystallisation at λ=2.0. The curves were slightly more complicated for the samples drawn at the highest rates, where the birefringence remained almost constant at longer times. This is similar to the behaviour observed for the samples drawn to λ=3.5, which have a higher crystallinity. It was probably due to the higher crystallinity of samples drawn at higher rates.

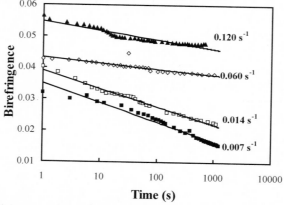

Figure 7.14 Birefringence measured during stress relaxation of PET drawn at 80°C and different strain rates to λ=2.0.

In calculating the crystalline fraction, the enthalpy of melting (ΔH_m) for fully crystallised PET was assumed to be 140 J/g [46].   The melting endotherms had a greater area at the higher draw ratio, indicating a higher crystalline fraction.   At both draw ratios, the relaxation produced an increase in the crystallinity, due to an annealing effect previously described. The DSC scans also showed the exotherm corresponding to the crystallisation of oriented amorphous material. This is less significant at the higher draw ratio because more of the oriented material was crystallised during drawing. At the higher draw ratio, the crystallisation occurred at a lower temperature because the energy required for crystallisation decreases as the chains become more oriented. Relaxation also lowers the crystallisation temperature and decreases the magnitude of crystallisation exotherm, which is clearer at $\lambda=2.0$. This suggests that oriented amorphous material was converted into crystalline material during the relaxation.

In Figure 7.15, the crystalline fraction is plotted against draw ratio, for PET drawn at 80°C and an initial strain rate of 0.017 s^{-1}. This shows that crystallinity increased at a higher rate for draw ratios above 2.3.   This is in line with previous work [40-45], showing that significant strain induced crystallisation only occurs at draw ratios above 2.3.   This also agrees with the stress and birefringence behaviour, described above. Using equatorial 2-theta X-ray scans on PET drawn at 80°C and 0.007 s^{-1}, it was found that at draw ratios below 2.0, there were no crystalline peaks. As the draw ratio increased, a shoulder formed on the amorphous pattern, which indicated that crystallites had formed. However, even at $\lambda=4.0$ the crystalline peaks are not well defined, indicating that the crystallites are small and imperfect.

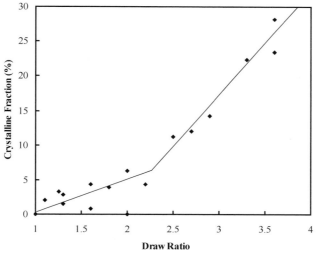

Figure 7.15 Draw ratio dependence of crystalline fraction for as-drawn PET drawn at 80°C and 0.007 s^{-1}.

Infrared dichroism measurements were made using front surface reflectance Fourier transform infrared (FTIR) spectroscopy [47-51, 38, 52-53]. Figure 7.16 shows the dependence on draw ratio of the three orientation functions, calculated from the polarizability function spectra. The orientation functions from the 1340 cm^{-1} and 1018 cm^{-1}

bands increased with draw ratio, but the orientation function for the 730 cm^{-1} band decreased. It has been shown [48, 54-55] that the 1340 cm^{-1} band corresponds to wagging of the CH$_2$ units in the trans conformation; the increase in this orientation function indicates greater orientation of trans units in the draw direction. The 1018 cm^{-1} and 730 cm^{-1} bands have both been assigned to motions of the hydrogens on the benzene ring [48, 54-55], in-plane and out-of-plane motions, respectively. During drawing, the polymer chains, and therefore the planes of the benzene rings, were oriented towards the draw direction. The 1018 cm^{-1} orientation function increased because the in-plane motions, responsible for the band, are also aligned with the draw direction. The 730 cm^{-1} orientation function decreased because the out of plane direction is increasingly oriented perpendicular to the draw direction. Up to a draw ratio of 1.5, the changes in the orientation functions are small but then the orientation increases more rapidly.

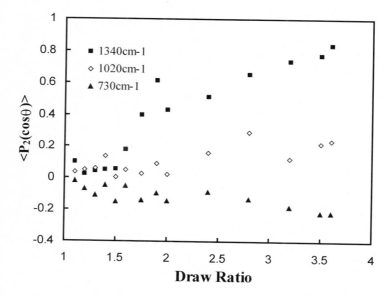

Figure 7.16 Draw ratio dependence of orientation functions calculated from polarizability function spectra for as-drawn PET drawn at 80°C and 0.007 s^{-1}

Orientation functions were also calculated using the ratio of the 1330-1240 cm^{-1} and 1729 cm^{-1} bands [53]. The draw ratio dependencies of these orientation functions, obtained from samples of as-drawn PET, are shown in Figure 7.17. The orientation increased in the draw direction but decreased in the transverse and normal directions. These samples were drawn uniaxially so the orientation functions, for the transverse and normal directions, would be expected to decrease by the same amount; this did not occur. The changes in orientation were small up to λ=2.0 but then the change was more rapid. This behaviour is similar to that observed in online birefringence measurements.

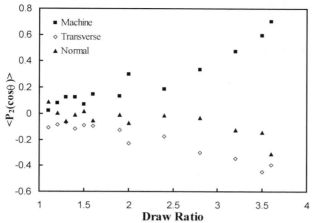

Figure 7.17 Draw ratio dependence of the orientation functions calculated from reflection spectra for as-drawn PET drawn at 80°C and 0.007 s^{-1}.

In Figure 7.18, the draw ratio dependence of orientation is shown for as-drawn and relaxed samples, where the orientation functions correspond to orientation in the draw direction. At low draw ratios, the orientation function was higher for the as-drawn samples, because the oriented chains tend to go back to the isotropic state during relaxation. At high draw ratios, however, the strain-induced crystallites hindered the motions of the chains reducing the amount of relaxation that could occur. The relaxation could also produce an increase in the orientation of the crystalline phase because movement of the chains within the crystallites could improve the packing. Even if there was some loss of orientation in the amorphous regions, the increased orientation in the crystalline regions would have increased the orientation function.

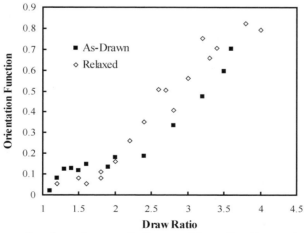

Figure 7.18 Draw ratio dependence of the machine direction orientation functions calculated from the reflectance spectra for as-drawn and relaxed PET drawn at 80°C and 0.007 s^{-1}.

7.3.2.2 Orientation and free relaxation

While many publications have dealt with the development of orientation in PET, there is a scarcity of information concerning the relaxation of orientation. A few publications have appeared discussing the results of shrinkage experiments on oriented PET [56-57]. These studies have demonstrated that beyond a certain degree of orientation, shrinkage is hindered. This observation is explained in terms of orientational relaxation, which is inhibited due to the development of crystallinity [56-58]. In order to study relaxation, IR transmission measurements were performed on previously drawn thin PET films, relaxed in a bath with a controlled temperature. The absorbencies were normalised to the area under the peak at 1410 cm^{-1}. The angle θ_m has been evaluated to be 21° for the peak at 1340 cm^{-1} and 34° for the peak at 970 cm^{-1} [59]. Using these values, we have calculated the orientation function based on IR dichroism as a function of draw ratio λ. The results are shown in Figures 7.19 and 7.20, together with the experimentally determined values of the dichroic ratio (D). It can be seen that a sigmoidal curve is obtained, the orientation function increasing slowly at low draw ratios and then more rapidly beyond $\lambda = 2$.

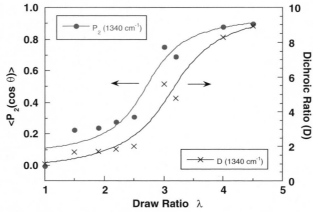

Figure 7.19 Draw ratio dependence of orientation function and dichroic ratio for the 1340 cm^{-1} peak.

Numerous PET IR peaks of exhibit the phenomenon of dichroism [38, 60]. The peaks at 1340 cm^{-1} and 970 cm^{-1} are highly dichroic and hence most suitable for monitoring changes in orientation, particularly at low draw ratios, where most other peaks have D values too close to unity to be measured accurately. These peaks also have the added advantage of being fairly weak in intensity and hence are less prone to saturation. This feature is especially important for transmission studies. A careful examination of the peaks at 1340 cm^{-1} and 970 cm^{-1} in transmission spectra of a film stretched to $\lambda = 2$ revealed a positive dichroism (i.e. the parallel peaks are more intense than the perpendicular peaks). When this film is kept at 70°C for a period of 2 minutes, the dichroism was found to disappear. The loss of molecular orientation as a function of time can be seen in Figure 7.21. It can be observed that after 60 seconds the level of orientation has been reduced by approximately one half. Figure 7.21 also shows the loss of orientation with time at 72°C for a film stretched to $\lambda = 2$. Once again, the relaxation is seen to be rapid. Approximately 40 seconds are required for the orientation function to be reduced to half of its initial value. If the temperature is increased further to 76°C, the relaxation is more rapid, as expected. In the same figure, it can be observed that only approximately 10 seconds are required for half the orientation to disappear.

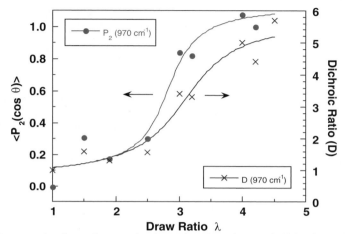

Figure 7.20 Draw ratio dependence of orientation function and dichroic ratio for the 970 cm^{-1} peak.

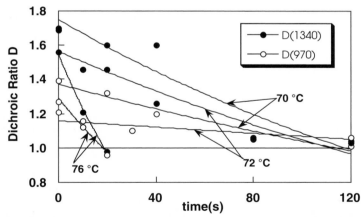

Figure 7.21 Time dependence of dichroic ratios at 70, 72 and 76°C for PET film with λ= 2.

The situation when λ is increased beyond 2 (to λ = 3 for example) is quite different. The dichroism was clearly visible at 1340 cm^{-1} and 970 cm^{-1}. While the orientation present in the film with λ = 2 relaxed after 120 seconds, there was no such relaxation evident. This observation is borne out by the data in Figure 7.22. Exactly the same behaviour was exhibited by a film with λ = 4, as shown in the same figure. The above-mentioned behaviour was observed at all temperatures examined. For example, at 80°C, the film with

$\lambda = 4$ showed no obvious signs of relaxation even after 10 minutes, as demonstrated in Figure 7.22.

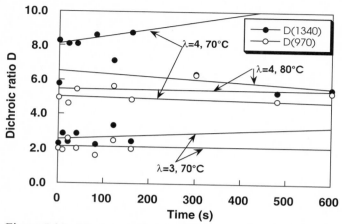

Figure 7.22   Time dependence of dichroic ratios for PET film

Terada *et al.* [57], using density measurements, observed a constant low level of crystallinity up to $\lambda = 2$ and a rapid increase between λ of 2 and 3. This has been confirmed by Dargent *et al.* [61] using birefringence measurements. These authors suggest the onset of strain-induced crystallisation occurs at $\lambda = 2.8$. Padibjo and Ward [62] also used density measurements and observed significant crystallisation beyond $\lambda = 2$. The most exhaustive study of this phenomenon is that of Salem [63] who investigated the effects of temperature and strain rate on crystallisation. The results for the films investigated here are consistent with these reports as shown in Figures 7.18 to 7.20. The onset of strain-induced crystallisation was found to occur between λ values of 2.2 and 2.5.

We postulate that the significant inhibition of orientational relaxation observed for $\lambda = 3$ and higher is due to the presence of strain-induced crystallisation. This is in accordance with the results of shrinkage experiments, as reported by Gupta *et al.* [56] and Terada *et al.* [57]. In both publications, shrinkage at temperatures between 80 and 90°C was found to increase with λ until a draw ratio is attained at which crystallinity increases. Beyond this point shrinkage is severely hindered. These authors attributed this behaviour to the development of crystallites which restrict relaxation in the amorphous zones, thus preventing these conformational segments from returning to an isotropic orientation distribution. Such a mechanism is often termed *pseudo*-crosslinking, to distinguish it from chemical crosslinking, as found in thermoset polymers. Matsuo *et al.* [64] have demonstrated by light scattering that PET crystallites formed upon strain-induced crystallisation are oriented with their c-axes in the draw direction.

There is an alternative explanation which may be considered. It is to be expected that any orientational relaxation will be confined to the *trans* conformers located in the amorphous regions, which according to some studies [61] constitute a mesophase. Therefore, an important issue is the distribution of *trans* content between the crystalline and amorphous regions.   If the amorphous *trans*   content is extremely small compared to the crystalline content, it is conceivable that any relaxation will be "masked" by the rigid crystalline *trans* conformers.

The effect of temperature, relative to T_g, on relaxation is illustrated in Figure 7.23, which shows the decay of dichroism for the 970 cm^{-1} peak at temperatures of 68 and 70°C.   The former temperature is below the beginning of the glass transition region and no relaxation is evident.   An increase of temperature to 70°C is sufficient to induce relaxation.   As reported by Dargent *et al.* [61], the latter temperature corresponds to the beginning of the glass transition region in PET.

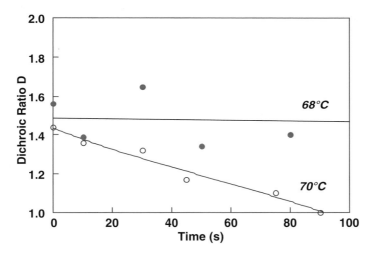

Figure 7.23 Time dependence at 68 and 70°C of dichroic ratio of peak at 970 cm^{-1} for films with $\lambda = 2$.

The result that strain-induced crystallisation inhibits orientational relaxation in the amorphous domains suggests that these zones should not be considered as discrete phases. Rather, as elongation progresses, a transformation of *gauche* to *trans* occurs [65], followed by orientation of *trans* segments.   A certain critical orientation of the *trans* segments is attained [58] beyond which these conformers pack together in fibrillar-type structures and orient parallel to the draw direction [64].   Concomitantly, the remaining conformational segments are trapped between these crystallites and can no longer relax to an isotropic orientation distribution.

7.4   Roll-drawing of PET

The roll-drawing of PET has been extensively studied using a complete pilot plant line as shown on Figure 7.24.   The line consists of a 63.5 mm single screw extruder equipped with a flat die to produce up to 1 cm thick by 10 cm wide profile which is cooled and formed in a calibrator.   The profile speed can be varied between 0.3 and 1.0 m/min. The profile is conditioned within tunnels 1 and 2 (Figure 7.24) to the test temperature. Rolling is performed using 4 rolling stations, separated by heated tunnels to keep the profile at the desired temperature. The driving motor power for the different rolling stations were as follow: 2.2 kW (3 hp) for the first station, 5.6 kW (7.5 hp) for the second and 3.7 kW (5 hp) for the two others. The roll speeds, roll gaps and the temperatures in all the sections are measured and stored on computer. The profile, preheated at the deformation temperature for 10 minutes, is fed into the first rolls where a slight compression was applied to pull the profile through. Then it passed to the second roll station where the main deformation was applied, the compressive force was also measured as illustrated in Figure 7.1.

Typically at start up, the gaps are open to the maximum and then lowered sequentially, one rolling station after the other. The top is lowered until in contact with the profile. Pressure is increased gradually and rolls speed adjusted to maintain the extrusion speed. Process data were acquired via real time data acquisition. The profiles were marked regularly between the calibrator and tunnel 1. The draw ratio in the length direction (λ_l) was calculated using these marks for length draw ratio. The thickness and width draw ratios were determined by measuring the initial and final thickness and width.

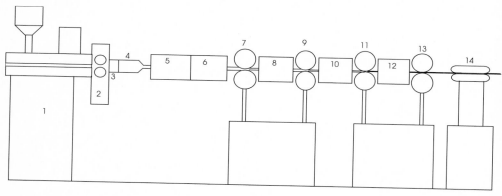

Figure 7.24 Schematic of pilot plant roll-drawing installation: 1-extruder, 2- gear pump, 3- die, 4- calibrator, 5, 6, 8, 10, 12- tunnels, 7- rolls I, 9- rolls II, 11- rolls III, 13- rolls IV and 14- puller.

7.4.1   Roll-drawing of amorphous PET

For the roll-drawing experiments, the amorphous sheet was prepared by extrusion, after drying the PET pellets, using a flat die with adjustable gap. The sheet was cast onto a chilled roll at 20°C at the exit of the die. The maximum thickness of amorphous sheet (crystallinity of less than 5% as determined by DSC) that could be obtained was 3 mm. The width of the sheet was 85 mm.

The crystallinity results as a function of draw ratio are shown in Figure 7.25 for amorphous roll-drawn samples. The roll-drawing speed varied from 5 to 20 cm/min. Crystallinity increased in a fashion similar to that observed for uniaxially drawn samples (Figure 7.10).

However, the stress-induced crystallinity onset occurs sooner (between 1.5 and 2). This is due to the higher drawing speeds compared to the tensile drawing. This behaviour has already been noted by Jabarin [66]. Crystallinity results obtained for the die-drawn amorphous samples are also shown on the same graph for comparison. These results are similar to those obtained for roll-drawn and tensile drawn samples for the same draw ratio, i.e. crystallinity around 30% for the draw ratio of 3 [36-37, 67].

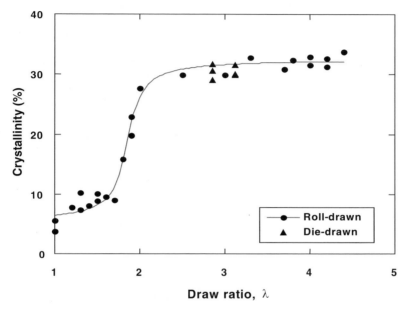

Figure 7.25 Crystalline content of roll-drawn initially amorphous sheets as a function of λ.

For roll-drawn amorphous sheets, constant increase in the conversion from *gauche* to *trans* conformers as drawing takes place was observed. The orientation of the trans conformers is comparable to that obtained for uniaxially tensile drawn films at draw ratios above 3. Below this draw ratio, the roll-drawn samples have a higher orientation, which is due to the higher deformation rate which induced higher crystallinity (and thus *trans* content). The orientation of the band related to the benzene ring C-H in-plane bending was representative of the overall orientation and was lower than that of the *trans* conformers [67].

Mechanical properties of the different amorphous roll-drawn PET samples were measured in tension. The results of the static mechanical tests in the tensile mode, in both the longitudinal and transverse properties, are presented in Figures 7.26 and 7.27 for the tensile modulus and strength respectively. For the longitudinal direction, an increase of the modulus and strength as a function of draw ratio was observed for all the samples. The transverse modulus and strength were both observed to decrease slightly with draw ratio. This is an indication of an essentially uniaxial deformation, despite the plane strain nature of the deformation.

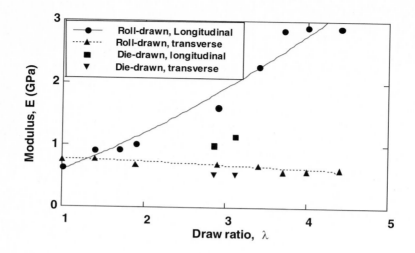

Figure 7.26 Longitudinal and transverse moduli of roll-drawn sheets as a function of draw ratio.

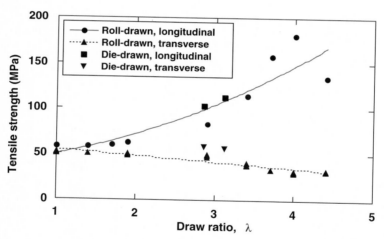

Figure 7.27 Longitudinal and transverse tensile strength of roll-drawn sheets as a function of draw ratio.

7.4.2 Roll-drawing of semi-crystalline PET

7.4.2.1 Effect of process parameters

In this section, we investigate the roll drawing process for the production of solid state oriented crystalline PET profiles. We will examine in particular a number of key process parameters such as the maximum draw ratio, the production rates, the effect of temperature, tension and roll gap.

Preliminary experiments were conducted to assess the evolution of the profile temperature between the rolls during rolling. The temperature inside the heated tunnels and the surface temperature of the rolls were set to 200°C. The temperature at the centre of the PET profile was measured by inserting a thermocouple in the profile at the exit of the extruder. The results of two tests are presented on Figure 7.28 for two different gaps. After about five minutes in the heating tunnels 1 & 2, the temperature of the profile reaches that of the tunnels and remains constant. For the gaps of 8 and 4 mm between the rolls of stations I and II respectively (Figure 7.28a), an increase of about 5 °C was observed when the profile passes between either pair of rolls. The second test was performed with a gap of 4 mm between the first rolls and 2 mm between the second rolls (Figure 7.28b). For rolls I, the same temperature increase as previously (5 °C) was observed, but the increase was higher in the second rolls, about 14 °C, because of the higher stress involved.

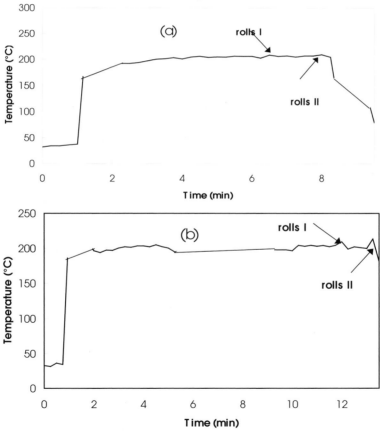

Figure 7.28 (a) Temperature in the centre of the profile for rolls I and II gaps of 8 and 4 mm respectively and (b) for rolls I and II gaps of 4 and 2 mm respectively.

Maximum draw ratio without tension

The first rolling trials were aimed at evaluating the maximum draw ratio that could be achieved at each pair of rolls. The initial tests were conducted with only the first pair of rolls by gradually decreasing the gap. The conditions for the maximum draw ratio without any fracture of the profile were obtained for a gap of 3.05 mm and was λ_1 of 2.2 for a profile speed of 80 cm/min and rolls speed of 194 cm/min. The surface of the rolled profiles was good and the maximum torque required when lowering the rolls was less than 3 kg.m (20 lb.ft). Further decreases of the gap caused either fracture of the profile or the stopping of the motor because the required power to deform the material was above its capacity.

The second pair of rolls was tested slightly differently. The speed, the torque necessary for closing the rolls and the draw ratios were measured. The speed of the PET profile, V_p, at the entrance of the first rolls, was kept at 80 cm/min, and a slight compression was applied at the first rolls in order to stabilise the profile. The gap in the first rolls was 6.35 mm and the thickness of the profile at the exit of these rolls was 7.9 mm. The resulting draw ratios were 1.10 in length and 1.15 in thickness. The rolls velocity was 92 cm/min. The tension was fixed to a minimum just to guide the profile.

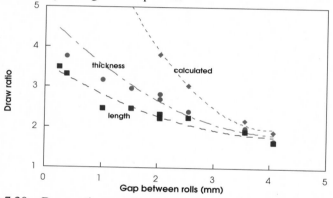

Figure7.29   Draw ratios as a function of rolls gap with minimum tension

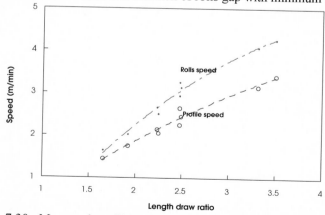

Figure 7.30   Measured profile and rolls speed as a function of draw ratio.

The results of this series of tests are presented on Figures 7.29 and 7.30 in terms of draw ratio and speeds respectively. Figure 7.29 shows the draw ratio as a function of the gap between rolls II. A difference is observed between λ_t and λ_l, particularly at high draw ratios. This difference is due, on the one hand, to changes in width, indicating some biaxial orientation, and to a lesser extent to thickness variations. The calculated results corresponding to a theoretical draw ratio computed as the ratio of initial profile thickness to the rolls gap is also reported. The difference between theoretical and measured draw ratios increases when the gap is lowered, indicating that as the deformation gets progressively larger, an increasing portion of it is lost when the material comes out of the gap. Indeed, the material shows a very large recovery, so large in fact that for small gaps, most of the deformation results in elastic energy stored during the passage through the rolls which snaps back afterward. For example, at a gap of 0.3 mm, the theoretical λ is around 30, whereas λ_l is experimentally around 3.3, i.e., an order of magnitude difference. This is an indication of the importance of the relaxation of the polymer at high deformation. Figure 7.30 shows the speed of the rolls and profile as a function of λ_l. The profile speed was lower than that of the rolls in all cases, which indicates slippage of the profile on the rolls. This difference in speed increased with draw ratio.

Effect of temperature and speed
The effect of temperature and speed on the final draw ratio was examined using two rolling stations, without applying significant tension on the profile. Figure 7.31 shows the results obtained for the draw ratio computed using length, thickness and width. Test conditions were: a temperature of 195°C and a speed of 40 cm/min. The profile dimensions were 8.5-9.3 mm in thickness and 94 mm in width. The gap of the first rolls was lowered gradually to about 1.8 mm to obtain a profile with a draw ratio of about 2 (1.99). The gap at the second rolling station was gradually lowered down to a final value of 0.9 mm. The maximum roll speed reached was 126 cm/min and the final λ was 3.5. From Figure 7.31, it is clearly seen that $\lambda_w > 1$ ($\lambda_t > \lambda_l$), which indicates some transverse orientation ($\lambda_w = 1.2$).

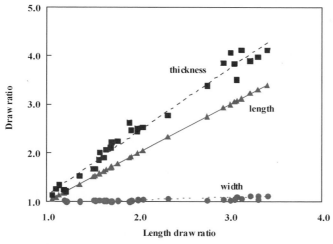

Figure 7.31 Draw ratios in the thickness and width directions as a function of that in the length direction for minimum tension.

The profile temperature was then set to 210°C while keeping the speed at 40 cm/min. While keeping all the other process variables relatively constant, the maximum draw ratio achieved was 3.3. Thus, increasing temperature reduced the maximum achieved draw ratio, which is due to the larger relaxation taking place at this temperature. Once again, some transverse orientation was observed, but to a lesser extent than for the previous temperature. For the third test, the profile speed was set to 80 cm/min and the temperature at 195°C. The minimum gap attained in the first rolls was 1.8 mm with rolls speed of 97 cm/min. When the gap between the second rolls was lowered to 1.3 mm, the profile surface began fracturing. Severe surface fracture was observed at a gap of 1 mm and rolls speed in this case was 250 cm/min. The maximum draw ratio obtained in this case was 2.8, which is much lower that for the other conditions above. This indicates that the deformation rate is a very important parameter. In fact, the relaxation will be much larger for this speed than for the previous one because of the larger energy stored at this deformation rate, and fracture of the polymer for the same gap indicate that at this deformation speed, tension must be applied to prevent relaxation. The biaxial orientation obtained in this case was also lower than that obtained in the previous cases. Table 7.2 summarises the main findings of these three tests. It can be seen that increasing the temperature or speed reduces the final deformation. Maximising the properties would then involve reducing production rate and temperature, and requiring larger forces to be applied.

Table 7.2: Draw ratios measured as a function of temperature and speed

Temperature (°C)	Profile speed (cm/min)	Final λ_1
195	40	3.5
210	40	3.3
195	80	2.8

Effect of tension

Roll-drawing was performed at four temperatures 190°C, 170°C, 150°C and 130°C, all at the same roll speed of 40 cm/min. The optimum process temperature could be determined by comparing the deformation obtained at each temperature. It was observed during the experiments that the highest tensions, applied by the final rolls, were measured while processing at 170°C. This temperature also produced the highest draw ratios, while maintaining constant width deformation. The discussion will therefore concentrate on the roll-drawing at 170°C, which give the widest range of deformation, with comparisons to the other temperatures.

Figure 7.32 shows the final draw ratio as a function of the tension applied to the profile for T= 170°C and two different gaps: 0.9 and 0.45 mm, which correspond to draw ratios of 3.3 and 6.6 respectively. For both roll gaps, the final draw ratio was linearly dependent on the tension. The 0.45 mm gap yielded higher draw ratios. The draw ratios, obtained from both gaps, converged at the highest tensions until failure occurred at a draw ratio of ~5.5. The width of the profile, processed at the 0.45 mm gap tended to be larger for a given draw ratio in the machine direction, suggesting that the small gap produced more biaxial orientation. The behaviour shown in Figure 7.32 was typical of that seen for all the temperatures. It should be noted, however, that the highest draw ratio obtained with the 0.9 mm gap is higher than the calculated maximum draw ratio, 3.3. This indicated significant tensile drawing, but the constant width nature suggests that it occurred mainly between the rolls.

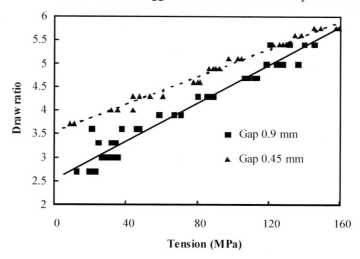

Figure 7.32 Dependence of final draw ratio on tension for roll-drawing at 170°C and 40 cm/min.

7.4.2.1 Structure development

Crystallinity
The results of the crystal fraction measurements, made using DSC, are shown in Figure 7.33. The crystal fraction showed the same trend for all the temperatures: crystallinity increased with draw ratio. However, the higher processing temperature produce a higher crystalline fraction for a given draw ratio, although there was some convergence at high draw ratios. Also, the figure shows that the roll gap has no effect on the crystal fraction. The same plot is shown for crystalline fraction values calculated from equatorial 2-theta X-ray scans in Figure 7.34. Although the absolute values were higher than those obtained from DSC, the temperature and draw ratio dependencies were similar. The absolute crystal fractions are different for X-ray and DSC measurements because the X-ray technique only takes into account the crystallites oriented in the draw direction. This is the orientation direction for the majority of crystallites, so it produces a crystal fraction value which is higher than the actual crystalline fraction of the sample. The DSC method, however, averages over all directions within the sample and gives a more accurate absolute value.

Increasing the processing temperature moved the melting peak, in the DSC scan, to higher temperatures, which indicated larger crystallites. This was supported by the 2-theta X-ray scans, where the crystalline peaks became narrower as the processing temperature increased. The crystallite size calculated from the (100) peak increased from 38 nm at 130°C to 53 nm at 190°C.

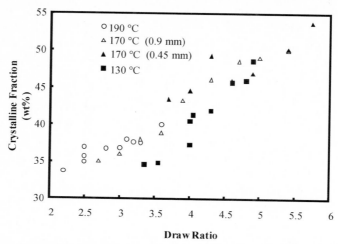

Figure 7.33 Draw ratio dependence of crystalline fraction (DSC) for PET roll-drawn at different temperatures and 40 cm/min.

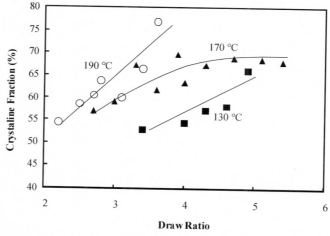

Figure 7.34 Draw ratio dependence of crystalline fraction calculated from the equatorial 2-θ scans for PET roll-drawn at different temperatures.

Birefringence

Figure 7.35 shows the birefringences corresponding to the axial directions Δn_{mn}, Δn_{mt} and Δn_{tn}, for PET roll-drawn at 170°C. If the orientation were uniaxial, the birefringences Δn_{mn} and Δn_{mt} would have been the same. The three birefringences were all different, so the orientation was biaxial in nature. The birefringence Δn_{mn} had the highest value so the level of orientation was greatest in the machine direction and lowest in the normal direction. This shows the constant width nature of the roll drawing process, which limits the reorientation of chains in the transverse direction. A similar behaviour was observed for processing at other temperatures but orientation was higher, for a given draw ratio, for lower processing temperatures as shown in Figure 7.36.

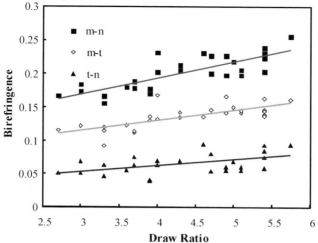

Figure 7.35 Draw ratio dependence of axial birefringences for PET roll-drawn at 170°C and 40 cm/min.

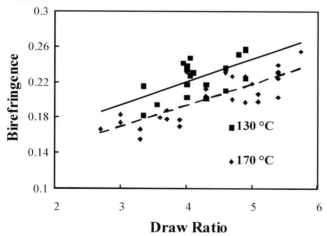

Figure 7.36 Draw ratio dependence of birefringence (machine-normal directions) for PET roll-drawn at 170°C, 130°C and 40 cm/min.

Orientation using infrared spectroscopy (FTIR)
Typical polarizability function spectra obtained from FTIR experiments, for polarisation's both parallel and perpendicular to the draw direction, are shown in Figure 7.37. In Figure 7.38, the dichroic ratios for the 1340 cm^{-1} band are plotted against the final draw ratio for material roll-drawn at 130°C, 170°C and 190°C (the data includes results for both roll gaps but they are not differentiated for overall clarity). The draw ratio dependence of the dichroic ratio was in agreement with the mechanical, birefringence and the crystalline fraction data: the dichroic ratio increased with the draw ratio. The scatter in the data, from samples

processed at 130°C makes determining the temperature dependence of the orientation difficult. If only the results for the data from material processed at 170°C and 190°C are compared, the orientation appeared to decrease as the processing temperature increased which is in agreement with the birefringence data.

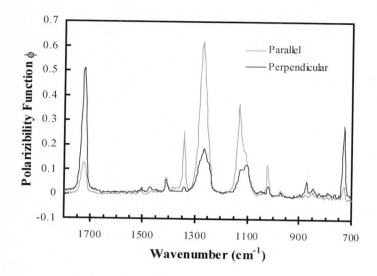

Figure 37   polarizability function spectra from PET roll-drawn at 190°C and 40 cm/min to λ=3.5.

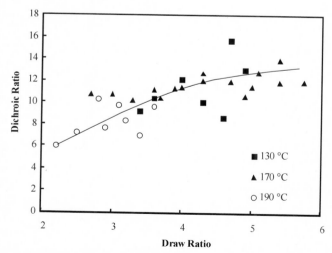

Figure 7.38   Draw ratio dependence of dichroic ratio from the 1340 cm^{-1} band for PET roll-drawn at different temperatures.

7.4.3   Properties

7.4.3.1 Mechanical properties

Figures 7.39 and 7.40 show, respectively, the tensile modulus and strength of the oriented profiles.   During the experiments, it was observed that samples with initial draw ratios below 4 failed after the formation of a neck whereas, at higher initial draw ratios the samples failed in a brittle manner.   Figure 7.39 indicates that the modulus is linearly dependent on the final draw ratio, which is consistent with previous work [51].   Neither the processing temperature nor the gap between the rolls had any significant effect on the modulus (the data for different roll gaps are not indicated for clarity).   Similarly, Figure 7.40 does not show any clear relationship between the tensile strength and the processing temperature.   However, it can be seen that the highest strength was obtained from

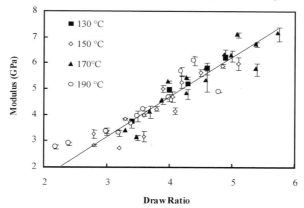

Figure 7.39 Draw ratio dependence of the modulus of PET roll-drawn at different temperatures and 40 cm/min.

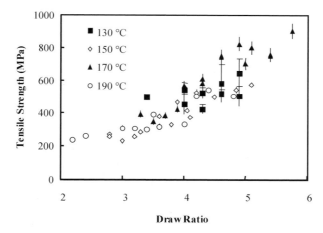

Figure 7.40 Draw ratio dependence of the yield or failure stress of PET roll-drawn at different temperatures and 40 cm/min.

processing at 170°C, because the highest draw ratios were obtained at this temperature. On the other hand, the yield strain decreased as draw ratio increased, up to $\lambda=4.0$. At higher draw ratios the samples failed in a brittle manner at a constant strain, around 0.2. Similar yield strain behavior has been observed for PET in previous works [40-41].

The most interesting result of the mechanical tests was that the mechanical properties appeared to be independent of the processing temperature. The crystal fraction and birefringence measurements may offer an explanation for this result. The most likely explanation of the behaviour is indicated by the difference in the relationships between crystalline fraction and orientation and the processing temperature. The crystalline fraction increased with the deformation temperature (Figures 7.33 and 7.34) but the orientation decreased (Figure 7.36), for a given draw ratio. The intrinsic birefringences of the crystalline and amorphous phases are approximately the same, so it is possible for the crystalline fraction to have increased while the birefringence, and therefore the orientation decreased. The mechanical properties are dependent on both crystalline content and orientation, so if one increased and the other decreased, the effects on the mechanical properties could cancel out. This would have produced the observed independence of the mechanical properties with respect to the processing temperature. These observations are also supported by FTIR orientation results.

7.4.3.2 Thermal conductivity and ultrasonic velocity

Thermal conductivity, k, is very sensitive to orientation, crystallinity and crystal perfection [68-69]. The results obtained for roll-drawn PET are presented in Figure 7.41-a in terms of the ratio of the thermal conductivities in the parallel and perpendicular directions, $k_{//}/k_{\perp}$, as a function of λ. A slight decrease was observed for k in the transverse direction: 0.220 W/mK for isotropic PET; 0.202 W/mK for $\lambda=1.7$ and 0.195 W/mK for $\lambda=3.2$. A large effect was observed for k in the direction of draw. It is also to be noted from Figure 7.41(a) that the ratio $k_{//}/k_{\perp}$ is nearly equal to draw ratio λ. A similar behaviour has been noted for rubbers up to λ of 4 [68-69]. In addition, the increase in $k_{//}/k_{\perp}$ as a function of λ does not level off, contrarily with other properties such as birefringence which shows a plateau at high draw ratios.

Different models have been used to relate the thermal conductivity of oriented polymers to properties of intrinsic constituents (crystalline and amorphous phases). An approach using the aggregate model, similarly to the models used for the elastic modulus E, has been examined, and k and E behaved similarly. A linear relationship was found for PE [68-69]. In Figure 7.41(b), the ratio $k_{//}/k_{\perp}$ is plotted as a function of the tensile modulus obtained for rolled PET. A linear relationship as already observed for other polymers [68-69] is clearly seen.

Ultrasonic velocity, v, measurements were carried out on three PET specimens. The results are presented in Table 7.3. The sound velocity is seen to increase with draw ratio, but to a lesser extent than the modulus and thermal conductivity. In the transverse direction, the ultrasonic velocity was the same for all the specimens, similar in that to the transverse modulus and thermal conductivity.

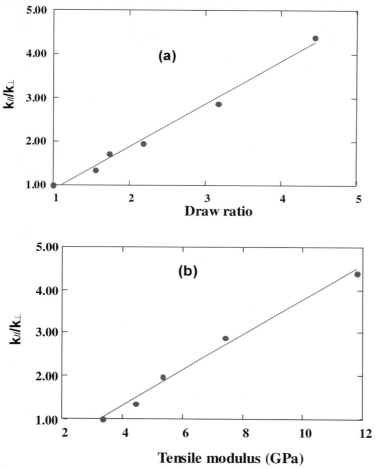

Figure 7.41 $k_{//}/k_\perp$ obtained for roll-drawn PET as a function of (a) draw ratio and (b) longitudinal tensile modulus.

Table 7.3: Modulus , ultrasonic velocity, thermal conductivity and birefringence for PET.

Draw ratio	Modulus (GPa)	$v_{//}$ (m/s)	$k_{//}/k_\perp$	Birefringence Δn_{MT}
1.00	3.35	2470	1.00	0.000
1.74	5.10	2770	1.72	0.047
3.17	7.40	3470	2.88	0.074

References

1. P.J. Lemstra and R. Kirschbaum, *Polymer*, **26**, 1372 (1985)
2. Proceedings of the Society of Plastic Engineers Regional Technical Meeting, RETEC 1987, "New Advances in Oriented Polymers", New Jersey, 1987.
3. I.M. Ward, *Adv. Polym. Sci.,* **70**, 1 (1985).
4. R.S. Porter, T. Kanamoto and A.E. Zachariades, *Polymer*, **35**, 4979 (1994).
5. I.M. Ward, *Macromol. Symp.*, **100**, 1 (1995).
6. P.D. Coates, G.R. Davies, R.A. Duckett, A.F. Johnson and I.M. Ward, *Trans I. Chem E.*, **73**, Part A, 753 (1995).
7. D.M. Bigg, *Polym. Eng. Sci.*, **28**, 830 (1988)
8. P.E. Burke, G.C. Watherly and R.T. Woodhams, *Polym Eng. Sci.*, **27**, 518 (1987).
9. H.J. Chen, M.T. Kortshot and K.G. Leewis, *Polym. Eng. Sci.*, **34**, 1016 (1994).
10. D.C. Sun, E.M. Berg and J.H. Magill, *Polym. Eng. Sci.*, **30**, 635 (1990).
11. J.S. Lee and M. Cakmak, *Polym. Eng. Sci.*, **33**, 1570 (1993).
12. A. Ajji, J. Dufour, N. Legros and M.M. Dumoulin, *J. Reinf. Plas. Comp.*, **15**, 652 (1996).
13. A. Richardson, F. Ania, D. R. Rueda, I. M. Ward and F. J. Balta Calleja, *Polym. Eng. Sci.*, **25**, 355 (1985).
14. T. Kunugi, T. Hayakawa and A. Mizushima, *Polymer*, 32, 808 (1990).
15. Y. Lee, J. Lefebvre and R. S. Porter, *J. Polym. Sci., Polym. Phys. Ed.*, **26**, 795 (1988).
16. Y. Lee and R. S. Porter, *Macromol.*, 24, 353 (1991).
17. A. Kaito and K. Nakayama, *J. Polym. Sci., Polym. Phys. Ed.*, 32, 691 (1994).
18. "The Strength and Stiffness of Polymers", A. E. Zachariades and R. S. Porter (Eds.), Marcel Dekker Publ., New York, 1983
19. "High Modulus Polymers", A. E. Zachariades and R. S. Porter (Eds.), Marcel Dekker Publ., New York, 1988
20. E. M. Berg, D. C. Sun, J. H. Magill, *Polym. Eng. Sci.*, **29**, 715 (1989).
21. J. Yang, C.E. Chaffey, G.J. Vancso, *Plast. Rubber and Composites Proc. and Appl.*, **21**, 201 (1994).
22. A. Kaito, K. Nakayama, H. Kanetsuna, *J. Appl. Polym. Sci.*, **30**, 4591 (1985).
23. Y. Higashida, K. Watanabe, T. Kikuma, *ISIJ International*, **31**, 655 (1991).
24. S. Bahadur and A. Henkin, *Polym. Eng. Sci.*, **6**, 422 (1973)
25. B.S. Thakkar and L.J. Broutman, *Polym. Eng. Sci.*, **21**, 155 (1981)
26. H.-Y. Oh and B.H. Kim, *Polym. Eng. Sci.*, **26**, 1290 (1986)
27. Shin-ichi Matsuoka, *J. Materials Processing Technol.*, **84**, 175 (1998)
28. M. P. Kubisiak, D. J. Quesnel, *Polym. Eng. Sci.*, **36**, 2253 (1996)
29. J. Radhakrishnan and V.B. Gupta, *J. Macromol. Sci.-Phys.*, **B32**, 243 (1993).
30. P. Lapersonne, J.F. Tassin, L. Monnerie and J. Beautemps, *Polymer*, **32**, 3331 (1991).
31. P. Lapersonne, J.F. Tassin and L. Monnerie, *Polymer*, **35**, 2192 (1994).
32. G. Lorentz and J.F. Tassin, *Polymer*, **35**, 3200 (1994).
33. B. Huang, M. Ito and T. Kanamoto, *Polymer*, **35**, 1210 (1994).
34. D.H. Gordon, R.A. Duckett and I.M. Ward, *Polymer*, **35**, 2554 (1994).
35. J. Guèvremont, A. Ajji, K.C. Cole and M.M. Dumoulin, *Polymer*, **36**, 3385 (1995).
36. A. Ajji, J. Brisson, K.C. Cole, M.M. Dumoulin, *Polymer*, **36**, 4023 (1995).

37.   A. Ajji, J. Guèvremont, K.C. Cole, M.M. Dumoulin, *Polymer*, **37**, 3707 (1996).
38.   K.C. Cole, J. Guèvremont, A. Ajji and M.M. Dumoulin, *Appl. Spectrosc.*, **48**, 1513 (1994).
39.   A. Ajji, J. Guèvremont, R.G. Matthews and M.M. Dumoulin, Society of Plastic Engineers Annual Technical Conference Proceedings, Atlanta, GA, USA, Vol. **2**, 1588 (1998).
40.   S.A. Jabarin, *Polym. Eng. Sci.*, **24**, 376 (1984).
41.   V. Pankaj, E.A. Lofgren and S.A. Jabarin, *Polym. Eng. Sci.*, **38**, 245 (1998).
42.   S. Fakirov, E.W. Fischer, R. Hoffman and G.F. Schmidt, *Polymer*, **18**, 1121 (1977).
43.   N.T. Wakelyn, *J. Appl. Polym. Sci.*, **28**, 3599 (1983).
44.   W.H. Yeh and R.J. Young, *J. Macromol. Sci., Phys. Ed.*, **B37**, 1, 83 (1998).
45.   R. Daubeny, C.W. Bunn and C.J. Brown, *Proc. Roy. Soc. London*, **A226**, 531 (1954).
46.   B. Wunderlich, *Polym. Eng. Sci.*, **18**, 431 (1978).
47.   D.J. Walls, *Appl. Spectrosc.*, **7**, 1193 (1991).
48.   W.W. Daniels and R.E. Kitson, *J. Polym. Sci.*, **33**, 161 (1958).
49.   S.K. Bahl, D.D. Cornell, F.J. Boerio and G.E. McGraw, *J. Polym. Sci., Polym. Lett. Ed.*, **12**, 13 (1974).
50.   F.J. Boerio, S.K. Bahl and G.E. McGraw, *J. Polym. Sci., Polym. Phys. Ed.*, **14**, 1029 (1976).
51.   I.M. Ward and M.A. Wilding, *Polymer*, **18**, 327 (1977).
52.   S. Krimm, *Adv. Polym. Sci.*, **2**, 51 (1960).
53.   K.C. Cole, A. Ajji and E. Pellerin, 11th International Conference on Fourier Transform Spectroscopy, Athens, Georgia, August 1997.
54.   D. Grime and I. M. Ward, Trans. Faraday Soc., 54, 959 (1958)
55.   I. M. Ward, Chem. and Ind., 905 (1956)
56.   V.B. Gupta, J. Radhakrishnan and S.K. Sett, *Polymer*, **34**, 3814 (1993).
57.   T. Terada, C. Sawatari, T. Chigono and M. Matsuo, *Macromol.*, **15**, 998 (1982).
58.   F.S. Smith and R.D. Steward, *Polymer*, **15**, 283 (1974).
59.   P. Spiby, M.A. O'Neill, R.A. Duckett and I.M. Ward, *Polymer*, **33**, 4479 (1992).
60.   M. Yazdanian, I.M. Ward and H. Brody, *Polymer*, **26**, 1779 (1985).
61.   E. Dargent, J. Grenet and X. Auvray, *J. Therm. Analys.*, **41**, 1409 (1994).
62.   S.R. Padibjo and I.M. Ward, *Polymer*, **24**, 1103 (1983).
63.   D.R. Salem, *Polymer*, **33**, 3182 (1992).
64.   M. Matsuo, M. Tamada, T. Terada, C. Sawatari and M. Niwa, *Macromol.*, **15**, 988 (1982).
65.   A. Cunningham, I.M. Ward, H.A. Willis and V. Zichy, *Polymer*, **15**, 749 (1974).
66.   S.A. Jabarin, *Polym. Eng. Sci.*, **32**, 1341 (1992).
67.   A. Ajji, K.C. Cole, M.M. Dumoulin and I.M. Ward, *Polym. Eng. Sci.*, **37**, 1801 (1997).
68.   D.B. Mergenthaler, M. Pietralla, S. Roy and H.G. Kilian, *Macromol.*, **25**, 3500 (1992).
69.   D. Greig, in "Development in Oriented Polymers-1", I.M. Ward Ed., Applied Science Publishers, Essex, UK, 1982.

8   Planar Deformation of Thermoplastics

S Osawa
Department of Materials Science and Engineering
Kanazawa Institute of Technology
Nonoichi, Ishikawa 921-8501, JAPAN

and

R S Porter
Department of Polymer Science and Engineering
University of Massachusetts
Amherst MA01003-4530, USA

8.1   Introduction

For over fifty years scientists have been keenly making and studying the property development resulting from oriented morphologies of thermoplastics. A particularly interesting morphological state is that for polymers oriented in a planar direction. The mechanical properties of tensile modulus, strength and impact particularly of semicrystalline thermoplastics are markedly increased by planar orientation. For the former the measurement and use properties are generally in tension, and for the last, impact in the transverse direction. The impact enhancement through thickness is most prominent even at a low draw. Optical clarity is also commonly enhanced by planar draw processes, while permeation through thickness can be changed dramatically.

Table 8.1 lists some of the main deformation processes for planar and biaxial draw. The widely-used processes of tentering and blowing for stretching thin films are inapplicable for the planar drawing of thick sections. The methods for thick sections in Table 8.1 also have the listed limitations.

Table 8.1   Processes for biaxial and planar draw.

PROCESS	CONSIDERATIONS
FOR THIN FILMS	
Tentering	only small extensions
Blowing	nearly uniplanar
FOR THICK SHEET	
Tube expansion	high pressures & cost
Cross rolling	step draw & recovery
Continuous forging	uniplanar, area limited

Planar deformation by forging has its intellectual roots in the thoughts of Turner Alfrey[1]. His 1950 process for films imparts biaxial orientation to obtain balanced mechanical properties in the machine and transverse directions. The process was referred at Dow as a "pancake stretcher". Figure 8.1 is a schematic representation of the stretching zone for a planar deformation between the die and a take-away [1]. Polymer is pumped radially outward through the circumference of a "pancake" die of radius R1. The film is extended in a 360° stretching zone with an outer boundary of R2. (The actual takeoff equipment consisted of eight draw rolls arranged in an octagon.) To obtain equal stretching rates, the radial velocity must conform to a highly specific pattern; i.e., all particles must move in diverging straight lines with velocities proportional to the distance from the convergence point (Turner Alfrey described this velocity distribution as "a 2-dimensional model of the expanding universe").

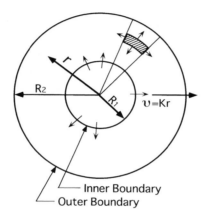

Figure 8.1. Schematic of the stretching zone of a "pancake" film stretcher. The actual device had eight draw rolls and winders arranged octagonally. u is velocity, r the radial position and K a constant.

The UMass planar deformation studies of Saraf[2~7], Osawa[8~11] and Lionti[12] with Porter are predicated on the octagon geometry. As such, the thick initial billet square is octagonally grooved. This diminishes the tremendous lateral strains on compression draw that develop without the groves. Billets are known to explode on forging due to hoop strains. As described following section, the hydrostatic pressure on forging has a synergistic effect on the draw efficiency. Rather than discussing the equibiaxial, solid-state deformation in general terms, it is instructive to describe the special case of the forging process for three reasons; (i) it is the simplest biaxial deformation geometry among the various biaxial processes; (ii) it is naturally equibiaxial in contrast to all the other processes mentioned above where special kinematics conditions are required to obtain equal strain (and strain-rates) in orthogonal directions; (iii) the process subjects the sample to hydrostatic pressure, an important variable to control the final morphology. The forging geometry and the possible resultant polymer orientation textures to be described in the following section, are shown in Figure 8.2.

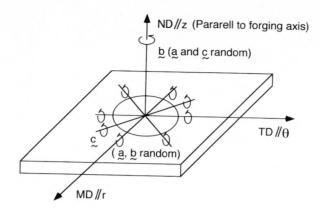

Fiber Texture: b // z   (the compression axis)

                  a, c random around z

                  a, c // r-θ plane

Planar Texture: c // r-θ plane

                  c random around z-axis

                  a, b random around c

Figure 8.2   The sample geometry of forging.   The two possible crystal orientation textures obtained when a semicrystalline polymer is equibiaxially stretched.

The fiber-axis direction, b in general may be <hk0> perpendicular to the major slip plane

(hk0).   For iPP, nylons and polyimides the fiber axis is b .

8.2   Concepts

8.2.1   Synergistic effect

Although the hydrostatic pressure of forging impedes chain mobility[13], it does improve draw efficiency and limits void formation, especially noted for the low temperature draw of single crystal mats[14]. Figures 8.3 and 8.4 compare the change in birefringence between the machine directions and the neutral direction for poly(ethylene terephthalate) (PET) biaxially deformed at 90°C by tenter-framing (free-drawing) and forging (hydrostatic)[5].   The refractive indices, Nx, NY, and Nz are for r, θ, and z directions in Figure 8.2, respectively. For a given draw ratio, the orientation is larger for deformation under hydrostatic pressure than without. This typifies the notion that hydrostatic pressure has a synergistic effect on the resultant morphology of a semicrystalline polymer subjected to any extensional deformation field.

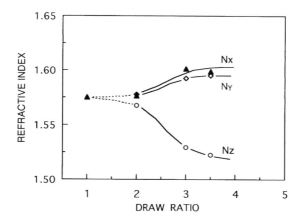

Figure 8.3   The change of refractive index of PET deformed at 90°C by tenter framing. Drawing speed was 45.0 cm/min, corresponding to initial strain rate 5.0 min^{-1}

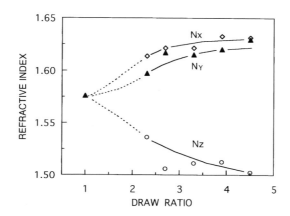

Figure 8.4   The change of refractive index of PET deformed at 90°C by forging. Compression speed was 0.51 cm/min, corresponding to an initial strain rate of 0.53 min^{-1}.

8.2.2   Orientation texture

Under any macroscopic (i.e., continuum) deformation field, the microscopic deformation in the crystal is always a shear process that involves specific crystallographic glide planes which translate along specific crystallographic directions[15]. Polymer crystals shear along their c-axis when subjected to mechanical strain, since it is the strongest direction. Such a shear process, referred to as c-slip mechanism, has been directly observed using a Transmission Electron Microscope (TEM) [16]. As a result, the crystal always orients with the chain axis (i.e., the c-axis) parallel to the (tensile) strain direction during uniaxial extension. For biaxial extension, the chains can align along two possible (principal) strain directions. This degeneracy forces a complicated shear field on the crystal, which may create a hierarchy of textures, depending on the original position of the crystal, chain mobility (related to temperature) and deformation dynamics. Two slip mechanisms, resulting in two distinct textures, are most likely to occur on biaxial deformation:

1. Pencil slip system: Here the slip-planes are more than one (hk0) planes and slip direction for each plane is <001>. The resultant texture would be a *planar-texture* where, the c-axis is parallel to the machine plane and the a- and b-axes are randomly oriented around the c-axis (Figure8.2).

2. Unique slip system: Here only one dominant (hk0)/<001> system dictates the microscopic shearing of the crystal. The resultant texture would be a *fiber-texture (uniplanar)* where, <hk0> (the direction normal to (hk0) shear plane) will be perpendicular to the machine plane and the c-axis is randomly oriented in the machine plane (see Figure 8.2).

A biaxially deformed polymer will contain fractions of both planar and fiber textures. As opposed to an ideal planar-texture, in an ideal fiber-texture all the slip planes are parallel to machine plane. As a result, the fiber-texture will strengthen the polymer in the machine plane, since there are no slip mechanism available for the crystal to (further) deform in this plane. Thus, a larger enhancement in mechanical property (in the machine plane) is expected for the fiber-texture than the planar texture. The selection criteria to identify those semicrystalline polymers, where the crystal shearing would occur by unique slip system leading to dominantly fiber texture, is as follows:

1. Polymers with crystals containing a unique (hk0) plane with specific, strong inter-chain interaction. An example would be nylons which have a (010) plane where the chains are hydrogen-bonded.

2. Polymers with no specific interaction but an (hk0) plane with an appreciably large density compared to other planes containing the c-axis. An example would be isotactic polypropylene (iPP) where the (040) plane has high density compared to other (hk0) planes.

Studies on texture analysis of biaxially deformed iPP indicate that the dominant texture is fiber-texture where the b-axis is perpendicular to the machine plane[4]. A similar result is obtained for Nylon-11, where, the (010) plane aligns with the machine plane on equibiaxial deformation[17].

8.2.3   Order-disorder transition

In certain polymers, a crystal-crystal phase transition is induced on large plastic deformation.[18] In certain cases the deformation induced crystal structure has a disordered inter-chain packing or a chain conformation with lower cross-section, making the subsequent structure easier to deform compared to the original crystal. Such a deformation induced transition can lead to in-situ 'plasticization' causing increased chain mobility and thereby improving draw efficiency.[18] The first studies demonstrating this principle were performed on equibiaxial, hydrostatic deformation of iPP where a lower activation energy to deform was observed in the presence of the order-disorder transition compared to conditions where this transformation was suppressed.[3,9,10] The disordered form is a draw-generated smectic phase on iPP forging. Furthermore, the order-disorder transition increases the relative amount of fiber-texture compared to the planar-texture[4].

On deformation, the increase in extension along the chain and decrease in coil cross-sectional area is consistent for all polymers, but the trend in average inter-chain distance (related to density and cross-sectional area) does not follow a monotonic trend. Further studies in this category would be fruitful to identify polymers where the principle of an order-disorder transition could be employed to produce extended morphology with low-energy processes.

8.3   Physical Properties Induced By Planar Deformation

8.3.1   Polyethylene

Major efforts have been made to attain high planar draw with polyethylene. The tensile properties[19,20], morphology[21], including draw of gels[22~24] and the mechanisms of planar deformation have been studied as a function of draw ratio and compared to the corresponding values on uniaxial draw. Figures 8.5 and 8.6 show the tensile modulus and strength of biaxially and uniaxially drawn films of ultra high molecular weight polyethylene (UHMW-PE). Here, the DR (draw ratio) = 10 for biaxially oriented film means a biaxial draw of 10x10. The modulus for uni- and bi-axially oriented films is seen to increase with draw. The tensile modulus of planar (biaxially) oriented film is 3/8 of uniaxial oriented film, applying laminate theory[20,25~27]. The modulus data in Figure 8.5, from several workers[19,20,24], approximates this relationship. A biaxially oriented film at DR = 15~20

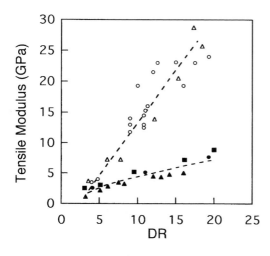

Figure 8.5   The relation between tensile modulus and draw ratio for uniaxially (O, Δ) and biaxially (●, ▲, ■) oriented UHMW-PE gel films. Data: O, ● from ref. 19, Δ, ▲ from ref. 20, and ■ from ref. 24. For biaxial draw, draw ratio (DR) =10 means 10x10.

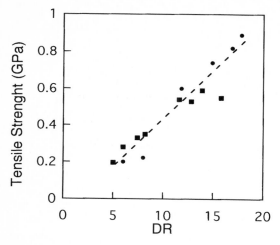

Figure 8.6 The relation between tensile strength and draw ratio for uniaxially (▲) and biaxially (●) oriented UHMW-PE gel films. Data from ref. 20.

has a tensile modulus of ~8 GPa. This value is about 3/8 of the uniaxially drawn films and 4 times higher than the modulus (only ~2 GPa)[19] of uniaxially oriented films for the transverse direction at a same DR. Thus, the significance of planar draw is the production of equal and enhanced properties in the planar direction. The elongation at break of planar oriented film is over 3 times higher than that of uniaxially oriented film[20]. The random orientation in the plane direction for biaxially drawn film changes to chain orientation along the stretching direction until the samples break[20]. Therefore, the tensile strength of bi- and uni-axially oriented film can be similar, see Figure 8.6. Above DR>10, there is a marked decrease in polyethylene draw efficiency, as evaluated by an elastic recovery test[28]. Kyu has also biaxially stretched a conventional morphology of polyethylene. In studying two molecular weights of UHMW-PE, 2 and 6 million, he obtained up to 10x10 draw, with moduli comparable to those in Figure 8.5, but with scatter in the corresponding strength values[29].

The development of multiaxially oriented films of lower molecular weight high density polyethylene with high mechanical properties in planar directions has been pursued by inducing fibrillar crystallization under curvilinear melt flow[30]. This has been done in a contained geometry using an extrudomolding process and by simulating similar crystallization conditions in an optical plate-plate rheometer. The schematic diagram of this study is shown in Figure 8.7a,b. The films are like the uniaxially drawn morphologies of the same molecular weight high density polyethylene made by solid-state extrusion. They have a high modulus (12-20 GPa) and some strength (250 MPa) along the flow lines, but they exhibited also a modulus enhancement (5 GPa) in the transverse direction as a result of the orientation gradient of the molecules in the thickness direction.

The polymer is processed in a contained geometry at a temperature near to, but below the isotropic crystalline melting point under curvilinear flow conditions generated by the combined effects of a compressive force and a rotational force perpendicular to the compressive force. The process has been demonstrated with thermoplastics using simple torsional flow[31], see Figure 8.7b.

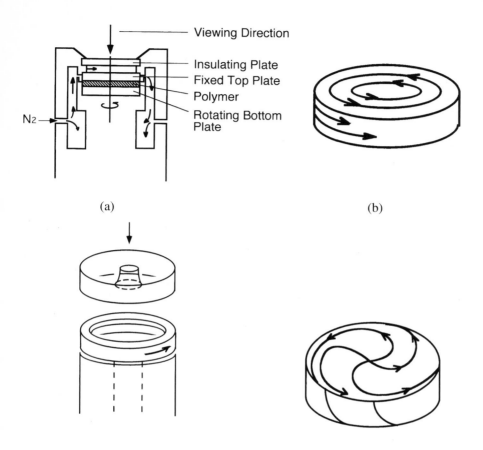

Figure 8.7   Schematic diagram of (a) the optical apparatus for the rheo-optical and melt flow crystallization studies; (b) the flow lines generated under the torsional flow conditions; (c) the cavity using in the rotational injection molding experiments, and (d) the chain orientation gradient in the plane and thickness directions of a molded sample under rotational injection molding conditions.

On injection molding, the polymer is oriented and extended in different directions by flow induced crystallization. One part of the mold cavity is rotated during the mold filling cycle to obtain a fibrillar product with enhanced planar mechanical properties[32]. A schematic diagram of a circular mold cavity for " rotational injection molding " is depicted in Figure 8.7c. The moving mold part could rotate, e.g. at 80rpm, as the mold is closed. The mold is then heated to a temperature close to, but below, the melting temperature of the polymer, e.g. polyethylene. A schematic diagram of the flow pattern generated under rotational injection molding, using a center-gated mold cavity, is shown in Figure 8.7d. In rotational injection molding, the polyethylene was injected into the mold cavity (< 4500 psi) and was deformed under the combined effects of injection pressure and the rotational force generated by the rotating part of the mold cavity. Multiaxially oriented polyethylene, in disc-shaped products obtained by rotational injection molding, has a flexural modulus of ~12GPa. The planar tensile modulus was 10 - 15 GPa along the flow lines (circular direction) and 3~4 GPa across (radial direction). Scanning electron microscopy showed that such products have a laminated fibrillar structure in which the orientation of the molecular chains differ in different layers, see the schematic in Figure 8.7d. The tensile strength of the fibrillar films along the flow lines was 0.15 - 0.20 GPa, which shows that tear strength was significantly enhanced in the radial direction.

The impact strength of conventionally injection molded and those generated under torsional flow conditions (see Figure 8.7) were compared in disc form for polyethylene and polypropylene. A falling steel ball test was also used to test the impact strength. It was found to be significantly higher for the samples processed under torsional flow conditions. For example, the impact strength of conventionally molded polypropylene (1.2mm thick) was 0.9 lb.in as compared to 10.5 lb.in. of products obtained under rotational injection molding[32].

8.3.2   Polypropylene

Figure 8.8 The Bethlehem Steel Corporation biaxially-oriented polypropylene (2.4 x 2.4) as impacted by a 35 mm bullet.

Isotactic polypropylene (iPP) has been long known for its special properties in uniaxial extension. These include the "living hinge" concept for developing reversible pliability and the fibrillation of uniaxially oriented sheet to produce fiber for indoor-outdoor carpet. On planar deformation also, remarkable achievements have been made with iPP. Even at low draw, < 3 x 3, a remarkable enhancement of impact properties has been observed. This includes products made from tubular expansion over a mandrel by hydrostatic extrusion. The expanded tubes can be slit and cold-rolled flat. The Bethlehem Steel process can be used to make 2.4x2.4 biaxially-oriented polypropylene sufficiently tough to stop a 35mm bullet (see Figure 8.8)[33]. Bonded cross-laminates of uniaxially drawn iPP have also exhibited improved impact toughness and ballistic energy absorption, with values for laminate that decrease above a draw ratio of 10[34]. Biaxially-rolled iPP has also been shown to exhibit large changes in mechanical properties[35], including high impact strength by the Izod test[36]. Under certain conditions, the impact strength is over 30 times that of the undrawn. These increases in property perpendicular to the plane direction are much higher than that along the plane direction.

The large enhancement of impact property is related to the layered structure that is developed as consequence of the planar chain orientation. The fractured samples show delamination into layers. As impact is applied normal to the planar draw direction, iPP tends to rupture in the transverse direction, i.e. along the plane direction due to the planar or fiber-texture. The impact energy is thus absorbed by the tearing apart of many thin layers along the plane[11]. The corresponding crystal structures in unidirectionally and cross-rolled iPP has been revealed by small angle x-ray tests[35]. The nominal long period decreases with draw (with decreasing thickness reduction t/to, where to is initial thickness), as shown in Figure 8.9. These changes are interpreted in terms of a model incorporating tilting both of lamellae and of chain stems within the lamellae, revealed by the SAXS paterms[35].

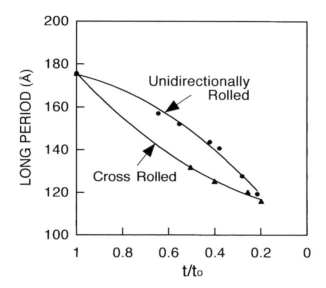

Figure 8.9   Effect of rolling on small angle X-ray crystal long period spacing of iPP on thickness reduction, t/to (to is initial thickness).

The impact strength is also directional dependent[11]. Figure 8.10 shows the impact strength of forged iPP tested for both parallel (//) and perpendicular (⊥) directions to the sample plane. The compression ratio, CR = do/dc, is defined by the ratio of thicknesses before (do) and after (dc) compression on forging. The sample geometry with the crystal orientation is shown in the Figure 8.11. The impact strength of (//) and (⊥) at CR of 30 are about 7 and 3 times of the original unoriented sample, respectively. Note that the value of (//) is about twice higher than that of (⊥) at a comparable CR. These are explained by the two energy absorption mechanism. One is the delamination of layers and one is an elongational break of the layers[11]. For (⊥), all lamellae or extended molecules must be pulled out of layer, since the forge direction is perpendicular to the molecular axis or 0k0 plane, leading to a break of layers at low elongation. For (//), on the other hand, the molecules are pulled within the plane direction, since the forge impacts parallel to the plane, leading to a high elongation at break. Major impact energy is absorbed by this mechanism.

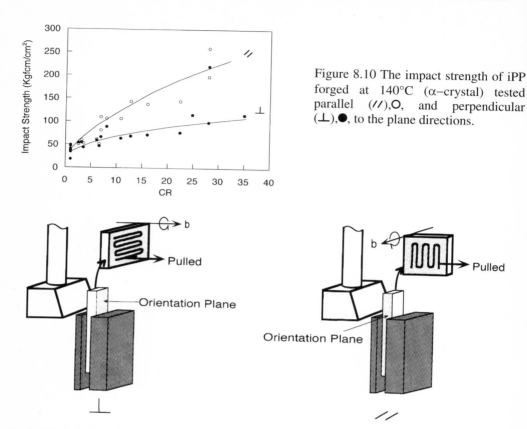

Figure 8.10 The impact strength of iPP forged at 140°C (α–crystal) tested parallel (//),O, and perpendicular (⊥),●, to the plane directions.

Figure 8.11   The sample geometry for impact tests with crystal orientation in the forged iPP for (⊥) and (//) directions.

A practical consequence for planar deformation is the enhancement of optical clarity through thickness. The optical clarity and haze of polymer film is a consequence of the light scattering that originates from both surface irregularities and bulk scattering. They are affected by the process conditions such as draw ratio and draw temperature[37]. On compression draw of iPP, the surface transmittance seems to be independent of compression ratio, CR, and draw temperature due to the tight contact between sample surface and the smooth cylinder surface during forging. This is in contrast to the major surface roughness caused by process variations of biaxial drawing of PET[38] and tubular extrusion of PE[37]. Figure 8.12 shows the turbidty, τ, of forged iPP as a function of CR[39]. The τ = -(logTb)/d, where Tb is bulk phase transmittance and d is the sample thickness. The sample prepared at 50°C includes both smectic phase and α–crystal (smectic sample), whereas the sample forged at 140°C includes only α–crystal (α–crystal sample). The τ of both smectic and α–crystal samples decrease rapidly with increasing CR. Further, the τ of smectic sample is lower than that of α–crystal sample at a comparable CR. These results suggest that at least two factors provide optical clarity for forged iPP. One is directly related to generation of the smectic phase and one is related to common structural changes in both samples induced by the forging.

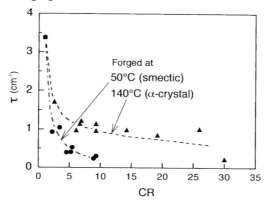

Figure 8.12   The turbidity, τ, of iPP forged at 50 (●) and 140°C (▲) as a function of CR.

In semicrystalline polymers, there are several factors that influence optical clarity, such as refractive index of the phases, and the shape and size of spherulites. Among the factors, the refractive index is a key to the difference between α–crystal and smectic samples of iPP. Figure 8.13 shows the refractive indices of amorphous, smectic, and α–crystal phases as a function of the phase density (data from ref.10) for iPP. Here, the density is proportional to the refractive index. By the forging, α–crystal transforms to smectic, which minimize the difference in the refractive indices between crystal (smectic is a disordered crystal form of iPP) and amorphous phases, leading to higher transparency. A large amount of α–crystal transforms to smectic phase by CR of 5. This trend follows the rapid decrease in τ for the smectic sample (forged at 50°C) in Figure 8.12. The amorphous density with both α–crystal and smectic samples is increased by the compression draw[9]. This also contributes to the increase in clarity of forged iPP. The crystallite size is also decreased by the forging process. As will be discussed later, the contribution of smectic phase fraction and planar deformation with CR can be estimated and compared by multiple regression analysis.

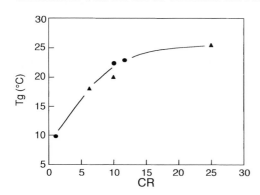

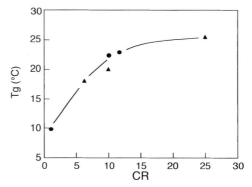

Figure   8.13   Relation   between   the refractive index of each phase and phase density for iPP.

Figure 8.14 Tg evaluated from the peak temperature of β relaxation as a function of CR   for samples compressed at 50°C (●) and 100°C (▲).

On forging of iPP, its glass transition temperature, Tg, (evaluated by dynamic mechanical analysis) is increased by the compression draw. From the β relaxation, Tg can be evaluated as a function of compression ratio (CR) as in Figure 8.14. Tg increases steeply from 9°C for the original undrawn sample to 23°C for a drawn sample at CR=10, followed by a gradual increase. The amorphous density increases with compression draw[10], the same trend in Figure 8.14. The increase in Tg appears to be associated with the increase in amorphous density on draw[10]. The Tg shift with increase in amorphous density can be roughly estimated by using a simple Simha and Boyer model[40,41]. The increase in Tg of iPP on forging is in contrast to the decrease noted on uniaxial deformation[42,43].

8.3.3   Higher poly-1-olefine

In contrast to the other poly-1-olefins, there exist unsurmounted limitations for attaining high uniaxial draw of poly-1-butene. Here the multiple crystal forms seem not to participate nor aid the draw process. Quite the reverse is true on draw of the highest available polyolefin, poly(4-methyl-1-pentene) (PMP). This polymer draws with great ease both uniaxially and on planar deformation (forging). This accomplishment is summarized in Table 8.2, which shows that under selected conditions, PMP has been planar drawn up to CR~240 times[44]. The draw efficiency of PMP (75%), as measured by elastic recovery, is lower than that for iPP. Nevertheless, the total and molecular planar draw ratios achieved for PMP are the largest reported for any polymer.

Table 8.2 Transition temperatures and Crmax of PMP and iPP for planar deformation

	Tg	Ti	Tc	Tm(peak)	CRmax
PMP	35	62	160	235	240 (at 180C)
iPP	0	77	120	162	50 (at 140C)

The PMP has the unusual property that the crystalline and amorphous phases have densities that are comparable (ρ_c = 0.83, ρ_a = 0.838 g/cm^3 at 20°C)[45], in contrast with other semicrystalline polymers where the crystals are always more dense than the amorphous phase. The high ductility of PMP is related to the loose packing and weak interaction of polymer chains in both the crystalline and amorphous phases. Rastogi with Keller[46] examined pressure-induced behavior of PMP. They suggest that the tetragonal crystal transforms to a liquid crystal and/or to an amorphous phase under the pressure above 2Kbar.

This order-to-disorder transition is consistent with the crystal density relation $\rho_c<\rho_a$ (in the same rare class as water). In the forging experiment, the compression pressure increases as the deformation proceeds and approaches 2 Kbar at high CR. The pressure-induced crystal to amorphous transition may be not only related to the high ductility, but also to the lower draw efficiency of PMP.

8.3.4   Poly(ethylene terephthalate)

Orientation of PET has been widely studied because of its application in the food and beverage container markets. Oriented PET (Mylar) tapes also have huge volume use.

The orientation and the structure in biaxially drawn PET (Intrinsic Viscosity 0.54) are found to be determined by the rates of draw, chain relaxation, and by crystallization[47]. When the draw rate is slow, a higher orientation is attained at lower draw temperatures due to slow chain relaxation. On initial deformation, draw speed and crystallization rate are important factors in determination of orientation. For example, higher orientation and crystallinity can be attained at 94°C than at 80°C, when the drawing speed is 50cm/s. Presumably this is because the stress-induced crystallization rate is greatest at 94°C. The overall increase in the trans isomer content of PET, as a consequence of draw, plays an important role in the determination of orientation and crystallization[47].

The incremental effect of hydrostatic pressure and process geometry on deformation have been considered[5]. Amorphous PET has been free drawn (i.e., no hydrostatic pressure) isothermally on a tenter frame device. The hydrostatic equibiaxial deformation was achieved by a forging process involving squeezing between two circular plates. These two processes have been carried out at 80-110°C at a series of deformation rates, with the three-dimensional refractive index measurements made on the PET using a modified Abbe refractometer, as per Samuels[48,49]. These reported trirefringence measurements provide a sensitive scale for the planar deformation. For PET, hydrostatic biaxial deformation is found to be more efficient than free drawing, with respect to: (a) more planar orientation and (b) more induced crystallinity in the former process. Forging is thus more effective than stretching in achieving stabilized planar draw under comparable processing conditions, as judged by stress-induced crystallization[50].

A detailed study has been made on the anisotropic optical properties in uni and simultaneous biaxially stretched PET films[51]. Cast amorphous sheets of PET were stretched to a series of extension ratios in two mutually perpendicular directions at 80 to 100°C. The changes in the principal refractive indices with the processing history were correlated with the orientation of PET chains and phenyl plane normals, which were determined independently by a wide angle X-ray pole figure technique[51].

PET has also been drawn in sequential steps[52]. In one process, the first step was to draw film to a ratio of 3.4 between rolls at 127°C and then transversally stretch it under controlled conditions. The results showed that the crystals were preferentially aligned with their chain axes along the first stretching direction, with breakdown during the second transverse draw. As a consequence, recrystallization takes place producing crystals aligned in the transverse direction[52].

8.3.5 Polyimide

The planar or fiber texture, as described in an early section, may also be obtained spontaneously without any deformation field. In polyimides, such as poly (pyromellitic dianhydride-oxydianiline) (PMDA–ODA) and poly(biphenyl dianhydride - p-phenylene diamine) (BPDA–PDA), the chains spontaneously tend to align parallel to the film plane during curing[53]. For PMDA–ODA, the planar-texture observed in the bulk is essentially a fiber-texture in the top from 10 nm the surface where the b-axis is perpendicular to the surface[54]. The reason for such a morphology for polyimides is not fully understood, nevertheless, the chain alignment and stretching at the interface has been demonstrated by computer simulation[55]. Two important applications of such spontaneous planar orientation in polyimides are in the liquid crystal (LC) display and electronics packaging technology. Spontaneous alignment of LC for display panel is achieved by contacting the LC to specially rubbed polyimide surface[56~58]. The effect is most pronounced in polyimides due to their planar texture. The planar orientation of polyimides leads to a small in-plane thermal expansion compared to an isotropic film[59]. The low in-plane thermal expansion reduces thermal stress at the polyimide/Si interface, essential for the reliability of integrated circuits[60].

8.3.6   Other thermoplastics

A range of additional polymers have been studied in planar and biaxial deformations. The structure and development in biaxially stretched polystyrene (PS) has been studied early[61] and later by White and Spruiell[62], other evaluations of oriented PS have been made by permeation studies[63,64]. A tubular film process to produce biaxially oriented poly(phenylene sulfide) has also been reported by White[65]. The results may be compared with uniaxial draw studied on these same two polymers[66]. The Nippon Steel group[67] has also reported on the cross rolling of ABS and PVC polymers. Sweeney and Ward have studied the planar and equibiaxial extension of PVC at temperatures just above and below Tg [68].

Cakmak and coworkers have performed impressive biaxial tests on additional thermoplastics including poly(ethylene 2,6 naphthalate)[38] and poly(ether ether ketone)[69]. Among the few biaxial studies of blends, draw of PVC plus a poly(methyl methacrylate coimide) has been reported[70]. The Dow scrapless molding process has been applied to most thermoplastics including multi component laminates[71]. The area of liquid crystal polymers hold great promise, with limited studies to date[12,72], for properties such as reduced permeation.

8.4   Analytical Approaches for Planar Deformation

8.4.1   Ductility and draw efficiency

High planar draw ratios, CR~50, have been attained for iPP[8], by a forging process and with a high draw efficiency of 85%. This high and efficient deformation is facilitated by generation of a smectic phase from the stable α–crystal. The conditions for achieving high draw in iPP are shown in Table 8.3. The density changes resulting from such deformation are shown in Figure 8.15. Here, Ts is the smectic phase upper temperature stability limit. Below Ts, the draw efficiency is high, ~100%, as defined by elastic recovery tests[73,74], under Elastic Recovery. The crystal deformation proceeds through the generation of the intermediate smectic phase (Process I) when the forging is carried out below Ts[3]. In the smectic phase, the lateral packing between the inter-chains (3/1 conformation) is disordered. A large fraction of the conventional α–crystal is transformed to smectic by a compression ratio (CR) of ~10[9]. In this process, the molecular chains are pulled out from the lamellae (lamellae unraveling). In addition to Process I, the smectic to amorphous transition (Process II) may also occur, but it is not major. At a deformation temperature (Td) >Ts, however, the smectic phase is not stable[3], therefore the pulled out chains fall back into their stable α–crystal form during the deformation, restricting the lamellae unraveling process. It can thus be concluded that Process I apparently facilitates the efficient draw of the α–crystal phase, as the mechanism to achieve high efficient draw.

Table 8.3 Polypropylene Compressive Draw Characteristics

Draw Temperature	Draw Efficiency	Ductility
Td<Ti	High	Low
Ti<Td<Ts	High	High
Ts<Td	Low	High

Ti: intercept temperature at zero yield energy (see Figure 8.20)
Ts: smectic phase stability limit temperature

Undeformed iPP generally consists of about half amorphous chains and half α– crystals. During forging deformation, amorphous chains are first extended by the compressive force. Saraf et al.[3] pointed out that amorphous region might be considered a rubbery network with entanglements as temporary cross-link junctions[75]. Furthermore, the network might behave like an ideal rubber for planar deformation up to a DR of ~4[3,76]. The amorphous density increases rapidly at a low compression ratio (CR<~5), caused by orientation via the compressive force. The increase in the smectic fraction and/or increase in amorphous density on compression draw, results in the increase in Tg, and the iPP also becomes transparent as described early[10].

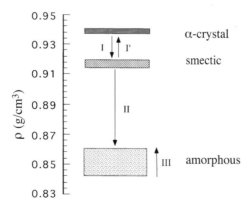

Figure 8.15   Schematic of density (ρ) scale for phase changes of iPP on compression draw.

8.4.2   Trirefringence

A technique for the determining the three principal refractive indices of anisotropic polymer films (trirefringence technique)[77,78] has been developed by Samuels et al.[48,49]. He has coined the word "trirefringence" for the measurement of the principal refractive indices. These measurements and their difference have made possible a wide range of optical criteria for evaluating property and processing variables. For planar (forging) and biaxial deformation, the Z-axis is designed normal to the plane direction and the X-Y plane is parallel to the sample surface (see Figure 8.2). The average refractive index and the three birefringences are expressed as follows:

$$\overline{N} = (Nx + Ny + Nz)/3 \qquad (1)$$

$$
\begin{aligned}
\bullet xz &= Nx - Nz \\
\bullet yz &= Ny - Nz \qquad (2) \\
\bullet xy &= Nx - Ny
\end{aligned}
$$

For equibiaxial deformation, Nx=Ny. This indicates orientation symmetry around the Z-direction. Further, Nx and Ny are greater than Nz (for the usual polymer), confirming that the chain orientation is in the plane normal to the Z-axis. To evaluate planar orientation, Equation (3) or (4) is chosen as an average birefringence normal to the plane or a total birefringence, respectively[79].

$$\bullet n = \tfrac{1}{2}(Nx + Ny) - Nz = \tfrac{1}{2}(Nx - Nz) + \tfrac{1}{2}(Ny - Nz) \qquad (3)$$

$$(\bullet n_{total})^2 = \tfrac{1}{2}(Nx - Ny)^2 + \tfrac{1}{2}(Ny - Nz)^2 + \tfrac{1}{2}(Nz - Nx)^2 \qquad (4)$$

In the equibiaxial process, ideally $\bullet xz = Nx - Nz$ and $\bullet yz = Ny - Nz$ should be same. For the ideal case, $\bullet n_{total}$ will equal $\bullet n$.

Taraiya and Ward examined the $\bullet n$ functions against the oxygen permeability coefficient of drawn iPP[79]. The through thickness permeability falls as $\bullet n$ increases, with both uniaxial and biaxial samples following a similar pattern. For uniaxially oriented iPP, Nx decreased and Nz and Ny increased almost symmetrically with draw[49,80] (where X in Figure 8.2 is the uniaxial draw direction).

8.4.3   X-ray analysis

Pole figure X-ray studies are widely used to characterize the crystal orientation with respect to the three perpendicular axes, for example MD (machine direction), TD (transverse direction) and ND (normal direction). The orientation parameters in terms of the second moments of intensity distribution, $<\cos^2\theta>$, in the three directions of the simultaneously biaxially drawn iPP have been evaluated by Taraiya et. al.[81,82]. The analyses showed that the c-axis orientation is nearly equal with respect to the MD and TD ($<\cos^2\theta c,MD>$ • $<\cos^2\theta c,TD>$) in the balanced biaxially drawn iPP. In the iPP films with variation of biaxial draw, there is a strong tendency for the b-axes of the crystallites to align normal to the plane of the films judged from $<\cos^2\theta b,ND>$ >0.5.

The amorphous orientation averages for the three directions can be also evaluated by combining the X-ray diffraction parameters and the refractive index measurements[81,82] as described earlier. It was assumed that the semicrystalline polymer was composed only of crystalline and amorphous phases (two phase model) to each of which can be attributed specific anisotropic, intrinsic properties. For both uni- and biaxially drawn iPP, a good correlation has been found between the gas permeability through the ND (thickness direction) and the amorphous orientation in terms of $<\cos^2\theta a,ND>$ (i.e. the permeability decreases monotonously with decreasing $<\cos^2\theta a,ND>$), see Figure 8.17 [81].

The crystal orientation of forged iPP (equibiaxial deformation achieved by uniaxial compression) has been studied[4]. This study shows equibiaxality of the deformation with the observation of two types of crystal textures referred to planar and fiber (b-axis orientation along normal direction). These textures are quantified by relating the intensity distribution for a (110) reflection to the orientation distribution of the crystals. The relative amount of the fiber-texture is higher if iPP is deformed via the order-disorder process at 60°C (in this case, smectic phase is generated during the deformation) and annealed at 140°C, rather than by deformation only at 140°C[4].

In thin sections, generated by the blown film process, small angle x-ray scattering (SAXS) has been of value to explain the mechanical properties of high density polyethylene film[83]. Study of x-ray scattering at angles <2°C revealed crystal structures of >50 Å size. The lamellar crystal structures of PE[84,85] and PET[86] developed by biaxial deformation have been studied by SAXS measurements.

8.4.4 Spectroscopy

Transmission infrared spectroscopy has been widely used to evaluate the effects of biaxial draw on molecular orientation. Combined with thermal analysis, the structure of PET has been investigated by using a (Fourier Transform) FTIR attenuated total reflection attachment for dichroism studies[87]. The angular profiles of IR absorbencies for the transverse electric and transverse magnetic polarization were collected at 16 different angles around a horizontal axis and used for the orientation analysis. The results show that the orientation and the structure of biaxially drawn PET is determined in a complex way by the draw rate, draw temperature, and draw ratio[47].

Coworkers with Ward have shown[82] that FTIR spectroscopy combined with X-ray diffraction and refractive index measurements can provide a much firmer basis for characterizing the molecular orientation distribution in uni- and biaxially drawn iPP film than can be provided by measurements using only two of the techniques. The 841 cm^{-1} IR peak gives information about the chain axes, which is similar to that given by the X-ray data for crystallites. The 973 cm^{-1} peak can give information about non-crystalline chains. The close agreement of orientation parameters derived from this peak with those deduced indirectly from refractive index data gives confidence in the values. It was concluded was that the 841 and 973 cm^{-1} bands of iPP form an alternative pair of peaks for the determination of crystalline and amorphous orientation to the peaks at 1220 and 2725 cm^{-1} used by Mirabella[88].

8.4.5   Neutron scattering

Small-angle neutron scattering (SANS) from blends of deuterated and hydrogenated polymer quantitatively determines over-all size (i.e., radius of gyration, Rg) of a chain in bulk[89]. For uniaxial deformation, the isotropic chain shape becomes highly anisotropic with a significantly larger Rg parallel to the draw direction[90]. A double tilt method[6] to measure the shape of iPP chain on equibiaxial deformation revealed that the chain has fiber symmetry in the machine plane but has a complex, non-ellipsoidal over-all shape[7]. In contrast, the uniaxially drawn and simple sheared polymer samples show ellipsoidal shape[7].

8.4.6   Elastic recovery

The extent of deformation can best be expressed by a molecular draw ratio, MDR. The advantages of this method have been described[73,74].

$$MDR = \frac{Le-Ls+Lo}{Lo} \qquad (5)$$

where Lo is the length before draw, Le and Ls are the length of drawn sample before and after shrinkage on melting or on heating above the Tg for an amorphous polymer, respectively. Draw efficiency may therefore be expressed as MDR/DR for each draw direction (Draw Ratio, DR=Le/Lo). In planar deformation (uniaxial compression draw), DR = CR$^{1/2}$ where CR is compression ratio, do/dc (do and dc are sample thicknesses before and after compression, respectively). The MDR parallel to the draw direction is as follows[8].

$$MDR = \frac{Le-Ls}{Le} CR^{1/2} + 1 \qquad (6)$$

Then draw efficiency in the planar direction is given by (MDR)^2/CR or MDR/CR$^{1/2}$.

A relation between draw efficiency and the formation of the smectic phase during planar deformation of iPP has been examined over a wide range of compression draw ratios(<50)[8]. Figure 8.16 shows the evaluation by elastic recovery of planar drawn iPP (using Eq.6). At a forging temperature below the smectic phase stability limit, the efficiency of draw, parallel to the planar direction, is high and independent of the compression ratio, pressure and draw temperature.

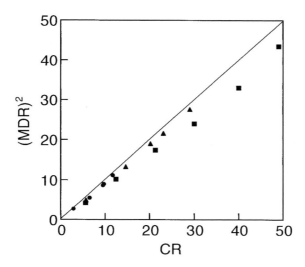

Figure 8.16   Molecular draw ratio as $(MDR)^2$ parallel to uniplanar direction vs. CR (compression ratio) for iPP forged at 50 (●), 100 (▲) and 140°C (■).

The molecular deformation of uni- and bi-axially drawn solution-crystallized PE, evaluated by the elastic recovery test, have been compared[28]. For uniaxially drawn tapes, a linear relationship with the macroscopic draw ratio has been observed. At low macroscopic draw ratios, biaxially drawn films show similar elastic recovery behavior to uniaxially drawn tapes. However, at higher draw ratios the molecular extension of a biaxially drawn film does not increase. Therefore it appears that the crystals deform without further chain extension due to transverse molecular displacement processes at higher biaxial strains. This is consistent with the ineffectiveness of high draw demonstrated by tensile properties[28].

Several specialized features of structure can also be revealed by elastic recovery. For PET differential draw has revealed the history of prior deformation[91]. For isotactic polystyrene, elastic recovery has identified the condition for the generation of continuous crystals induced by draw. For isolated crystals, elastic recovery occurs above Tg, for continuous crystals, dimensions are retained below the melting point.

8.4.7   Gas permeation

The common gases are known to permeate only the amorphous regions of semicrystalline polymers[92~94]. The exception is PMP which has crystal densities in the range of the amorphous phase and indeed the crystalline phases are known to be permeated by the common gases[95,96]. For amorphous phase characterization, gas permeation is not only a valuable tool, but also a property of utility, such as for CO_2 retention in carbonated beverage containers, as with PET bottles[97], or for retaining the flavor of beer by oxygen exclusion.

Two published studies by Ward and coworkers have provided insight on amorphous orientation in iPP via oxygen permeability tests. They also measured the refractive indices, x-ray pole figures and density [79,81]. Oxygen permeability was found to decrease with increasing uniaxial and biaxial draw. Figure 8.17 shows oxygen permeability as a function of the amorphous orientation, $<\cos^2\phi_n>$, perpendicular to the draw directions, where ϕ_n is the angle between amorphous chain and the normal direction. A remarkable correlation was developed that showed a reduced oxygen permeation for both uni and biaxially-oriented iPP which correlates with the amorphous orientation.

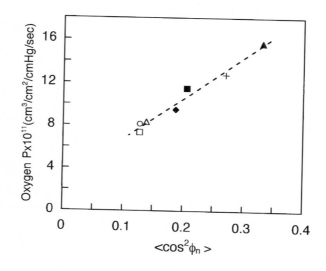

Figure 8.17   Effect of amorphous orientation of iPP in the normal direction on oxygen permeability. (▲) Isotropic; Uniaxially drawn: (△) DR = 10.5 and (□) DR = 12.0; constant-width drawn DR= 7.0: (O) 118 and (■) 145°C; simultaneously biaxially drawn: (+) DR = 3x3 and (□) DR = 8x8.

The permeability (P) and diffusion coefficients (D) for several gases in biaxially oriented polystyrene film has been reported[63,64]. Both P and D increase when the specimen is stretched in simple tension and decrease when the strain is then held constant. These effects were attributed to an increase in free volume with strain and to subsequent densification at constant strain. The strain dependence at small strain, of P and D for Ar, Kr, N2, CO2, and Xe at 1 bar and 50°C shows that the size distribution of free volume elements is not distorted when polystyrene is stretched. At a constant strain of 1.8% at 50°C, P and D of Xe increases about 13.8 and 11.8%, respectively, per decade of time, 2-3 fold faster than for

CO2. These results and those obtained for Ar, whose molecular diameter is smaller than that for Xe, suggest that the larger free volume elements decrease in size faster than the smaller ones as volume recovery progresses after strain.

The CO2 permeability of forged iPP as a function of pressure is unusual. Although the molecular chains orient in the plane with draw, the permeability through thickness direction of an α–crystal sample forged at 140°C is increased. A reduction of permeability with draw was observed only on the sample with smectic phase. The permeability of the forged iPP seems to increase with pressure, but the effect is minor, see Figure 8.18. This is attributed to a decrease in crystallite size on forging, which provides a new diffusive pathway for CO2 through the thickness direction[98]. In the smectic samples, the α–crystal has been transformed to a smectic phase. This phase is a laterally disordered form of the α–crystal and is deformed easily by the compression force, providing no diffusive pathway in the smectic regions. Further, the crystallites are pan-caked which also increases the tortuosity. These structural change in the crystal region as well as the increase in amorphous density on planar orientation reduce the permeability of only forged samples that develop the smectic phase.

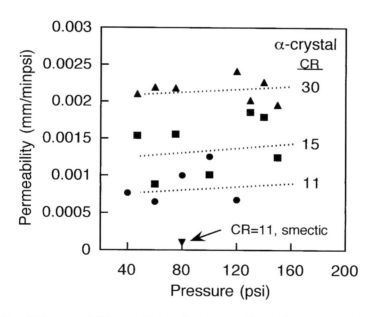

Figure 8.18   CO2 permeability coefficient for planar oriented iPP as a function of CR.

8.4.8   Mechanical tests

Mechanical tests such as tensile, dynamic, indentation, yield, creep and impact on planar oriented thermoplastics have been carried out, not only for evaluation of the properties, but also for investigation of the structures produced by draw.

Among the processes listed for biaxial and planar draw in Table 8.1, only continuous forging involved measuring stress and strain during draw. The compressive stress-strain behavior of a commercial polystyrene has been studied and the effects of deformation temperature on modulus, yield stress, percent yield strain and yield energy have been determined[99]. Yield energy is a parameter that changes regularly with temperature. The energy extrapolates to zero at the Tg of PS in compressive and tensile modes of deformation. The nearly coincident occurrence of a change in slope of the both activation energy and yield stress with the polymer β transition suggest that the main chain motion is affected by the β transition for polystyrene. Data indicate that there is a correlation between the temperature at which yield stress changes slope and the β transition temperature (evaluated from dynamic mechanical test at 1 Hz) for a number of glassy polymers[99].

The crystal-crystal transition temperatures of semicrystalline iPP[3] and PMP[44] on compression draw have been also evaluated by an Arrhenius plot of compressive yield stress measured during deformation. The slope of log(yield stress) vs. reciprocal absolute temperature changes at the transition temperature (see Figure 8.19). The transition temperatures (Tc) are affected by the compression speed and pressure.

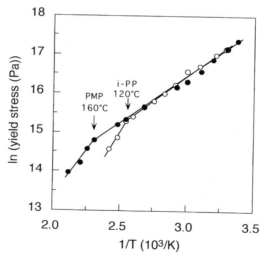

Figure 8.19   The Arrhenius plots of ln (yield stress) vs. reciprocal draw temperature for PMP (O) and iPP (●).

In the compression deformation for uniplanar draw, the yield energy (Ey), ie. The area under the compressive stress-strain curve up to the yield point, is found to consist of two linear components of different slope, as a function of draw temperature (Td)[8,44]. Figure 8.20 shows the Ey versus Td at two compression speeds. For each speed, the relation between Ey and Td consists of two linear components. These two linear relations arise from the glassy and crystalline phases of the semicrystalline iPP. Ti in Table 8.3 is an intercept temperature at zero yield energy for the glassy phase. For tensile deformation, Ti agrees with the measured Tg[100,101]. However, for the compression deformation, the Ti is higher than the ambient Tg due to the high pressure and strain rate history under compression[8,99,102].   The attainable compression draw ratio without sample rupture, increases regularly on increasing forging temperature between Ti and Tm. Performing the deformation in this temperature range results in extremes of uniplanar draw and with high draw efficiency.   It may also noted that both Ti and Tc go up with increasing the compression speed.

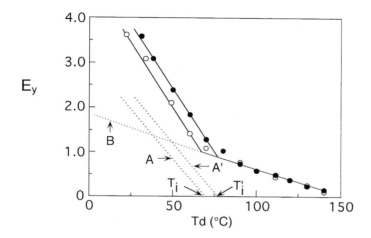

Figure 8.20   Ey vs. Td for two compression speeds; 0.254 (●) and 0.0254 cm/min (O). The dotted line A (or A') and B represent the each component of amorphous and crystal phases, respectively.   A and A' are for higher and lower compression speed, respectively. Ti and Ti ' are the intercept temperature at zero yield energy for A and A'.

On tensile strain, planar oriented iPP breaks into a layered structure. This feature is most prominent for highly compression-drawn samples.   This behavior is confirmed by a discontinuous stress-strain curve[10]. As shown in a former section on iPP, the layered structure has significance for impact toughness[11]. Such a layer structure, caused by the planar deformation, is also observed for PMP, high density polyethylene, as well as atactic polystyrene as induced in forging experiment[10].

Indentation tests provide an evaluation of variation in surface mechanical properties of polymers, and as affected by changes in processing, heat treatment, microstructure and aging[103].   Indentation hardness testing is also used in the investigation of the microstructure of semicrystalline polymers at several morphological levels[104~106].

Anisotropy in mechanical properties of forged iPP is readily observed by the micro indentation hardness test, see Figures 8.21a and 8.21b. As described earlier, iPP forged at 50°C includes draw generated smectic phase, on the other hand samples forged at 140°C consist of α–crystal plus amorphous. The anisotropy is sensitive to the phase structure. For samples forged at 140°C (resulting in α–crystal only plus amorphous), the indentation hardness perpendicular to plane direction, H(⊥), slightly increased with draw.   The hardness tested parallel to film surface, H(//), decreased rapidly with draw. For the sample forged at 50°C (containing draw-generated smectic), both H(//) and H(⊥) decreased with draw, suggesting the softness of the smectic phase. The anisotropy, H(⊥)/H(//), for samples forged at 50°C is higher than that for samples forged at 140°C at a comparable compression ratio. The decrease of H(//) is due to the planar chain orientation and b-axis orientation normal to plane direction (i.e. 0k0 plane orient parallel to plane direction). Above a CR of 10, it is difficult to measure H(//) because the crack along the film plane direction takes place on indentation. This is thus an easy break from the edge, as discussed earlier for iPP.

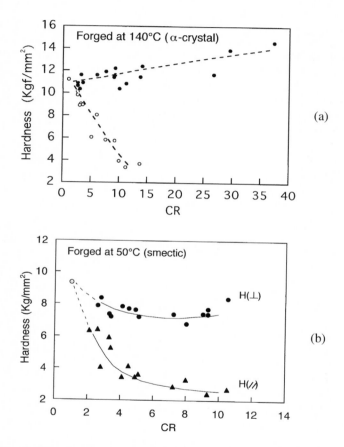

(a)

(b)

Figure 8.21   The indentation hardness of planar oriented iPP (a) forged at 140°C as a function of compression ratio, CR: indentation for perpendicular (●) and parallel (O) to the plane direction; (b) forged at 50°C as a function of CR: indentation for perpendicular (●) and parallel (▲) to the plane direction.

The dynamic mechanical properties and the creep rate of biaxially drawn films of UHMW-PE dried gels have been studied and compared with the results of uniaxially drawn films[24]. The tensile modulus of biaxially drawn film is much smaller than that of uniaxially drawn films, yet their temperature dependence is similar. The creep strain rate in a 16x16 film is over two orders of magnitude greater than those of x10 and x20 uniaxially drawn films. This large creep reflects the morphology of the biaxial drawn films characterized by microfibrils randomly orienting in the film plane: the creep strain of biaxially drawn film is attributed mainly to the orientation of fibrils in the straining direction, accompanying the straightening of bent fibrils and the rotation as well as the slippage among the fibrils[24].

The crack growth behavior and analysis of the notch damage zone in cross-rolled iPP has been studied[107~109]. Unoriented iPP formed a narrow wedge-shaped damage zone that grows with increasing stress until catastrophic brittle fracture occurs. The 50% oriented (50 % thickness reduction by the rolling) material initially developed a wedge-shaped damage zone that grew wider as loading was increased. The specimen fractured with stable crack growth in a ductile manner, showing a large resistance to crack growth. The 80% oriented (more oriented than 50%) material had a circular damage zone that consisted of many delamination crazes. These crazes grew by splitting the specimen in the thickness direction. Stable crack growth dominated the final failure process with the 80% oriented material showing nearly three times the toughness of the unoriented material[107].

Also in tests on biaxial iPP, the creep[110] and the ultrasonic velocity and attenuation have been reported[111].

8.4.9   Multiple regression analysis

In any solid-state deformation of polymers, the resultant physical property is correlated to morphologies such as molecular orientations, conformations, and phase structures. The processing variables such as draw ratio, compression ratio and draw temperature also affect the morphologies. Therefore, it is important to understand the contributions of each morphology and the processing variable to the resultant property. The multiple regression analysis is one of the statistical methods for describing the characteristic of the subject as a function of many factors[112]. The method is used to describe analyses of data that are multivariant in the sense that numerous observations or variables are obtained for each individual or unit[113,114]. This method can be applied to understanding property-structure relationships of planar deformation of iPP as an example[115].

The basic equation for linear multiple regression analysis is

$$Y = a_0 + \sum_j a_j X_j \tag{7}$$

where Y is a criterion variable, X1, X2....Xj are explanatory variables, a_0 is a constant and $a_1, a_2...a_j$ are the regression coefficients, which mean the rate of contribution of each variable. In the section of *polypropylene*, see Figure 8.12, the turbidity τ seems to be as functions of smectic phase fraction (induced by the draw) , compression ratio CR, as well as the diameter of spherulite. Let τ be the criterion variable Y. Then the phase fraction of smectic, Xs, amorphous, Xa, and CR^{-1} are taken for this analysis as explanatory variables of Xj. The sum of phase fractions, Xs + Xa + Xc (crystallinity) =1, Therefore, two of the three fractions are chosen for the analysis (The fraction data is from ref. 115).   To compare each contribution to the explanatory variable to τ, by comparison of each regression coefficient, data must be normalized as X_i' .

$$x_i' = \frac{x_i - \overline{x}_i}{\sqrt{Sxx}}$$

(8)

where x_i is data of each variable, $\overline{x}_i$ is the average of x_i, and Sxx is a variance of Xj. Then the following equation for τ as a function of Xs, Xa, CR^{-1} is obtained.

$$\tau = -0.0002 - 0.310Xs - 0.087Xa + 0.893CR^{-1}$$

(9)

This result demonstrates that the contribution of CR^{-1} (aCR^{-1} =+0.893) and Xs (aXs=-0.310) to τ are large and that of Xa is minor. The adjusted R-square for the calculation is 0.994, indicating τ can be well described by sum of contributions of the three variables. (Adjusted R-square of 1 means that the criterion variable Y is completely expressed by the sum of contribution of each explanatory variable.) The minus and plus for regression coefficients of Xs and CR^{-1} mean to reduce the τ and to increase τ , respectively. Namely, increase in CR reduce the τ. The smectic contribution is about 1/3 of draw as revealed by CR^{-1}. When CR is taken as a explanatory variable for this analysis, the adjusted R-square drops to 0.64. This means that the CR^{-1} is an appropriate variable to describe the optical clarity of iPP rather than CR. This also confirm that the increase in clarity is directly related to the decrease of spherulite diameter, D, since it correlates linearly to CR^{-1} (see Figure 8.22).

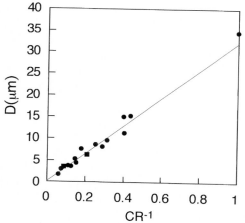

Figure 8.22   Relation between spherulite diameter, D and CR^{-1}.

Another application of this method is for the tensile properties of forged iPP. Figures 8.23 and 8.24 show the tensile strength and modulus of forged iPP as a function of CR. Both tensile modulus and strength parallel to the planar direction increased with CR. The smectic phase generated by forging at 50°C, however, reduced the tensile modulus with no significant effect on tensile strength. In planar deformation, draw ratio, DR, for the planar direction equals $CR^{1/2}$ [8].

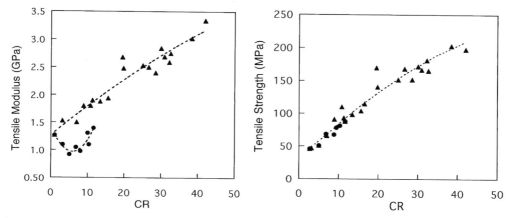

Figure 8.23 The tensile modulus vs. CR for samples prepared at 50°C (●) and 140°C (▲).

Figure 8.24 The tensile strength vs. CR for samples prepared at 50°C (●) and 140°C (▲).

Therefore, the $CR^{1/2}$ is taken as a explanatory variable in this analysis. The analytical equation obtained from the normalized data of Xs, Xa and $CR^{1/2}$ for strength and modulus are expressed by Equations (10) and (11), respectively

$$\text{Modulus} = 0.00001 - 0.509Xs - 0.009Xa + 0.728CR^{1/2} \qquad (10)$$

$$\text{Strength} = 0.00002 - 0.043Xs - 0.056Xa + 0.953CR^{1/2} \qquad (11)$$

The adjusted R-squares are 0.982 and 0.995 for Equations (10) and (11), respectively. The contribution (refer to the regression coefficient) of the smectic fraction Xs is higher in modulus than that in strength. Equations. (10) and (11) indicate that $CR^{1/2}$ contributes strongly to tensile properties. When CR is taken as a explanatory variable in place of $CR^{1/2}$, the R-square is not changed significantly, suggesting the CR is also a reasonable variable to describe the tensile properties. The multiple regression analysis is thus useful to understand the property-structure-processing variable relationships for any kind of solid state deformation.

References

1. W. J. Schrenk, J. Polym. Sci., Polym. Symp., **72**, 307 (1985).
2. R. F. Saraf and R. S. Porter, J. Rheol., **31**, 59 (1987).
3. R. F. Saraf and R. S. Porter, Polym. Eng. Sci., **28**, 842 (1988).
4. R. F. Saraf, Polymer, **35**, 1359 (1994).
5. J. Y. Guan, R. F. Saraf and R. S. Porter, J. Appl. Polym. Sci., **33**, 1517 (1987).
6. R. F. Saraf, Macromolecules, **22**, 675 (1989).
7. R. F. Saraf, J. M. Lefebvre, R.S. Porter, and G.D.Wignall, Mat. Res. Symp. Proc., **79**, 263 (1987).
8. S. Osawa and R. S. Porter, Polymer, **35,** 540 (1994).
9. S. Osawa and R. S. Porter, Polymer, 35, 545 (1994).
10. S. Osawa, R. S. Porter and M. Ito, Polymer, 35, 551 (1994).
11. S. Osawa and R. S. Porter, Polymer, 37, 2095 (1996).
12. Lionti and R. S. Porter, Polymer, 31, 1228 (1990).
13. C. Nakafuku and T. Takemura, Mem. Fac. Eng., Kyushu Univ., 34, 11 (1974).
14. T. Kanamoto, A. Tsuruta, K. Tanaka, M. Takeda, and R.S . Porter, Macromolecules, 21, 470 (1988).
15. J. P. Hirth and J. Lother, "Theory of Dislocation" Second Edition, McGraw-HillInc., New York (1982).
16. C. J. Speerchneider and C. H. Li, J. Appl. Phys., 33, 1871 (1962):
    P.H. Geil, J. Polym. Sci. **A2**, 3813 (1965);
    Ingram and A. Peterlin, J. Polym. Sci., **B2**, 736 (1964);
    H. Kiho, A. Peterlin and P.H. Geil, J. Polym. Sci.. **B3**, 257 (1965);
    M. Kojima, J. Polym. Sci. Polym. Phys., **5**, 597 (1967);
    J. Petermann and H. Gleiter, J. Polym. Sci. Polym. Phys., **10**, 2333 (1972);
    R. M. Gohil and J. Petermann, J. Polym. Sci. Polym. Phys., **17**, 525 (1979).
17. Nylon-11 Biaxial Deformation, J.P. Autran Ph.D thesis, University of Massachusetts.
18. R. F. Saraf and R.S. Porter, J. Polym. Sci. Polym. Phys. Ed., 26, 1049 (1988).
19. S. Minami and K. Itoyama, Am. Chem. Soc., Polym. Prepr. , 26, 245 (1985).
20. N. S. J. A. Gerrits, R. J. Young and P. J. Lemstra, Polymer, 31, 231 (1990).
21. N. S. J. A. Gerrits and P. J. Lemstra, Polymer, 32, 1770 (1991).
22. Y. Sakai,and K. Miyasaka, Polymer, 29, 1608 (1988).
23. Y. Sakai and K. Miyasaka, Polymer, 31, 51 (1990).
24. Y. Sakai, K. Umetsu and K. Miyasaka, Polymer, 34, 318 (1993).
25. C. W. M. Bastiaansen, P. J. R. Leblans and P. Smith, Macromolecules., 23, 2365 (1990).
26. H. L. Cox, Brit. J. Appl. Phys., 3, 72, (1952).
27. D. Hull, 'An Introduction to Composite Materials', Cambridge, 1981, P. 81.
28. N. S. J. A. Gerrits and R. J. Young, J. Polym. Sci. Part B: Polym. Phys., 29, 825 (1991).
29. T. Kyu, University of Akron, personal communication,1989.
30. A. E. Zachariades, J. Appl. Polym. Sci., 29, 867 (1984).
31. A. E. Zachariades and J. Economy, Polym. Eng. Sci., 23, 266 (1983).
32. A. E. Zachariades and B. Chung, Adv. Polym. Technology, 7, 397 (1987).
33. A. R. Austen and D. V. Humphries, SPE ANTEC Tech. Papers, 837 (1982).
34. H. J. Chen, M. T. Kortschot and K. G. Leewis, Polym. Eng. Sci., 34, 1016 (1994).

35. V. J. Dhingra, J. E. Spruiell and E. S. Clark, Polym. Eng. Sci., 21, 1063 (1981).
36. K. Watanabe, T. Kanda, Y. Higashida and T. Kikuma, Nippon Steel Co., private commun.
37. H. Ashizawa, J. E. Spruiell and J. L. White, Polym. Eng. Sci., 24, 1035 (1984).
38. M. Cakmak, Y. D. Wang and M. Simhambhatla, Polym. Eng. Sci., 30, 721 (1990).
39. S. Osawa and R. S. Porter, Rept. Progr. Polym. Phys. Jpn., 38, 431 (1995).
40. R. Simha, and R. F. Boyer, J. Chem. Phys., 37, 1003 (1962).
41. *Properties of Polymer, Their estimation and correlation with chemical structure*, by D. W. Van Krevelen and P. J. Hoftyzer, Elsevir Scientific Publishing Company, Amsterdam, Oxford, New York, 1976.
42. S. K. Roy, T. Kyu, and R. St. J. Manley, , Macromol., 21, 499 (1988).
43. M. Matsuo, C. Sawatari, and T. Nakano, Polym. J.,18, 759 (1986).
44. S. Osawa and R. S. Porter, J. Polym. Sci., Polym. Phys., 32, 535 (1994).
45. J. H. Griffith and B. Randy, J. Polym. Sci., 44, 369 (1960).
46. S. Rastogi, M. Newman, and A. Keller, J. Polym. Sci. Polym. Phys., 31, 125 (1993).
47. K. H. Lee and C. S. P. Sung, Macromolecules, 26, 3289 (1993).
48. R. J. Samuels, J. Polym. Sci. Part □□2, 7, 1197 (1969).
49. R. J. Samuels, J. Appl. Polym. Sci., 26, 1383 (1981).
50. J. Y. Guan, L-H. Wang and R. S. Porter, J. Polym. Sci. Polym. Phys., 30, 687 (1992).
51. M. Cakmak, J. L. White and J. E. Spruiell, Polym. Eng. Sci. 29, 1534 (1989).
52. J. B. Faisant de Champchesnel, D. I. Bower, I. M. Ward, J. F. Tassin, and G. Lorentz, Polymer, 34, 3763 (1993).
53. T.P. Russell, H. Gugger, and J.D. Swalen, J. Polym. Sci., Polym. Phys. Ed., 21, 1745 (1983).
54. R. F. Saraf, C. Dimitrakopoulos, M.F. Toney, and S.P. Kowalczyk, Langmuir, to be published.
55. D.N. Theodorou, Macromolecules, 21, 1391 (1988); 21, 1400 (1988).
56. Mauguin, Bull. Soc. Miner., 34, 71 (1911).
57. D.W. Berreman, Phy. Rev. Lett., 28, 1683 (1972).
58. S.W. Depp, and W. E. Howard, Scient. Am., 268, 90 (1993).
59. S. Numata, and K. Fujisaki, N. Kinjo, Polymer, 28, 2282 (1987).
60. R.R. Tummala, and E.J. Rymaszewski, Eds., "Microelectronics Packaging Handbook" Van Nostrand Reinhold, New York, (1989).
61. L. E. Nielsen "The Mechanical Properties of Polymers" Ch 10, Reinhold New York (1962).
62. K-J. Choi, J. E. Spruiell, and J. L. White, Polym. Eng. Sci., 29,1516,1524 (1989).
63. G. Levita and T. L. Smith, Polym. Eng. Sci., 21, 936 (1981).
64. T. L. Smith, W. O. Oppermann, A. H. Chan and G. Levita, Am. Chem. Soc. Polym. Prepr., 24, 83 (1983).
65. J. L. White, personal communication, 10/10/92.
66. L. H. Wang and R. S. Porter, J. Macromol. Sci. Review, C35, 63 (1995).
67. Y. Higashida, K. Watanabe and T. Kikuma, ISIJ. Int. 31, 655 (1991).
68. J. Sweeney and I. M. Ward, Polymer, 34, 299 (1995).
69. M. Simhambhatla and M. Cakmak, Polym. Eng. Sci., accepted 1994.
70. M. A. Kotnis and I. E. Spruiel, Polym. Eng. Sci., 29, 1528(1989).
71. Alcan assigned U.S. Pat. to Symplastics, 5,169,587 and 5,169,589 (1992), 5,204,045(1993).
72. S. M. Hong and J. Economy, Macromolecules, 28, 6481 (1995).

73. M. P.C. Watts, A. E. Zachariades, and R. S. Porter, J. Mater. Sci., 15, 426 (1980).
74. R. S. Porter, M. Daniels, M. P. C. Watts, J. R. C. Pereira, S. J. DeTeresa, and A. E. Zachariades, J. Mater. Sci., 16, 1134 (1981).
75. W. W. Graessley, Adv. Polym. Sci., 16, 1 (1974) and 47, 67 (1982).
76. L. R. G. Treloar, Trans. Faraday Soc., 42, 83 (1946).
77. S. Okajima and Y. Koizumi, Kogyo Kagaku Zasshi, 42, 810 (1939).
78. G. W. Scheal, J. Appl. Polym. Sci., 8, 2717 (1964).
79. K. Taraiya, G. A. J. Orchard and I. M. Ward, J. Appl. Polym. Sci., 41, 1659 (1990).
80. C. Cha, S. Moghazy and R. J. Samuels, Polym. Eng. Sci., 32, 1358 (1992).
81. K. Taraiya, G. A. J. Orchard and I. M. Ward, J. Polym. Sci., Polym. Phys., 31, 641 (1993).
82. A. K Karacan,. D. I. Taraiya, Bower and I. M. Ward, Polymer, 34, 2691 (1993).
83. A. Gupta, D. M. Simpson and I. R. Harrison, Plast. Eng. Nov., 1993, P.33, and J. Appl. Polym. Sci., 50, 2085 (1993).
84. K-J. Choi, J. E. Spruiell and J. L. White, J. Polym. Sci. Polym. Phys. Edn., 20, 27 (1982): Polym. Eng. Sci., 29, 463 (1989).
85. F. Ania, F. J. Balta Calleja, and R. K. Bayer, Polymer, 33, 233 (1992).
86. M. Cakmak, J. E. Spruiell, J. L. White and J. S. Lin, Polym. Eng. Sci., 27, 893 (1987).
87. E. A. Lofgren and S. A. Jabarin, J. Appl. Polym. Sci, 51, 1251 (1994).
88. F. M. Mirabella, J. Polym. Sci. Polym. Phys., 25, 591 (1987).
89. G.D. Wignall, Encyclopedia of Polymer Science and Engineering, vol. 10, 2nd ed., Wiley & Sons, NY (1987), p. 112.
90. G. Hadziioannou, L.-H. Wang, R.S. Stein, and R.S. Porter, Macromolecules, 15, 880 (1982).
91. M. F. Vallat and D. V. Plazek, J. Polym. Sci. Polym. Phys., 24, 545 (1986).
92. A. S. Michaels, W. R. Vieth and J. A. Barrie, J. Appl. Phys., 34, 1 (1963).
93. A. S. Michaels and R. B. Parker, Jr, J. Polym. Sci. 41, 53 (1959).
94. A. S. Michaels and H. J. Bixler, J. Polym. Sci., 50, 393 (1961).
95. P. K. Puleo, D. R. Paul and P. K. Wong, Polymer, 30, 1357 (1989).
96. J. H. Petropoulos, J. Polym. Sci., Polym. Phys., 23, 1309 (1985).
97. N. C. Wyeth and R. W. Roseveare, U.S. Pats., 3,733,309 (1973), and 3,849,530 (1973).
98. S. Osawa, S. Davis and R. S. Porter, Rept. Progr. Polym. Phys. Jpn., 38, 287 (1995).
99. L. Beatty and J. L. Weaver, Polym. Eng. Sci., 18, 1109 (1978).
100. W. Macosko and G. J. Brand, Polym. Eng. Sci., 12, 444 (1972).
101. B. Hartmann and R. F. Cole, Jr., Polym. Eng. Sci., 23, 13 (1983).
102. P. Zoller, J. Appl. Polym. Sci., 21, 3129 (1977).
103. F. J. Balta Calleja, Adv. Polym. Sci., 66, 117 (1985).
104. F. J. Balta Calleja, J. Martinez Salazar and T. Asano, J. Mater. Sci. Let., 7, 165 (1988).
105. F. J. Balta Calleja and H. G. Kilian, Colloid Polym. Sci., 263, 697 (1985).
106. J. Martinez Salazar, J. Garcia and F. J. Balta Calleja, Polym. Commun. 26, 57 (1985).
107. J. Snyder, A. Hiltner, and E. Baer, J. Appl. Polym. Sci., 52, 217 (1994).
108. J. Snyder, A. Hiltner, and E. Baer, Polym. Eng. Sci., 34, 269 (1994).
109. J. Snyder, A. Hiltner, and E. Baer, J. Appl. Polym. Sci., 52, 231 (1994).
110. I. X. Li and W. L. Cheung, J. Appl. Polym. Sci., 56, 881 (1995).
111. S. Malinarie, J. Appl. Polym. Sci. 53, 1405 (1994).
112. A.A. Afifi, "Computer-aided Multivariate Analysis", Lifetime Learning, Belmont, CA,1984.

113.T. Ogawa and T. Yamada, J. Appl. Polym. Sci., 53, 1663 (1994).
114.T. Ogawa, Y. Fukushima, S. Osawa and K. Kimura "Numerical prediction of weatherability for polymers" Proc. 5th Japanese International SAMPE Symposium, Oct. 28-23, 37 (1997)
115.S. Osawa, H. Mukai, T. Ogawa and R. S. Porter, J. Appl. Polym. Sci., 68, 1297 (1998)

9   Solid State Extrusion and Die Drawing

I M Ward*A K Taraiya* and P D Coates+
IRC in Polymer Science & Technology,
*University of Leeds & +University of Bradford, UK

9.1   Ram Extrusion

A very simple method of producing an oriented rod of a thermoplastic such as polyethylene and polypropylene is to apply pressure to a piston which pushes an isotropic plug of polymer through a die of reducing cross-section, most simply a conical cross-section. The first reports of the application of this technique to polymers came from Takayanagi and co-workers [1] in Japan, but it has also been used extensively by Porter and co-workers [2] in the USA, in some cases with spectacular results in terms of the high stiffness and strength of the oriented extrudates. As in the case of tensile drawing the deformation is most effective at comparatively high temperatures (~100°C for polyethylene) and there is a similar relationship between modulus and deformation ratio to that observed for tensile drawing.

In terms of deformation ratios and the Young's modulus of the extrudates, the results obtained by Takayanagi and his co-workers for polyethylene were interesting but not outstanding, with a Young's modulus of 10GPa obtained for a deformation ratio of 16, and a significantly high degree of crystalline orientation.

The Takayanagi group also undertook a quantitative analysis of the mechanics of the process [1] following the classic Hoffman-Sachs lower bound analysis [3] originally proposed for metals, which sets up the stress equilibrium equation for deforming a slice of material in the conical die. The analysis did take into account strain hardening, using an empirical relationship between the true stress σ and the plastic strain ε where

$$\log(\sigma / \overset{*}{\sigma})\log\left(\varepsilon / \overset{*}{\varepsilon}\right) = -\overset{*}{c}$$

where $\overset{*}{\sigma}$, $\overset{*}{\varepsilon}$ and $\overset{*}{c}$ are constants.

This theory does not accurately model the very large rise in pressure for high extrusion ratios because it does not take into account the major effects of strain rate and pressure on the flow stress.

The principal thrust of the research on ram extrusion by Porter and his co-workers was very different. The Porter group focussed almost entirely on the production of very highly oriented polymers. In the first experiments on polyethylene [4] ram extrusion was undertaken using a capillary rheometer, so that the polymer was molten at the entrance to the die. As the polymer became oriented in the die it was also crystallising from the melt, and very fine strands of highly oriented polyethylene were produced. Although material with a comparatively low overall modulus of only 6.6 GPa (at 110Hz) was obtained, DSC measurements showed a high melting component which could be attributed to chain extended material. This work of Southern and Porter [4] was later repeated by Keller and co-workers [5], who showed that it was possible to find optimal conditions where a highly oriented plug of polyethylene could be obtained with a modulus of 70GPa. It was clear that in these experiments molten polymer entered the die under high pressure and that as the polymer was cooled from the molten state it formed a unique structure of interlocking lamellae, which was revealed by transmission electron microscopy. The lamellar thickness was found to be about 300 Å, consistent with a very clear two point SAXS pattern and a comparatively low melting point. It was proposed that the high Young's modulus was consistent with a parallel lamellae structure where the deformation of the amorphous regions is constrained by the lamellar surfaces, analogous to the constraints imposed on the thin rubber sheets where sheets of steel are laminated with thin layers of an incompressible rubber.

In further work, Porter and his co-workers [6] invented a novel variant of ram extrusion where a sheet of polymer is extruded between a split billet consisting of two hemi cylinders of a second polymer. There appear to be two advantage of the die drawing process by this co-extrusion technique. First it was found that much faster extrusion rates could be obtained. For example, in the first experiments on polyethylene extrusion rates of 10cm/min were obtained for draw ratios of 30 with a Young's modulus of 30GPa. Secondly, the co-extrusion technique could be used to obtain very high draw ratios for polymer materials which could not be easily deformed, such as ultra high molecular weight PE, solution grown mats of PE and amorphous polymers.

The co-extrusion technique cannot be considered to be a practical engineering method for the production of high modulus and high strength oriented polymers. Results obtained by Porter and his colleagues are, however, so remarkable that no account of solid phase extrusion would be complete without a summary of what was achieved.

In the case of polyethylene, the highest moduli were obtained by a two stage process [7], where solution grown mats of ultra high molecular weight PE (Hizex 240M, $M_v = 2 \times 10^6$) were first co-extruded between two split billet halves of lower molecular weight PE ($M_w = 6.7 \times 10^4$), and then drawn in a conventional tensile drawing technique. An extreme value of 250 for the tensile draw ratio was achieved by drawing to an extrusion draw ratio 6 sample at 115°C. A tensile modulus of 222GPa was reported which is approaching the range of values of 240-320GPa reported from theoretical calculations and experimental results obtained from X-ray crystal strain and neutron diffraction measurements. Correspondingly high values of tensile strength (up to 6GPa) were also obtained.

Porter and his co-workers also used the two stage drawing (co-extrusion followed by tensile drawing) with success for PE reactor powders, obtaining a tensile modulus of 130 GPa for a total draw ratio of 85. Other polymers for which spectacular results were obtained include polypropylene [8] and poly tetra fluoroethylene (PTFE) [9]. In the case of PTFE, ultra high molecular weight polymer $M_n \sim 10^7$ was first co-extruded and then drawn rapidly over a heated metal cylinder to a very high draw ratio at a temperature just above the polymer melting point. At a maximum total draw ratio of 140, a tensile modulus and strength at 24°C of 82GPa and 1.2GPa were reported. The modulus corresponds to 55% of the crystal modulus of 158 GPa obtained by X-ray strain measurements.

Other polymers were successfully drawn using the technique of solid state co-extrusion, although the absolute values of tensile modulus and strength were not exceptional. These included polystyrene [10], polyethylene terephthalate (starting with both amorphous PET [11] and crystalline PET [12]). Poly (4-methyl pentene 1) [13] nylons 6 and 11 plasticized with anhydrous ammonia [14], and poly (1-butene) [15].

9.2   Hydrostatic Extrusion

9.2.1   Introduction

The ram extrusion technique has two principal limitations. First, there is high friction between the polymer and the metal die, which still applies in the case of split billet extrusion, and limits the process to very low extrusion rates. Many of Porter's most spectacular results were obtained in cases where it took several minutes or even hours to extrude a few millimetres of product. Secondly, this high friction can give rise to a significant variation in structure across the extrudate section, as shown by Farrell and Keller [16], due to the very high deformation in the polymer close to the polymer/die interface.

In hydrostatic extrusion, the piston is replaced by fluid under high pressure, and the billet is clear of the walls of the pressure vessel. A schematic diagram of a typical experimental set up is shown in Figure 9.1. The pressure transmitting fluid (typically castor oil) acts as a lubricant between the polymer and the die wall so that the deformation in the case of conical billet is essentially plug flow (i.e. plane sections remain plane and the deformation is one of pure homogeneous strain, identical to tensile drawing with a free surface. The resulting extrudates are therefore homogeneously oriented across the section to a very good approximation. The haul-off load shown in Figure 9.1 is very small, only sufficient to ensure that straight extrudates are produced, and to provide additional control of the extrusion process.

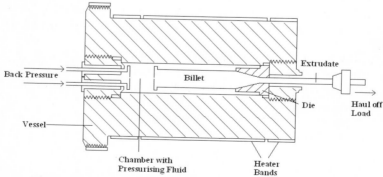

Figure 9.1.   Apparatus for hydrostatic extrusion of polymers

First reports of the hydrostatic extrusion of polymers were given by Buckley and Long [17], Alexander and Wormell [18] and Nakajima and Kanetsuma [19], but the results were not very interesting because of the rather low orientation and hence very moderate mechanical properties of the extrudates.

The first hydrostatic extrusion results to be of interest with regard to enhancement of polymer properties come from the ICI group at the Petrochemical and Polymers Laboratory at Runcorn, Cheshire. Williams [20] reported that polypropylene rods could be produced with a room temperature Young's modulus of 16GPa. In 1972 Gibson and Ward, who had been working at ICI, transferred the equipment to Leeds University and in collaboration with Cole and Parsons [21], produced very highly oriented polyethylene rods with a room temperature Young's modulus of 60 GPa. This research then continued in parallel with the related attempts to produce high modulus polyethylene fibres by tensile drawing which have been reviewed in chapter 5.

Many of the guidelines established for the successful production of high modulus polyethylene tapes and fibres were soon shown to be relevant to hydrostatic extrusion. First, it was clear that the modulus of the extrudates related primarily to the actual extrusion ratio R_A, based on the ratio of the initial and final billet cross-section (assuming deformation occurs at constant volume). This parameter is equivalent to the draw ratio of tensile drawing, but may not be exactly the same as the nominal extrusion ratio R_N, the ratio in the billet cross-sectional area to that of the die exit. In general, there is very little die-swell in hydrostatic extrusion so that $R_A \approx R_N$, for all except the lowest deformation values, where some recovery from the imposed deformation can occur.

Secondly, both hydrostatic extrusion and tensile drawing are limited by the strain hardening behaviour of the polymer, hence the importance of the chemical composition of the polymer, molecular weight and molecular weight distribution, initial billet morphology, etc. Ward and co-workers [22] proposed that the hydrostatic extrusion behaviour could be understood in terms of two basic principles. First, the properties of the extrudate are determined by the total plastic strain imposed i.e. the extrusion ratio, for a given initial isotropic billet where the chemical composition and initial morphology are fixed. Secondly, the processing behaviour in terms of factors such as extrusion pressure and extrusion rate, are determined by a mechanical equation of state, the true stress-strain curve, which is a function of strain rate, temperature and pressure (for a given initial isotropic billet).

The first principle is analogous to that established for tensile drawing, where the modulus of the drawn polymer was shown to relate to the draw ratio. Figure 9.2 shows results for the Young's modulus of a range of polyethylene extrudates [23]. Perhaps surprisingly there is a unique relationship between modulus and draw ratio irrespective of the inclusion of polymers of very different molecular weight, which is similar to the results shown for tensile drawing in chapter 5.

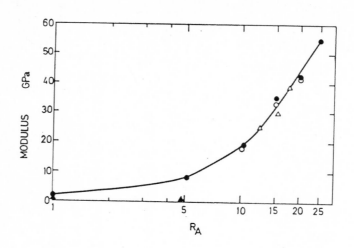

Figure. 9.2   Axial Young's moduli (10s flexural creep at 20°C) of 2.5 mm LPE extrudates: ▲ HGUR, ●R50, ○R25, ΔR140 (Ref. 23). © John Wiley & Sons. Reproduced by permission.

It was, however, also shown that ultra high molecular weight polyethylene did not conform to this unique relationship. The differences have been attributed to major structural differences between UHMPE and lower molecular weight polymers, relating to much lower degrees of crystallinity. For the lower molecular weight polyethylenes, structural studies principally by WAXS and SAXS, show that the structures of the hydrostatically extruded materials are very similar to those obtained by tensile drawing. The relationships between structure and properties are therefore analogous to those of the high draw ratio fibres and films described in detail in chapter 5 above.

9.2.2   The mechanics of the hydrostatic extrusion process

The mechanics of the hydrostatic extrusion process has been considered by Ward and co-workers [24,25] in two publications. The approach was to combine four elements in a comprehensive analysis.

(1)   The lower bound analysis of Hoffman and Sachs [26] for the extrusion of metals.
(2)   The Avitzur analysis [27] for the strain rate field in the die.
(3)   The dependence of the flow stress on plastic strain, strain and rate based on experimental data obtained in the Leeds Laboratories.
(4)   The dependence of the flow stress on pressure.

The analysis starts by considering the force balance on a thin, parallel sided element in the deformation zone of a conical die (Figure 9.3) which is given by

$$\frac{d\sigma_x}{d\varepsilon} = \sigma_x - \sigma_y \left(1 + \mu \cot \alpha\right) \tag{1}$$

where σ_x and σ_y are the stresses parallel and perpendicular to the extrusion direction, ε is the strain, μ is the billet-die friction coefficient and α is the die semi-angle.

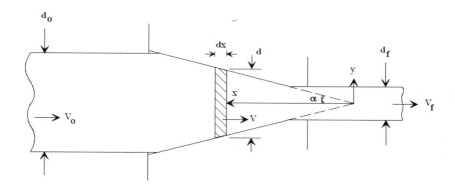

Figure 9.3. Co-ordinate system used to represent the deformation zone in conical die extrusion.

σ_x and σ_y are related through the yield or flow stress criteria. Accepting that the flow stress σ_f is dependent on the plastic strain ε (strain hardening), strain rate $\dot{\varepsilon}$ (strain rate sensitivity) and pressure P, we have

$$\sigma_x - \sigma_y = \sigma_f \left(\varepsilon, \dot{\varepsilon}, P\right) \tag{2}$$

The effect of pressure can be considered either in terms of a linear, additive pressure effect where

$$\sigma_x - \sigma_y = \sigma_f \left(\varepsilon, \dot{\varepsilon}\right) + \gamma' p \tag{3}$$

where γ' is a pressure coefficient, or more conveniently in terms of a Coulomb yield criterion where

$$\sigma_x - \sigma_y = \sigma_f \left(\varepsilon, \dot{\varepsilon}\right) + \gamma' \sigma_y \tag{4}$$

and the effect of pressure is represented by the effect of the normal stress σ_y.

Equation (1) can now be written as

$$\frac{d\sigma_x}{d\varepsilon} = \sigma_x - \left[\frac{\sigma_x - \sigma_f \left(\varepsilon, \dot{\varepsilon}\right)}{1 + \gamma'}\right] \left(1 + \mu \cot \alpha\right) \tag{5}$$

There is good experimental evidence from the observation of the deformation of split billets where grid lines have been printed, that plane sections remain plane during the deformation in the die and plug flow. The strain ε_x is therefore given in terms of the instantaneous deformation ratio R at a distance x from the die exit as

$$\varepsilon_x = lnR \tag{6}$$

and, following Avitzur, the strain rate at the point x is given by

$$\dot{\varepsilon}_x = \frac{4V_f}{d_f}\left(\frac{R}{R_N}\right)^{\frac{3}{2}} \tan\alpha \tag{7}$$

where V_f is the die exit velocity and d_f is the exit die diameter.

The dependence of the flow stress on ε and $\dot{\varepsilon}$ is determined experimentally, following the assumption that it depends on the total plastic strain and the current strain rate i.e. the assumption of a true stress, true strain, strain rate relationship. At a fixed temperature, the flow stress can be given a three-dimensional representation as shown in Figure 9.4 for Rigidex 50 linear polyethylene at 100°C.

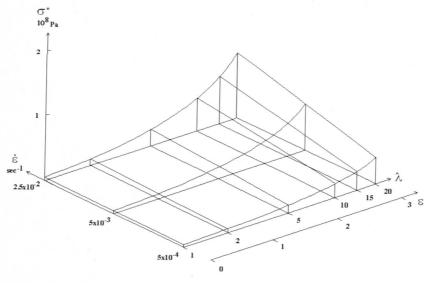

Figure 9.4.   Three-dimensional representation of the stress-strain rate relationship for Rigidex 50 LPE at 100°C.

Several approaches have been given for the effect of pressure. An interesting possibility, which reduces the number of independent variables in equation, is to assume that the pressure coefficient γ' can be related to the friction coefficient μ, as proposed by Briscoe and Tabor [28].

We then have

$$\gamma'(\varepsilon) = 2\mu(\varepsilon) \tag{8}$$

where both γ' and μ can be strain dependent.
Equation (5) then reduces to

$$\frac{d\sigma_x}{d\varepsilon} = \sigma_x \left[\frac{\sigma_\chi - \sigma_f(\varepsilon, \dot{\varepsilon})}{1 + \gamma'(\varepsilon)} \right] \left[1 + \frac{\gamma'(\varepsilon)\cot\alpha}{2} \right] \qquad (9)$$

The analysis presented above was used to compare experimental and numerically calculated extrusion pressures for linear polyethylene (LPE), polyoxymethylene (POM) and polymethylmethacrylate (PMMA). Figure 9.5 shows some collected results from which it can be seen that the observed very rapid rise in pressure at high R_N is reproduced theoretically. There are a number of points to be made about the analysis.

First, a numerical procedure was adopted, based on the finite difference method, connecting the force balance of equation (9) for successive sections throughout the conical die. In the die, the strain and strain rate were determined from equations (6) and (7). The flow stress was then determined for each section from experimental data, such as those illustrated in Figure 9.4. It was shown that the redundant work due to shear at the die entry and exit has a negligible effect on the final computed extrusion pressures and the redundant shear strain has a negligible effect on the flow stress. However, the redundant strains do modify the strain boundary conditions significantly, and should therefore be taken into account.

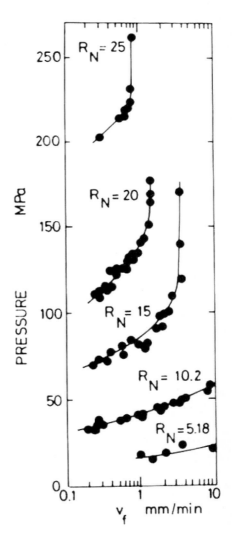

Figure 9.5   Best analytical fits to experimental extrusion pressure – extrusion ratio data [25]   © Chapman & Hall. Reproduced by permission.

Secondly, it is clear that the rapid rise in pressure arises from three effects:

(1)   Strain hardening, because in all polymers the flow stress increases with molecular orientation (with extrusion ratio) i.e. towards the die exit.

(2)   Strain rate sensitivity, because in polymers there is increasing strain rate sensitivity of the flow stress with increasing orientation i.e. extrusion ratio.

(3)   Pressure dependence of the flow stress. Because there is very likely a link between the friction and pressure coefficients as proposed by Briscoe and Tabor [28], these are both very significant at the die exit where boundary lubrication occurs. (It is likely that hydrodynamic lubrication occurs at the die entrance giving lower friction).

These effects are very well illustrated by Figure 9.6 which shows the results obtained for the hydrostatic pressure required to give a given extrudate velocity v_f for a series of imposed deformation ratios R_N , for a low molecular weight polyethylene (BP Rigidex 50, $M_w\sim700,000$). In all cases the die semi angle was 15° (a very usual choice) and R_N determined by changing only the initial diameter of the billet, keeping the die and die exit diameter constant. For low R_N, there is a linear relationship between pressure and extrudate velocity, implying that the extrusion process is akin to that for a polymer of constant viscosity. For $R_N=15$ and greater, a very rapid rise in pressure is observed, especially if it is attempted to increase the extrudate velocity beyond a critical value. The general increase in pressure with increasing R_N is due to the increasing strain hardening and strain rate sensitivity, as illustrated in Figure 9.4. The very rapid upturn in pressure can be attributed to the pressure dependence of the flow stress which means that any attempt to reach high extrudate velocities by increasing the pressure becomes counter productive.

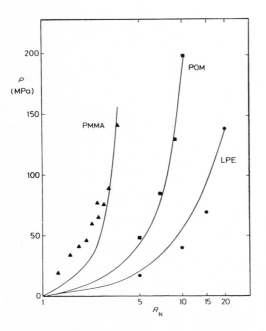

Figure 9.6.   Relationship between pressure and extrusion velocity at 100°C for 2.5 mm R50 LPE extrusions at different values of R_N [23]. © Chapman & Hall. Reproduced by permission.

Gupta and McCormick [29], in an alternative approach, have presented an upper bound analysis of the hydrostatic extrusion of linear polyethylene. They showed that a completely analytical solution can be obtained on the basis of results presented by Coates and Ward[24] where the dependence of the flow stress on strain and strain rate is represented by the Eyring equation for a thermally activated process. The strain rate $\dot{\varepsilon}$ is given by

$$\dot{\varepsilon} = \dot{\varepsilon}_o \exp-\left(\frac{\Delta U - \tau_f v + P\Omega}{kT}\right) \qquad (10)$$

where
$\dot{\varepsilon}$ is a pre-exponential factor,
ΔU is the activation energy,
τ_f is the shear flow stress
v is the shear activation volume
P is the hydrostatic pressure
Ω is the pressure activation volume
k is Boltzmann's constant
and
T is absolute temperature.

As discussed by Hope and Ward [25], equation (10) is a very useful way of representing the flow stress behaviour, providing that the increase in strain rate sensitivity with strain is taken into account by making v dependent on strain. In linear polyethylene, for R_N up to 9, v can be considered to decrease exponentially with strain, so that an analytical solution to equation (7) can be obtained. Gupta and McCormick showed that this assumption gave a good first order fit to hydrostatic extrusion data for LPE.

For PMMA, Inoue and co-workers [30] showed that it was adequate to assume that there is negligible strain hardening and strain rate sensitivity up to $R_N = 2.5$, and that a constant pressure coefficient and constant friction coefficient could also be assumed. The lower bound approach then produced a reasonable fit to the hydrostatic extrusion data.

9.2.3 Hydrostatic extrusion as a possible engineering operation

Hydrostatic extrusion has two major limitations in terms of developing a cost-effective process for oriented polymers. First, it is a batch process, so that the production rates are much reduced over those for continuous processes such as tensile drawing or die drawing (to be discussed). Secondly, the analysis of the mechanics of the hydrostatic extrusion process shows that under isothermal conditions small diameter material can only be produced very slowly (rates ~1cm/min) if significant draw ratios are to be obtained. For non-circular sections such as I-beam sections or tubes, the processing conditions are less favourable than for circular rods, due to higher strain rates in the deformation zone and increased die friction. There are, however, two possible advantages for large scale hydrostatic extrusion at higher extrusion rates. First, for a given strain rate, the exit velocity is proportional to the exit die diameter. Secondly, as shown by Hope and Parsons [31], stable hydrostatic extrusion can be achieved at rates ~50cm/min in an adiabatic regime, because the extrusion pressure is markedly reduced by adiabatic heating of the extruding billet. Similar results were obtained by Inoue and co-workers [32]. The results obtained by Hope and Parsons and by Inoue are shown in Figure 9.7.

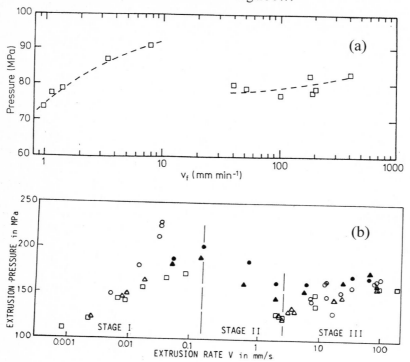

Figure 9.7. (a) Extrusion pressure versus extrudate velocity for polyethylene – large scale R40 polyethylene extrusion (T_N = 90°C, product diameter 15.5 mm). I.M. Ward, Adv.Polym.Sci. 70, 1 (1985). © Springer - Verlag. Reproduced by permission. (b) Relationship between extrusion pressure and extrusion rate for hydrostatic and ram extrusion of LPE with R_N = 7 at room temperature. Hydrostatic Extrusion: ○ steel die, Δ polycarbonate die, □ nylon die. Ram Extrusion: ● steel die, ▲ polycarbonate die (Ref. 32). © Chapman & Hall. Reproduced by permission.

9.2.4 Hydrostatic extrusion of pressure annealed polyethylene

An interesting scientific development, which could have some practical applications, is the production of high modulus polyethylenes by the hydrostatic extrusion of polyethylenes which have been annealed under high pressures. In a series of publications [32-35], Ward and co-workers showed that the optimum results were obtained when samples are annealed to produce substantial increases in lamellar thickness but without too great a reduction in the integrity of the molecular network [32]. For these optimum conditions the crystallites are rapidly aligned by hydrostatic extrusion, following close to the pseudo-affine deformation scheme, with the c-axes of the crystallites (i.e. the chain axes) rotating towards the deformation direction like lines joining pairs of points on the macroscopic billet. This means that high orientation and high moduli are achieved for comparatively low deformation ratios ($R_N \sim 10$ compared with the very high deformation ratios $R_N \sim 30$ required for deformation of a conventional melt crystallized polymer).

Collaborative research between Ward and co-workers and Bassett and co-workers enabled these ideas to be quantified [34,35].   Four different grades of polyethylene with molecular weights $\overline{M}_w$ in the range 100,000 - 300,000 were examined as a function of the morphology produced by annealing in the vicinity of the orthorhombic - hexagonal phase boundary.   It was shown that annealing within the hexagonal phase to produce a chain-extended morphology, is appropriate for high molecular weight i.e. a very high degree of lamellar thickening can be produced without reducing the effectiveness of the hydrostatic extrusion step. For low molecular weight, better overall results are obtained by annealing in the transition region between the hexagonal phase and the orthorhombic phase, because although the lamellar size can be increased by annealing in the hexagonal phase, this occurs at the expense of reducing the integrity of the molecular network, so that for high lamellar thicknesses the modulus falls and eventually the samples lose their mechanical coherence. For the optimum results there has to be a balance between increasing crystal size and loss of network integrity. It appears that the critical parameter is the ratio of the lamellar size to the number average molecular chain length. As shown in Figure 9.8, the optimum value for this parameter is about 0.5, presumably because this marks the limit where the molecules are incapable of completing two crystalline transverses and the crystalline regions no longer provide the required network coherence.

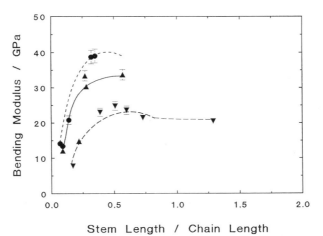

Figure 9.8   Variation in bending modulus of polyethylene extrudates as a function of the ratio L/M_N. The different grades have M_w x 10^{-3} of ▲ 135, ● 312, ▼ 742.   All samples annealed at 234°C for 1 hour before extruded to a fixed deformation ratio of 7 [35].
© Chapman & Hall. Reproduced by permission.

The hydrostatic extrusion of these chain extended polyethylenes is well described by a simple aggregate model of rotating crystalline units [36].   Figure 9.9 confirms that the crystal orientation is close to the pseudo affine deformation scheme. Figure 9.10 shows that the mechanical behaviour is well described by the supplied results form of the aggregate model, where the relationship between the tensile E and show moduli G is given by

$$\frac{1}{E} = \frac{1}{G} \langle \sin^2 \theta \rangle + \frac{1}{E_c}$$

where θ is the angle between the longitudinal direction of the aggregate unit and the extrusion direction, and E_c is the tensile modulus of the aggregate unit. Values of $\langle \sin^2 \theta \rangle$ can be obtained from plots such as Figure 9.10, and compared with values obtained by wide angle X-ray diffraction, (e.g. Table 9.1). The values are generally within experimental error, with the WAXS values systematically lower, which can be attributed to a small spread of crystallite orientation within the chain extended crystalline units.

It is important to note that the single phase aggregate model did not apply to the extruded high molecular weight samples. In these samples two distinct phases exist and a more appropriate model is the fibre composite model, which was shown to provide a reasonable fit to the data [36].

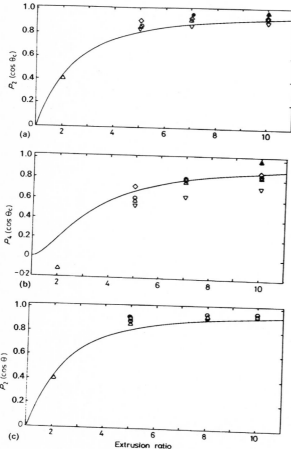

Fig. 9.9   Relationship between orientation function and extrusion ratio for various extrudates. (a) $\langle P_2(\cos \theta_c) \rangle$ relationship with extrusion ratio. (b) $\langle P_4(\cos \theta_c) \rangle$ relationship with extrusion ratio. (c) The effect of annealing time on the $\langle P_2(\cos \theta_c) \rangle$ relationship with extrusion ratio for R006-60.

(– pseudo affine, ● R006-60, 234°C, 5 mins, 450 MPa, △ R006-60, 234°C, 15 mins, 450 MPa, ○ R006-60, 234°C, 60 mins, 450 MPa, ◊ H020-54P, 234°C, 5 mins, 450 MPa, ▽ R006-60, 236°C, 60 mins, 450 MPa, □ R006-60, 234°C, 120 mins, 450 MPa, ▲ Conventional melt-crystallized polyethylene.) A.K. Powell, G. Craggs and I.M. Ward, J.Mater.Sci. 25, 3990 (1990).
© Chapman & Hall. Reproduced by permission.

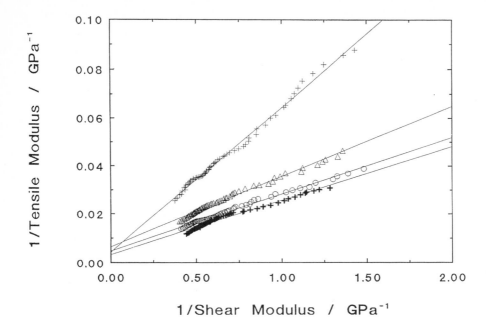

Figure 9.10 Application of aggregate model to oriented samples of polyethylene, Grade A. All samples annealed at 500 MPa and 234°C for 1 h before being extruded to a deformation level of +, 5; Δ, 7; ○, 10, and +, 14 (Ref. 36). © Elsevier Science Ltd. Reproduced by permission.

Table 9.1   Orientation moment averages, $<P_2(\cos \theta)>$, for extruded samples.   All samples annealed at 500 MPa, 234°C for 1 hour.

Extrusion ratio	WAXS	Aggregate model
5	0.85 ± 0.02	0.86 ± 0.03
7	0.90 ± 0.02	0.92 ± 0.03
10	0.91 ± 0.02	0.94 ± 0.03
14	0.92 ± 0.02	0.95 ± 0.03

9.2.5   Hydrostatic extrusion of filled polymers

9.2.5.1 Glass filled polyoxymethylene

A major advantage for the hydrostatic extrusion of filled polymers is that the maintenance of a positive hydrostatic pressure during the deformation ensures that there is no reason for the breakdown of bonding between the filler and the matrix due to the presence of a tensile stress [37].   A second important advantage in the case of glass-fibre filled polymers is that a high degree of alignment of the glass fibres can be achieved. For short glass fibre filled polyoxymethylene (POM) , Curtis et al [38] showed that the alignment of the glass fibres followed the pseudo-affine deformation scheme and that the development of mechanical stiffness could be described to a good approximation by the aggregate model. As proposed by Brody and Ward [39] for an oriented fibre composite it is necessary to consider the aggregate unit as a fibre surrounded by the polymer matrix. Recently this approach has been developed further in extensive studies at Leeds University [40,41]. Fibre orientations were measured using an image analysis method developed by Clarke and co-workers [42], and the matrix orientation by WAXS.   It was confirmed that the development of fibre orientation was close to the pseudo-affine deformation scheme, although the fibre orientation was greater than predicted at low draw ratio and slightly less at the highest draw ratio [40]. As found previously by Curtis et al [38] the development of orientation in the crystalline regions of the POM matrix was always significantly greater than that predicted by the pseudo-affine scheme. A further point noted by the image analysis was that the average fibre length was reduced from 150μm to 133μm by the hydrostatic extrusion process.

In a further publication Hine et al [41] compared a complete set of elastic constants for the extrudates, obtained by the measurement of ultrasonic velocities, with those predicted by a modification of a theory due to Wilczynski where the sub units of the aggregate model consist of fibres are surrounded by an oriented matrix phase, together with the Ward aggregate model. With these rather sophisticated modifications to the original Brody and Ward theory, excellent agreement could be obtained between the measured and predicted elastic constants (Table 9.2).

Table 9.2   Calculation of the Elastic Properties of the Most Highly Aligned Matrix

Elastic constant	Misaligned Properties (Measured)	Fully Aligned Properties		
		Lower Bound	Upper Bound	Average
E_{33}	6.62	6.68	7.12	6.90 ± 0.22
E_{11}	4.62	4.55	4.65	4.60 ± 0.05
v_{13}	0.406	0.405	0.415	0.410 ± 0.05
v_{12}	0.415	0.423	0.427	0.425 ± 0.002
G_{13}	1.83	1.78	1.82	1.80 ± 0.02

The fracture toughness of these oriented glass-filled POM samples was also studied by Hine et al. [43].   Unfilled and glass-filled POM extrudates were compared for a nominal extrusion ratio of 8, by undertaking fracture tests for samples with the compact tension geometry Samples were loaded parallel and perpendicular to the extrusion direction, and tests were carried out at +20 and -50°C. The glass-filled extrudate showed significantly higher fracture toughnesses than the unfilled extrudate, and stable rather than unstable crack propagation even at -50°C. The values of K_c were in the range 5.3 - 6.4 Mnm$^{-3/2}$ for the filled extrudate compared with 1.7 - 5.4 Mnm$^{3/2}$ for the unfilled extrudate.

9.2.5.2 Polyethylene filled with hydroxyapatite

Polyethylene reinforced with hydroxyapatite (HA) has been pioneered as a bone substitute by Bonfield and co-workers [44,45]. An optimum composition for mechanical and biological performance is obtained by an HA content of 40vol%, and this has been commercialised under the name HAPEX™ by Smith and Nephew, Richards USA. Clinical applications have been developed for middle ear implants and orbital floor reconstructions [45]. Although HAPEX™ does not have the stiffness and strength to be used in major load-bearing applications, it has been shown that very significant improvements can be produced by hydrostatic extrusion of the isotropic material.

In a collaborative project between the IRC in Polymer Science and Technology, University of Leeds and the IRC in Biomedical Materials, Queen Mary and Westfield College, University of London, several routes have been developed for the production of load-bearing materials based on PE and HA. The simplest route is the extrusion of PE/HA mixtures, including the HAPEX™ 40 vol % composition. Most usually the initial cylindrical billets were compression moulded, heated to 175°C and then cooled slowly under gradually reducing pressure. The hydrostatic extrusion of such billets was similar to that observed for unfilled polyethylene, and it was generally desirable to encapsulate the billets in a polypropylene sleeve to eliminate stress-cracking. A full account of the procedures, together with details of the flexural properties of the extrudates has been given by Ladizesky et al [48]. Extruded products with flexural moduli of 10GPa and flexural strengths of 90MPa were produced, which can be considered as candidates for load-bearing substitute materials.

Table 9.3   HA/PE systems for bone substitute material

No Hydrostatic Extrusion	Material		Flexural Properties		
			Modulus (GPa)	Strength (GPa)	D(%)
Melt Extruded	40 HA/60 PE		4	32	1.6
Multiple Sandwich	15 woven layers/(2 mm) + 60 HA/40 PE		14	107	4.2
Cortical Bone			7-30	50-150	0.5-3.0

With Hydrostatic Extrusion	HA Content Vol %	Extrusion Ratio	Flexural Properties		
			Modulus (GPa)	Strength (GPa)	D(%)
Melt Extruded	40	5:1	7	76	6.2
		8:1	9	83	6.0
		11:1	11	75	3.8
HA/Chopped Fibre	30	7:1	16	104	3.0
		11:1	20	117	2.8
HA/Chopped Fibre	50	7:1	12	72	2.8
		11:1	14	69	1.6

In a more elaborate procedure [49], Ladizesky et al showed that hydrostatic extrusion of HA/high modulus PE fibre composites could give even higher mechanical properties i.e. flexural moduli approaching 20GPa and flexural strengths greater than 100MPa. Table 9.3 summarises the results for all the bone substitute materials.

At present, the exploitation of these interesting materials awaits the results of current biocompatibility trials.

9.2.6   Other properties of hydrostatically extruded materials

9.2.6.1 Thermal conductivity

The thermal conductivity of hydrostatically extruded PE has been studied by Gibson, Greig and colleagues [50]. There is a very significant increase in the thermal conductivity parallel to the extrusion direction (Figure 9.11), which is similar to the increase in Young's modulus, and has been shown to relate to the degree of crystal continuity i.e. a simple Takayanagi model can provide a quantitative description of the thermal conductivity as in the case of the modulus.

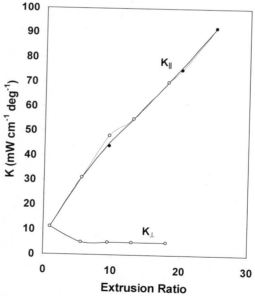

Figure 9.11 Variation of thermal conductivity of hydrostatically extruded R50 LPE samples at 100 K for various values of extrusion ratio.

In more physical terms, the thermal conductivity in the extrusion direction relates to the phonon velocity, which increases with increasing Young's modulus and to the phonon mean free path, which is increased due to the increase in crystal length. The thermal conductivity perpendicular to the extrusion direction is little changed, as would be anticipated on these simple arguments.

9.2.6.2 Thermal expansion

There is a large anisotropy in the thermal expansion behaviour of the oriented extrudates. For unannealed PE and PP the coefficient of linear thermal expansion in the extrusion direction α_{11} is small and negative $\sim -10^{-5} \text{K}^{-1}$ for significant draw ratios (~ 10), but that in the direction perpendicular to the extrusion direction $\alpha_{\perp}$ remains similar to that for isotropic material i.e. $+10^{-4} \text{K}^{-1}$.

The negative value has been explained by Orchard et al [51] on the basis that the structure of these oriented materials consists of two components acting mechanically in parallel. The first component is responsible for the stiffness of the polymer and consists essentially of the various elements of the crystalline bridge model, lamellae and long crystals or crystalline bridges. The second component is a large scale stretched molecular network which has been shown to give rise to a very significant internal stress [52]. With increasing temperature the first component loses its stiffness and the retractive force of the network becomes more important. Orchard et al set up a qualitative model to show that

$$\alpha_{11} \propto \frac{1}{E}\left[\frac{T}{E}\frac{dE}{dT} - 1\right]$$

where E is the modulus and T is absolute temperature. With falling temperature the values of α_{11} for all high draw samples converge to the c-axis expansion value of $-12 \times 10^{-6} \text{K}^{-1}$.

9.2.6.3 Barrier properties

Measurements of gas barrier properties show that there are significant reductions in the permeability for oriented polymers, especially for uniaxial orientation. Results have been obtained for PE [53], PP [54] and PET [55,56]. This also applies to equilibrium sorption and diffusion of solvents [57]. Chemical resistance to acids and alkalis is also excellent [58].

9.3   Die-drawing

9.3.1   The die-drawing process

The die-drawing process is a technique which combines the best features of free tensile drawing and hydrostatic extrusion enabling large section products to be made at the high modulus levels obtainable by free drawing. Die-drawing has several advantages over hydrostatic extrusion. The main disadvantage of hydrostatic extrusion through a conical die is that the polymer experiences the highest strain rates at the exit of the die where the plastic strain is greatest. Coates and Ward [59] have shown that the strain-rate sensitivity of flow stress in solid-state extrusion increases rapidly with plastic strain, a situation which incurs very high flow stresses as the polymer reaches the die exit, high extrusion pressures therefore being required. [By contrast, in die-drawing the polymer necks down and follows an optimal strain rate field.] This strain rate field is such that the highest strain rates are encountered at low levels of plastic deformation, consistent with a constant load at all sections of the final product.] Only the polymer in contact with the die wall surface suffers a strain rate field imposed by the die geometry. In the die wall contact area the material is of low deformation ratio; hence the strain rate effects on the flow stress are comparatively small and not detrimental to the process.

It follows from the above considerations that production rates in hydrostatic extrusion decrease with increasing deformation ratio, whereas in die-drawing the higher deformation ratio products are obtained at higher draw speeds.

The die-drawing process is shown schematically in Figure 9.12. A heated polymer billet is drawn through a heated conical die by applying a pulling force on the billet at the exit side of the die. The polymer billet is free to neck down and follows an optimum strain and strain rate path through the die, leaving the die wall at an appropriate point, hence reducing the required flow stress considerably. This leads to three distinct deformation regimes:

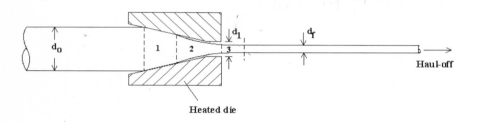

Figure 12   Schematic diagram of the die-drawing process: 1, 2, 3 are the drawing zones

Zone 1: Isothermal conical die flow; the initial region where contact with the die wall is maintained.

Zone 2: Free drawing within the die, at approximately the die temperature, but some adiabatic heating is likely.

Zone 3: In this region continuous deformation takes place outside the die until the polymer cools and reaches its final deformation ratio.

The exact positions of the boundaries between the zones are difficult to define in practice. In particular, the position at which the polymer separates from the die wall depends on the value of nominal draw ratio, R_N, of billet, the die temperature, and the imposed haul-off speed.

The degree of deformation imparted to the material by the die-drawing process is characterised by the actual draw ratio, R_A, which is expressed in terms of the initial and final cross-sectional dimensions of the product.

R_A = (Initial billet cross-sectional area) / (Final product cross-sectional area)

For circular cross-section rods,

$$R_A = (d_o / d_f)^2$$

where d_o = billet diameter
d_f = final product diameter.

The initial billet size is characterised by R_N, the nominal draw ratio, which is defined as the ratio of the billet cross-sectional area to the die exit:

$$R_N = (d_o / d_1)^2$$

where d_1 = die exit diameter.

The actual deformation ratio is always greater than the nominal deformation ratio since the polymer draws away from the die wall during the process.

This very simple process was first demonstrated successfully on the large scale for polyoxymethylene, then on the small scale for polypropylene [60] with the equipment which was designed for use on an Instron tensile testing machine. This die-drawing apparatus is shown in Figure 9.13.

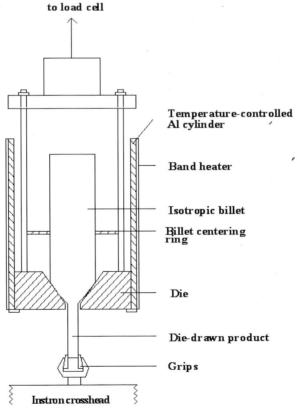

Figure 9.13. Schematic diagram of small-scale die-drawing assembly

A conical steel die of semi-angle 15° was used. The die was housed in an aluminium block fitted with a band heater and a thermocouple for temperature control. The aluminium block containing the die extended for a sufficient distance upstream to ensure the uniform temperature of material reaching the die. On the Instron tensile testing machine, jaws of serrated self-gripping type were used to draw the product through the die. A continuously variable haul-off speed over the range of 0.5 to 50 cm/min was available. Draw loads were measured using the load cell on which the die holder block was mounted.

Two key results for the die-drawing process are illustrated in Figures 9.14 and 9.15 for polypropylene [60]. In Figure 9.14 it can be seen that as the imposed draw speed is increased from ~ 10 mm/min to 50 cm/min, the actual deformation ratio increases. In the case of the die designed to give $R_N = 7$, the actual deformation ratio R_A rises to 20. The second key result is shown in Figure 9.15, where it can be seen that to a very good approximation, there is a unique relationship between the axial Young's modulus and the actual deformation ratio which is identical for die-drawing, hydrostatic extrusion and die-drawing.

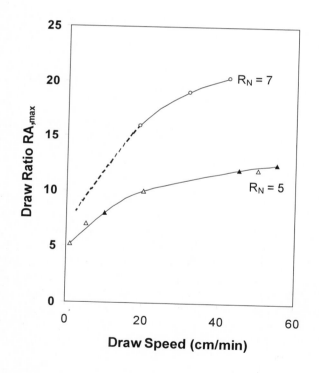

Figure 9.14. Dependence of maximum steady deformation ratio R_A on imposed draw speed for PP copolymer die drawn at a nominal temperature of 110°C. (○, $R_N = 7$, 15.5 mm die, △, $R_N = 5$, 15.5 mm die, ▲, $R_N = 5$, 7 mm die).

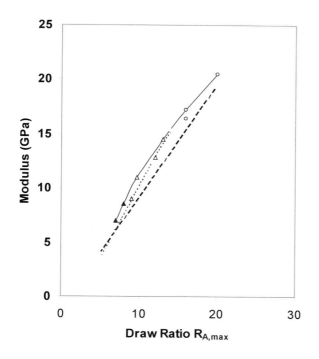

Figure 9.15. Axial Young's modulus versus draw ratio for PP copolymer die drawn at 100°C. (o, $R_N = 7$, 15.5 mm die, Δ, $R_N = 5$, 15.5 mm die, $\blacktriangle$, $R_N = 5$, 7 mm die). Tensile drawing results for PP homopolymer fibres (dashed line), hydrostatic extrusion of PP homopolymers (dotted line).

These interesting results on polypropylene together with similar results obtained for the die-drawing of polyethylene [61] were sufficiently encouraging to ensure the construction of a large scale die-drawing rig [62] capable of drawing large rod, sheet and tube cross-sections, in addition to thick monofilament, in lengths up to 11m. In this batch process polymer billets are heated to the required drawing temperature in a 1.8m long tubular heating chamber. The steel die is held in an independently heated block located in the die holder at one end of the heating chamber. A wide variety of die shapes and sizes are available, together with suitable mandrels for tube drawing. The product is drawn through the die gripped by a variety of devices, depending on the cross-section shape and size. The gripper unit is held in a load cell, and all of this is carried on a trolley mounted on ball bearings which locate in guide rails on a draw bench capable of giving a draw travel of 11m. The trolley is pulled by a chain which is hauled over the sprocket on the output shaft of a 1015:1, 1.5 HP reduction gearing box. The haul-off speeds typically range from 0.5 cm/min for start up, when the billet cone must be allowed to strain harden slowly, to about 2 m/min. Higher speeds have been obtained by simple gearing changes. A photograph of the die-drawing rig is shown in Figure 9.16.

Initial research on this large scale batch die-drawing rig was directed at obtaining a comprehensive picture of the die-drawing behaviour of a wide range of polymers by producing families of curves of the types illustrated in Figures 9.14 and 9.15 i.e. both R_A versus draw speeds and modulus versus R_A plots at a range of draw temperatures. Extensive studies of polyethylenes [62] (both homopolymers and copolymers), polypropylene [63], polyoxymethylene [64], polyvinylchloride, polyvinylidene fluoride [65], polyethylene terephthalate and polyar25yletherketone [66] demonstrated the viability of the process for a very wide range of polymers.

Figure 9.16   Large scale die-drawing machine; general view of tube drawing in progress.

The die-drawing rig can be readily adopted to draw sheet products from rectangular cross-section billets. Although this can be undertaken using a fixed geometry slit die, it has also been achieved by drawing through the nip of a pair of heated 65 mm diameter steel rollers. Such die-drawn sheets exhibit excellent mechanical properties in terms of fracture toughness and impact strength. They are also readily post formed in an appropriate temperature range well below the polymer melting point so that the structure and properties are retained. Charpy impact test results using flat unnotched specimens for polyoxymethylene (POM), linear polyethylene LPE and a high modulus PE fibre/epoxy composite are shown in Table 9.4.

Table 9.4 Sheet energy absorption over six consecutive impacts, Charpy test

Sample	Absorbed energy per unit area (kJ m^{-2}) Test No.					
	1	2	3	4	5	6
POM, R_A=7.5	152	152	152	152	152	152
LPE, R_A=14.7	159	99	84	76	71	68
Composite	170	72	51	44	39	37

Fracture toughness measurements have also been undertaken on die-drawn polyethylene sheets, selecting for study homopolymers of different molecular weight and molecular weight distribution and copolymers, such as that used for pipe production [67]. These sheets are biaxially oriented (effectively drawing takes place at constant width) and for the copolymer and a homopolymer with a bimodal molecular weight distribution showed values for both K_C^{long} (applied stress parallel to draw direction) and K_C^{trans} (applied stress perpendicular to draw direction) which were substantially greater than those for isotropic material (Table 9.5).

Table 9.5 Summary of the fracture test results

Grade	Average K_{IC} (MN m$^{-3/2}$)		
	Isotropic	Transverse ~ 9.5:1	Longitudinal ~ 9.5:1
Rigidex 006-60	1.8 ± 0.24	1.2 ± 0.16	8.0 ± 1.7
Rigidex 002-40	2.1 ± 0.17	5.0 ± 0.4	12.0 ± 1.3
Hizex 7000F	1.5 ± 0.1	4.6 ± 0.15	14.0 ± 1.0

9.3.2 Die-drawing of tube

The first experience with tube drawing used the batch rig [68] and initially the two different tooling configurations were used as shown in Figures 9.17(a) and 9.17(b).

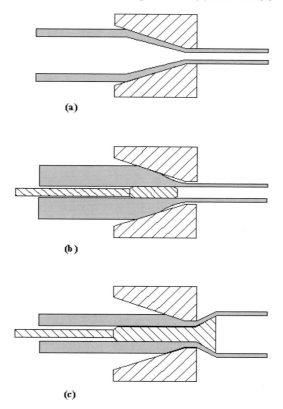

Figure 9.17. Tooling configuration for die-drawing tube: (a) drawing to reduced bore; (b) drawing to constant bore; (c) drawing to increased bore.

The simplest configuration, Figure 9.17(a), is similar to the die-sinking process normally applied to metals and involves drawing the tubular billet through a conical die of much smaller bore than the outside diameter of the billet. The second configuration (Figure 9.17(b)) is one where the tube is drawn over a fixed mandrel having nominally the same diameter as the bore of the tubular billet, similar to the die-drawing of sheet through a slot die.

It was soon found that there were substantial benefits to be gained by die-drawing to a greatly increased bore and the configuration for this is shown in Figure 9.17(c). In the case of expanded bore tube drawing, the hoop draw ratio of the product varies from the inner to the outer wall. These draw ratios are defined as:

$$R_{hoop,inner} = \frac{product\, bore}{billet\, bore}$$

$$R_{hoop,outer} = \frac{product\, outer\, diameter}{billet\, outer\, diameter}$$

The axial draw ratio of the drawn tube is defined as the ratio of the cross-sectional area of the tubular billet to that of the drawn tube, that is,

$$R_{axial} = \frac{(billet\, outer\, diameter)^2 - (billet\, inner\, diameter)^2}{(produt\, outer\, diameter)^2 - (product\, inner\, diameter)^2}$$

As the tube is to be hauled-off over the expanded mandrel, it is necessary to open out the leading end of the tubular billet beforehand. This is done by heating the tubular billet end to be deformed to the draw temperature and pulling into it a hot metal plug in the shape of a tapered-nose cylinder. The billet end and the plug are heated using a cartridge heater; this arrangement conveniently ensures that the body of the billet remains cool and stiff while the material at the end is soft. After a sufficient length of the billet has been deformed the end is cooled and the plug is removed.

In order to start the drawing, the tubular billet is loaded into the heating chamber through the die. The mandrel is screwed to one end of a long rod and the other end of the rod is fixed by the rear end cap closure of the billet heating chamber. The neck of the mandrel is a smooth fit in the bore of the tube which, in turn, is a smooth fit in the bore of the die. The die and billet chamber are heated to the desired drawing temperature. Billets are allowed sufficient time to attain thermal equilibrium before starting the drawing. The protruding end of the tubular billet is gripped by the jaws of the drawing winch. Initially the tube is drawn very slowly, a few millimeters per minute, until some oriented, hence, strain hardened material emerges. If necessary, the grip is then reattached to this stronger material and drawing is continued, initially at a slow speed, but then increased to the desired rate. The drawing load is monitored throughout the run.

Early biaxially drawn tubes produced on the batch die-drawing rig were often curved along their length and non-round along their cross-section. The production of biaxially oriented tube of commercial quality depends on a number of parameters in the drawing process being carefully controlled. The most important influences on the tube quality are those of dimensional accuracy of the feed stock and uniformity of temperature during the drawing operation. The isotropic tubular billet is heated in a chamber prior to being drawn over a conical mandrel. Variations in the temperature of the material as it is being drawn produce an uneven drawing force around the mandrel circumference resulting in variations in the wall thickness and curvature of the drawn tube. The modifications described below are designed to improve the temperature control of the drawing operation. The result of good temperature control is an improvement in the straightness and uniformity in wall thickness of the drawn product.

9.3.2.1 Rotating mandrel assembly

Figure 9.18 shows a longitudinal-section through the draw bench. The isotropic tube to be drawn is fitted over a rod and mandrel and supported in a heated tubular chamber. When the tube reaches the drawing temperature it is pulled over the conical mandrel by the haul-off mechanism. Heating to the chamber is provided by simple band heaters which fit around the circumference of the chamber. Tests have shown that there is variation in temperature within the chamber causing uneven drawing. The effect of this temperature gradient can be eliminated by arranging for the mandrel, mandrel rod, billet and drawn tube to rotate about the long axis of the machine during the drawing operation. This is achieved by two thrust bearings, one at the rear end of the mandrel support, and one in the clamp of the haul-off device. Rotation is provided by a slow speed electric motor coupled to the rear of the mandrel assembly. During the drawing operation, rotation ensures an even temperature throughout the material and perfectly straight and uniform tubes may be produced.

9.3.2.2 Heated mandrel

It is normal practice to preheat the conical mandrel prior to the commencement of the drawing operation. While this is normally satisfactory for the thick walled billets it is beneficial to be able to accurately control the mandrel temperature during the drawing operation especially when drawing thin walled tube. Figure 9.18 shows a section through the mandrel assembly. The conical mandrel is machined so that the cone is in two parts, an

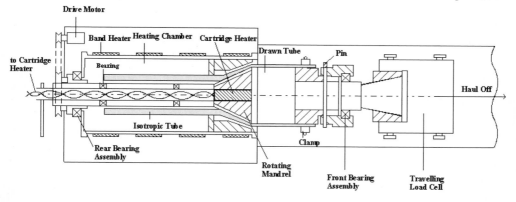

Figure 9.18. Modified batch die-drawing rig with rotating mandrel assembly

inner stationary cone and an outer rotating cone. The inner cone is machined to take a small cartridge heater and thermocouple. The wires from these pass back through a hollow shaft to the rear of the drawing bench. Miniature bearings support this tube within the outer stationary mandrel tube. The outer cone and tube are rotated in the same manner as described previously. Contact between the inner and outer cone is maintained by the axial force from a light spring fitted to the rear of the tubes.

9.3.2.3 Rotating mandrel and die assembly

It is normally the practice during the drawing operation to draw the material over the conical mandrel without any contact with the outer die assembly. Occasionally when high axial draw ratios are desired the mandrel can be drawn back against an outer ring squeezing the outer surface of the polymer against the cone. Figure 9.19 shows this technique adapted for use with a rotating assembly. The diameter of the stationary heating chamber has been enlarged to enable a bearing to be located inside. Attached to this bearing is a ring which can rotate freely. This ring is used to squeeze the polymer as it is being drawn over the mandrel. The mandrel assembly is rotating and when squeeze is applied the ring will also rotate, although remaining stationary with respect to the drawn product.

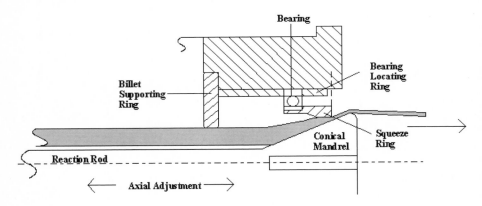

Figure 9.19   Die-drawing set-up for tubes with rotating squeeze ring.

9.3.2.4 Post-mandrel cooling

When the tubes are drawn at higher speeds the hoop draw ratio decreases as the tube draws down after leaving the mandrel. This loss of hoop expansion can be minimised by cooling the product as it leaves the mandrel. For this purpose a vacuum operated water cooler has been made and is shown in Figure 9.20. This water cooler is placed next to the mandrel end and the product is drawn through it. The cooling water is circulated through the cooler at room temperature at the rate of 2 litres/min. Drawn thick products have been made up to a draw speed of 2 m/min.

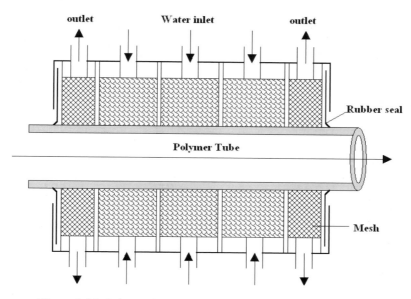

Figure 9.20. Schematic diagram of multi stage water cooler.

9.3.2.5 No contact air-bearing mandrel

In the simple tube expanding process, the feedstock is pulled over a conical mandrel, and the hot polymer slides directly over the metal surface. For gas and water pipes drawn from polyethylene a good quality internal surface arises naturally. But when PET is drawn over a solid mandrel the surface of the product is slightly marked by its contact with the metal.

A mandrel was developed which incorporates an air bearing which 'floats' the polyester tube over the mandrel without contact, so as to yield a glass-clear product. This is possible in this particular instance because the draw stress of PET is low, and the product is thin walled. It was found that an air pressure 50 psi is sufficient to give complete separation between the mandrel and the tube being drawn over it. Figure 9.21 shows the air-bearing mandrel. Air at 50 psi and 100°C is fed down the centre tube and out through the annular slot of width 50 - 100μ. The mandrel is approximately hemispherical in shape so that the tube being drawn is everywhere pulled onto the surface of the mandrel by the drawing tension.

Figure 9.22 shows the drawing equipment whereby glass clear PET tube may be made. A die is shown but it functions mainly as a centralising and supporting device. The squeeze between the die and the mandrel is quite low, only enough to stop the lubricating air from flowing upstream. A cooling device is shown, to prevent the freshly drawn hot tube losing diameter as it leaves the mandrel. The starting material in this example is quenched bottle grade PET tube 25mm OD and 16mm bore. The product is a tube 75mm diameter with a wall thickness of about 0.4mm.

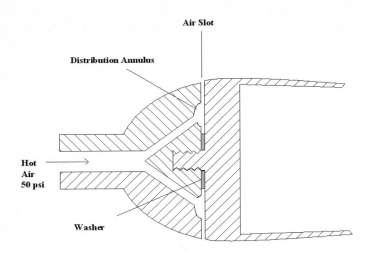

Figure 9.21   Mandrel with air-bearing.

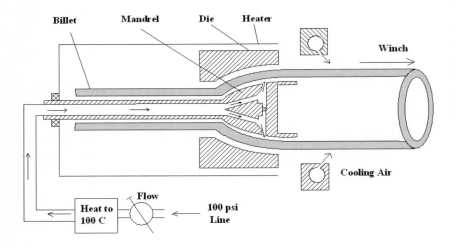

Figure 9.22   Drawing set-up for glass clear PET tube

9.3.2.6 Drawing square and rectangular profiles

It is known from experience on die-drawing of tubes that most of the materials draw down after leaving the mandrel at higher drawing speeds. This fact can be exploited to shape the drawn product by drawing over an extension of that shape (convex polygonal) beyond the mandrel.

Commercially available PVC tubes of OD 42mm and bore 32mm were used. The belling and start-up procedure is same as in simple biaxial drawing of tubes. PVC tube was drawn at 100°C over a 60mm mandrel. Drawing was started very slowly so that the tube will not draw down much. After drawing about 30cm long product, drawing was stopped and the product was cut near the grips. The square extension (40x40x200mm) was inserted and screwed in position on the mandrel. The extension was hollow and an insulating disc was placed between mandrel and extension. The leading end of the product was regripped and drawing commenced. The drawing speed was increased slowly so that the round product draws down over the extension and takes a square shape. The square tube was cooled over the extension end to retain the shape. A long length of square tube was successfully made. The drawing load increases from 210Kgf to 350Kgf on drawing over the extension, due to friction. To reduce the friction an extension can be made with concave surfaces.

Alternatively biaxially oriented round PVC tubes can be heat shrunk over convex polygonal shapes (as long as the cross-section area of the convex polygonal is greater than that of the isotropic billet). Similar work has been carried out with polypropylene and polyethylene tubes.

9.3.3   Development of the continuous die-drawing process

Although the batch die-drawing rig is capable of producing drawn lengths up to 11m long at drawing speeds of 2 m/min and drawing force of 2 tonnes, to realise the full potential of the die-drawing process it was appreciated that it was necessary to construct and put into operation a continuous process [60,69]. It was also decided that it would be most useful if this facility could be designed to produce tube, in addition to rod and sheet, which would clearly be simpler operationally.

The prototype facility developed at Leeds University is shown in Figure 9.23. The extruded polymer tube is cooled in a cooling bath and drawn in the die-drawing chamber between two caterpillars. A mandrel is located in the centre of the die to expand the tube in the hoop

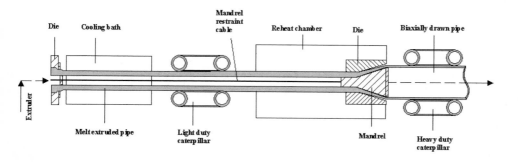

Figure 9.23. Continuous die-drawing line for biaxially drawn tube

direction producing hoop orientation, while the axial haul-off produces axial extension and axial orientation, i.e., a biaxially oriented product is made.

A prototype line for continuous controlled production of specified dimension, oriented polymeric rods and monofilaments has also been developed. This was based upon a combined single screw melt extrusion and solid phase die-drawing facility. Temperature profiles in the extruded rod, prior to the die-drawing operation, which are important with respect to the effectiveness, stability and control of the process, were predicted by finite difference modelling. These results were used to help optimise the thermal design of the process.

The cooling, solidification and uniform reheating of the extruded rod prior to the die-drawing determines the length of the entire die-drawing line for a given production rate. A finite difference method was used to predict the temperature profile of the cross-section of the rod through these phases. The latent heat of solidification was taken into account but the temperature dependence of density, specific heat and thermal conductivity were not. The temperature profile of a 18mm diameter rod with a 1mm diameter bore is shown in Figure 9.24. On the basis of these results the lengths of the water cooling bath and the hot air oven were decided.

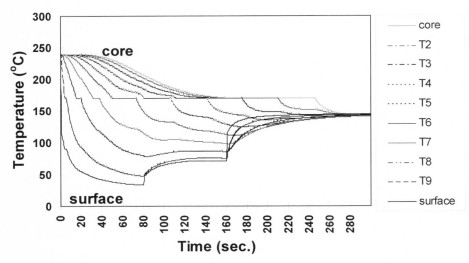

Figure 9.24. Temperature profile of the 18 mm diameter polypropylene rod. Extrusion temperature 240°C. Water bath temperature 20°C, residence time 80 seconds. Ambient temperature 20°C, residence time 80 seconds. Hot air oven temperature 150°C.

In the conventional die-drawing process the actual draw ratio is usually greater than the nominal draw ratio set by the die as the polymer draws away from the die wall. Further drawing occurs beyond the die exit as well. This causes the lack of control over the dimensions of the drawn product. To overcome this problem the drawing was carried out in two stages. A schematic diagram of this set-up is shown in Figure 9.25. In the first stage the conventional die-drawing is carried out and the die and the billet are heated to the desired drawing temperature. As the hot drawn rod leaves the first die it passes through the second die which is attached at the front of a water cooling bath which shapes the product to right size and stop any further drawing. The nominal draw ratio imposed at the first die is roughly half the final desired draw ratio. The gap between the two dies can be adjusted during the drawing so that the second die imposes the minimum required deformation to maintain the desired size at the high drawing speeds of up to 20m/min.

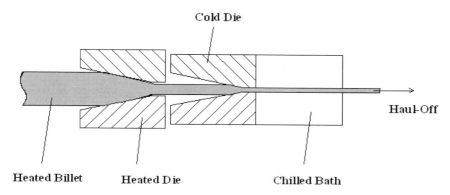

Figure 9.25 Schematic diagram of die-drawing and shaping process

A prototype line for a continuous die-drawing process [70] was designed and built for Bridon plc. at Doncaster. The line layout is shown in the Figure 9.26. Melt extruded rod is fed into a cooling bath, before being continuously hauled off by a light duty caterpillar and fed into a heating chamber. The heated rod is first drawn through a die which is fixed at the exit end of the heating chamber and then passes through a second shaping die which is attached at the front of a chilled water bath. The drawn product is continuously pulled by a second heavy duty caterpillar.

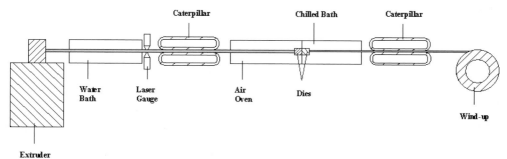

Figure 9.26   Schematic layout of the continuous die-drawing plant

The process parameters are monitored and stored for analysis throughout the experiment on the computer. These parameters include the temperatures of extruder zones and die, die pressure, rod diameter and ovality by a laser gauge, first caterpillar speed, temperature of heating chamber, temperature of chilled bath, die pull, product diameter and final haul-off speed of the second caterpillar.

The extruder was heated for two hours to achieve constant temperatures throughout. On starting the extruder the molten polymer was drawn by hand and linked to the first caterpillar. Initially the caterpillar was set at higher speed so that the rod diameter was smaller than the die-drawing die. Once the line was linked through the dies to the second caterpillar, the speed of the first caterpillar was reduced to make the desired diameter rod and hence die-drawing commenced. The volume of polymer output passing through each caterpillar was carefully matched by controlling their speeds.

Extruded rod of 11.5mm diameter was drawn at 150^0C through a first die of 8mm diameter and then drawn through a second die of 6mm diameter attached at the front of cold water bath. For comparison, the 11.5mm diameter rod was also drawn directly through a 6mm diameter die at 150^0C and no cooling on the product was applied. The variation of the product diameter with the draw speed is shown in Figure 9.27. It can be seen that by using the two dies arrangement with the cold second die leading the drawn product directly into a cold bath reduces the amount of draw down substantially the product diameter does not vary with the draw speed.

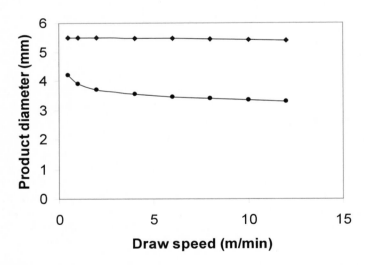

Figure 9.27. The variation of the product diameter with the draw speed. (●) conventional die-drawing, (♦) two-dies.

Monofilament of 2mm diameter at draw ratio 9 was produced at a drawing speed of 22 m/min. Over the long production run the variation in the product diameter was within ± 3%. Rod of 6mm diameter at draw ratio 6 was produced at a draw speed of 10m/min. The product quality was very good and the diameter variation was within ± 1.5% over the five hours long production run. The Young's modulus and the tensile strength of these drawn samples increased with the draw ratio. An isotropic sample had modulus of 1 GPa and a tensile strength of 30 MPa. A sample of draw ratio 9 gave a modulus of 6 GPa and a tensile strength of 350 MPa.

9.3.4 Mechanics of the die-drawing process

The mechanics of the die-drawing process has been analysed for rod and for tube. The approach in all cases was similar to that described for hydrostatic extrusion, and start by analysing the stresses in a small element of material in the die. An equation similar to Eq. (5) above was obtained with

$$\frac{d\sigma_x}{dD} = \frac{2}{D}[\sigma_x(1-B)+\sigma_f B] \tag{11}$$

where σ_x is again the axial stress, D, is the diameter of billet at the point x, σ_f is the flow stress and

$$B = \frac{1+\mu\cot\alpha}{1-\mu\cot\alpha} \tag{12}$$

This approach differs slightly from Equation (2), where the approximation that the normal pressure is exactly σ_y meant that B = 1 + μ cot α. Equation (11) was derived on the assumption that σ_x - σ_y = σ_f, i.e., a simple Von Mises or Tresca yield criterion.

It was shown by Motashar et al. [71] that for die-drawing, hydrostatic pressure effects are negligible. The strain and strain rate at any point in the die can be obtained from geometrical considerations, with relationships exactly analogous to Eqs (6) and (7) above. The stress-strain-strain rate characteristics of the polymer were deduced from uniaxial tensile data obtained at the same temperature as the die-drawing temperature. The axial strain along the die was determined using the "frozen" grid pattern on the flat faces of a split billet on which parallel grid lines 2 mm apart had been scribed, filled with ink, and dried. The two halves of the billet were fastened at these ends before drawing through the die, after which the billet was cooled and the "frozen" grid pattern observed. To a very good approximation plug flow was obtained. The axial stress distribution throughout the die was determined by successive reductions of the length of the die, by slicing material from the die at its exit, and determining the axial draw stress at the die exit for different lengths of die. This is equivalent to determining the axial stress at different points in the original die. Comparison of experimental results and the analytical predictions showed good agreement.

In a further analysis of the die-drawing of rod, Kukereka et al. [72] have taken into account thermal effects due to the heat generated by plastic deformation. In the absence of heat conduction the temperature rises even for moderate haul-off speeds ($\approx$ 10 mm/min for initial billet diameters of 40 mm) can be significant ($\approx$ 10°C), but these are substantially reduced when heat conduction is taken into account. Good agreement was obtained between the theoretically predicted drawing stresses and those determined experimentally for a range of drawing speeds.

The mechanics of the die-drawing process have also been analysed for biaxially oriented tube. Craggs [73] developed the force balance approach for this case also and showed that for PET a satisfactory analysis could be based on the Tresca yield criterion using uniaxial stress-strain data for amorphous PET produced by Foster and Heap [74]. A similar analysis has been undertaken by Kakadjian et al. [75] for biaxially oriented PVC tube. In this case the biaxial deformation was modelled by the Ogden equations for hyperelastic behaviour, which had been shown by Sweeney and Ward [76] to give a good representation of stress-strain behaviour of PVC at temperatures above the glass transition.

9.3.5 Properties of die-drawn products

The die-drawing process improves the mechanical properties of most thermoplastics. The improvements in the modulus of die-drawn rods and sheets compared with the various undrawn polymers are given in Table 9.6.

Table 9.6 Improvements in modulus of die-drawn rods and sheets* compared with various undrawn polymers

Material	Uniaxial Draw Ratio	Axial Youngs Modulus (GPa)	
		Isotropic	Drawn
Polyethylene HDPE	20	1.0	20
Polypropylene	20	1.5	20
Polyoxymethylene	16	3.0	25
PET*	4	3.0	10
PEEK	4	3.7	11
PVC*	3	2.5	6
PVDF	6	2.0	4

In addition to enhancement of Young's modulus and strength, molecular orientation gives rise to changes in other physical properties. These properties include:

- Low creep
- Low axial coefficient of thermal expansion
- Low permeability of gases and fluids
- Improved resistance to chemical attack

Uniform biaxially oriented tubes of polyethylene [77], polypropylene [78], PET [79] and PVC [80] of various draw ratios and thickness are produced by the die-drawing process. The drawing conditions for some of these samples are given in Table 9.7.

Table 9.7 Typical die-drawing conditions for tubes

Material	Billet Size (mm)		Draw Temperature (^{o}C)	Draw Ratio	
	OD	ID		Axial	Av. Hoop
Polyethylene Rigidex 002-40	63.0	25.0	115	4.0	2.0
	32.0	25.0	115	4.4	2.2
Polypropylene GSE 108	32.0	25.0	150	6.2	2.8
	32.0	25.0	150	8.5	2.2
PVC	42.0	32.0	100	1.9	1.75
	42.0	32.0	100	1.5	1.85
PET Eastman 9221	25.0	17.0	95	3.6	3.6

A specimen drawn using the rotating system showed a marked improvement in both straightness of tube and uniformity of wall thickness when compared with the product drawn without rotation. Compared with isotropic material, biaxially oriented samples have a higher tensile strength and modulus in both directions. The impact strength of biaxially drawn samples is much better than that of the isotropic material. On impact the isotropic uPVC tubes showed brittle fracture and shattered completely. The unbalanced biaxially drawn tubes of axial draw ratio 1.2 and inner hoop draw ratio 1.9, outer hoop draw ratio 1.6, gave a brittle fracture spiralling around the hoop. In the nearly balanced biaxially drawn tubes the fracture was localized and the rest of the tube remained intact. These fractured samples are shown in Figure 9.28.

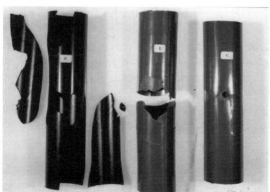

Figure 9.28   Impact tested samples of PVC tubes. (a) Isotropic,   (b) Biaxially drawn: axial draw ratio = 1.2, hoop inner = 1.9, hoop outer = 1.6,   (c) Biaxially drawn: axial draw ratio = 2.0, hoop inner = 1.9, hoop outer = 1.6

Table 9.8 Mechanical properties of isotropic and die-drawn tubes

Material	Draw Ratio		Tensile Strength (MPa)		Modulus (GPa)		Relative Impact Strength
	Axial	Hoop	Axial	Hoop	Axial	Hoop	
Polyethylene	1.0	1.0	22	22	0.5	0.5	1.0
	4.4	2.2	99	44	1.2	0.9	3.0
Polypropylene	1.0	1.0	25	25	1.0	1.0	1.0
	6.2	2.8	218	45	3.7	1.5	3.0
PVC	1.0	1.0	50	50	3.0	3.0	1.0
	1.9	1.75	110	80	3.1	3.1	2.0
PET	1.0	1.0	50	50	2.4	2.4	1.0
	3.6	3.6	93	85	3.1	3.1	3.0

Table 9.8 shows impact results for various biaxially drawn materials. High performance plastic pipes manufactured by this process have improved toughness and bursting strength. Biaxially drawn polyethylene tube (axial draw ratio 4 and hoop draw ratio 2) burst in a ductile fashion at a pressure level twice that for the undrawn tube. Die-drawn pipes can be welded to produce a pipe-line system and also work well with electro-fusion couplings [80]. A thin-walled highly transparent 'plastic can' in PET has been made for use in the food, beverage and other packaging industries. These plastic cans have improved gas barrier properties, increased toughness, and are resistant to crushing. The structure and properties of die-drawn materials have been discussed in several publications.

9.3.6   Applications of die-drawn materials

Die-drawn rods, monofilaments, sheets, tubes and profiles have found many commercially viable applications. Some of the current and potential applications are listed in Table 9.9:

Table 9.9 Some applications of die drawn materials

Application Area	Benefits	Material
Rods and Monofilaments		
Central core for wire rope	Improved service life	PP, PE
Packaging ties	Non-metallic	PE, POM
Strength members for Fibre Optic Cables	High specific strength and stiffness	POM
Pipes and Tubes		
Gas & Water pipe	Higher pressures	PE
Sewage & Drainage pipe	High impact and hoop strength	PE PVC
Food Containers	Transparent Low permeability	PET
Storage Drums	Improved stacking and crush strength	PE PP
Medical Applications		
Artificial joints	Increased resistance to wear	PE
Bio-resorbable Implants	High strength and Stiffness	Special polymers

References

1. K. Imada, Y. Yamamoto, K. Shigematsu and M. Takayanagi, J. Mater. Sci., **6**, 537 (1971).
2. N.E. Weeks and R.S. Porter, J. Polym. Sci. Polym. Phys., Ed. **12**, 635 (1974).
3. O. Hoffman & G. Sachs, Introduction to the Theory of Plasticity New York: McGraw Hill, 1953.
4. J.H. Southern & R.S. Porter, J. Appls. Polym. Sci., **14**, 2305 (1970).
5. C.J. Farrell & A.Keller, J. Mater. Sci., **12**, 966 (1977).
6. P.D. Griswold, A.E. Zachariades & R.S. Porter, Polym. Eng. Sci., **18**, 861 (1978).
7. T. Kanamoto, A. Tsuruta, K. Tanaka, M. Takeda & R.S. Porter, Polym. J. **15**, 327 (1983).
8. R.S. Porter, T. Kanamoto & A.E. Zachariades, Polymer, **35**, 4979 (1994).
9. R. Endo, T. Kanamoto & R.S. Porter, J. Polym. Sci., Polym. Physics, **36**, 1419 (1998).
10. B. Appelt & R.S. Porter, J. Macromol. Sci. Phys., **B20**, 21 (1981).
11. J.R.C. Pereira & R.S. Porter, J. Polym. Sci., Polym. Phys. Edn. **21**, 1133 (1983).
12. J.R.C. Pereira & R.S. Porter, J. Polym. Sci., Polym. Phys. Edn. **21** 1147, (1983).
13. T. He and R.S. Porter, Polymer, **28**, 946 (1987).
14. A.E. Zachariades & R.S. Porter, J. Appl. Polym. Sci., **24**, 1371 (1979).
15. R. Ball & R.S. Porter, J. Polym. Sci., Polym. Letters Edn., **15**, 519 (1977).
16. C.J. Farrell and A. Keller, J. Mater. Sci., **12**, 966 (1977).
17. A. Buckley and H.A. Long, Polym. Eng. Sci., **91**, 115 (1969).
18. J.M. Alexander and P.J. Wormell, Ann. C.I.R.P., **19**, 28, (1971).
19. Nakayama K. & Kanetsuma H., Eng. Edn., **3**, 1489, (1974).
20. T. Williams, J. Mater. Sci., **8**, 59 (1973).
21. A.G. Gibson, I.M. Ward, B.N. Cole & B. Parsons, J. Mater. Sci., **9**, 1193 (1974).
22. B. Parsons & I.M. Ward, Plast. Rubber. Proc. Applns., **2**, 215, (1982).
23. A.G. Gibson & I.M. Ward, J. Polym. Sci., Polym. Phys. Ed., **16**, 2015, (1978).
24. P.D. Coates, A.G. Gibson and I.M. Ward, J. Mater. Sci., **15**, 359, (1980).
25. P.S. Hope & I.M. Ward, J. Mater. Sci., **16**, 1511 (1981).
26. O. Hoffmann & G. Sachs, Introduction to the Theory of Plasticity, New York: McGraw Hill, 1953.
27. B. Azitzur, Metal Forming: Process & Analysis, McGraw Hill, New York, 1968.
28. B.J. Briscoe & D. Tabor, Polymer Surfaces: D.J. Clark and W.J. Feast (Eds) New York, Wiley, Chapter 1.
29. R. Gupta & P.G. McCormick, J. Mater. Sci, **15**, 619, (1980).
30. N. Inoue, T. Nakayama & T. Ariyama, J. Macromol. Sci. Phys., **B, 19**, 543 (1981).
31. P.S. Hope & B. Parsons, Polym. Eng. Sci., **20**, 589, (1980).
32. N. Inoue, Hydrostatic Extrusion: Theory & Applications: N. Inoue, M. Nishihara (Eds.) London: Applied Sci. Publ., (1985).
33. A.S. Maxwell, A.P. Unwin & I.M. Ward, Polymer **37**, 3293 (1996).
34. M.M. Shahin, R.H. Olley, D.C. Bassett, A.S. Maxwell, A.P. Unwin & I.M. Ward, J. Mater. Sci., **31**, 5541, (1996).
35. A.S. Maxwell, A.P. Unwin, I.M. Ward, M.I. Abo El Maaty, M.M. Shahin, R.H. Olley & D.C. Bassett, J. Mater. Sci., **32**, 567, (1997).

36. A.S. Maxwell, A.P. Unwin, I.M. Ward, Polymer, **37**, 3283 (1996).
37. P.S. Hope, A. Richardson & I.M. Ward, Polym. Eng. Sci., **22**, 307, (1982).
38. A.C. Curtis, P.S. Hope & I.M. Ward, Polymer Composites, **3**, 138 (1982).
39. H. Brody & I.M. Ward, Polym. Eng. Sci., **11**, 139, (1971).
40. P.J. Hine, N. Davidson, R.A. Duckett, A.R. Clarke & I.M. Ward, Polym. Composites, **17**, 720 (1996).
41. P.J. Hine, S. Wire, R.A. Duckett & I.M. Ward, Polym. Composites, **18**, 634, (1997).
42. A.R. Clarke, N. Davidson & G. Archenhold, Trans. Royal Microscopical Soc., **1**, 305, (1990).
43. P.J. Hine, R.A. Duckett & I.M. Ward, Composites, **24**, 643, (1993).
44. W. Bonfield, J.A. Bowman & M.D. Grynpas, U.K. Patent GB 2085461 B (1984).
45. W. Bonfield, J. Biomech. Eng., **10**, 522, (1988).
46. K.E. Tanner, R.N. Downes & W. Bonfield, Brit. Cer. Trans., **93**, 104 (1994).
47. I.M. Ward, W. Bonfield & N.H. Ladizesky, Polymer International, **43**, 333, (1997).
48. N.H. Ladizesky, I.M. Ward & W. Bonfield, Polym. Adv. Tech., **8**, 496, (1997).
49. N.H. Ladizesky, I.M. Ward & W. Bonfield, J. Appl. Polym. Sci., **65**, 1865, (1997).
50. A.G. Gibson, D. Greig, M. Sahota & I.M. Ward, J. Polym. Sci., Polym. Lett. Edu., **15**, 183 (1977).
51. G.A.J. Orchard, G.R. Davies & I.M. Ward, Polym., **25**, 1203 (1984).
52. G. Capaccio & I.M. Ward, Colloid Polym. Sci., **260**, 46 (1982).
53. P. Holden, G.A.J. Orchard & I.M. Ward, J. Polym. Sci., Polym. Phys. Edn., **28**, 709 (1985).
54. A.K. Taraiya, G.A.J. Orchard, I.M. Ward, J. Appl. Polym. Sci., **41**, 1659 (1990).
55. J.A. Slee, G.A.J. Orchard, D.I. Bower & I. M. Ward, J. Polym. Sci., Polym. Phys., **27**, 71, (1989).
56. G.A.J. Orchard, P. Spiby & I.M. Ward, J. Polym. Sci., B., Polym. Phys., **28**, 603, (1990).
57. J.M. Marshall, P.S. Hope & I.M. Ward, Polymer, **23**, 142 (1982).
58. G. Capaccio & I.M. Ward, J. Polym. Sci., Polym. Phys. Edn., **19**, 667, (1981).
59. P.D. Coates and I.M. Ward, J.Mater.Sci., **13**, 1957 (1978).
60. P.D. Coates and I.M. Ward, Polymer, **20**, 1553 (1979).
61. A.G. Gibson and I.M. Ward, J.Mater.Sci., **15**, 979 (1980).
62. A. Richardson, B. Parsons and I.M. Ward, Plast.Rubb.Proc.Appl., **6**, 347   (1986).
63. A.K. Taraiya, A. Richardson and I.M. Ward, J.Appl.Polym.Sci., **33**, 2559   (1987).
64. P.S. Hope, A. Richardson and I.M. Ward, J.Appl.Polym.Sci., **26**, 2879 (1981).
65. A. Richardson, P.S. Hope and I.M. Ward, J.Polym.Sci.,Polym.Phys.Edn., **21**, 2525 (1983).
66. A. Richardson, F. Ania, D.R. Rueda, I.M. Ward and F.J. Balta Calleja, Polym.Eng.Sci, **25**, 355 (1985).
67. S-W. Tsui, R.A. Duckett, I.M. Ward, D.C. Bassett, R.H. Olley and A. Vaughan, Polymer, **33**, 4527 (1992).
68. A. Selwood, I.M. Ward and B. Parsons, Plast.Rubb.Proc.Appl., **8**, 49 (1987).
69. C.C. Morath, A.K. Taraiya, A. Richardson, G. Craggs and I.M. Ward, Plast.Rubb.Comp.Proc.Appl., **19**, 55 (1993).

70. J Sweeney, M Nugent, A K Taraiya, I M Ward and P D Coates, Plast.Rubb.Composites, **29**, No.1, (2000), in press.
71. F.A. Motashar, A.P. Unwin, G. Craggs and I.M. Ward, Polym.Eng.Sci., **33**, 1288 (1993).
72. S.N. Kukereka, G. Craggs and I.M. Ward, J.Mater. Sci., **27**, 3379 (1992).
73. G. Craggs, Proc.Inst.Mech.Eng., **204**, 43 (1990).
74. E.L. Foster and H. Heap, Br.J.Appl.Phys., **8**, 400 (1957).
75. S. Kakadjian, G. Craggs and I.M. Ward, Proc.Inst.Mech.Eng., **210**, 65 (1996).
76. J. Sweeney and I.M. Ward, Trans.Inst.Chem.Eng., **A17**, 232 (1993).
77. A.K. Taraiya and I.M. Ward, J.Appl.Polym.Sci., **59**, 627 (1996).
78. A.K. Taraiya and I.M. Ward, Plast.Rubb.Comp.Proc.Appl., **15**, 5 (1991).
79. A. Selwood, I.M. Ward and G. Craggs, Plast.Rubb.Proc.Appl., **10**, 93 (1988).
80. A. Selwood, A.K. Taraiya, I.M. Ward and R.A. Chivers, Plast.Rubb.Proc.Appl., **10**, 85 (1988).

10. Mathematical Modelling

J Sweeney* P D Coates* and I M Ward+,
IRC in Polymer Science and Technology
*University of Bradford and +University of Leeds, UK

10.1 Constitutive Equations

An adequate constitutive equation to describe the solid phase processing behaviour of polymers requires the following ingredients:

(1)   Recognition that finite strains are involved.

(2)   Incorporation of the effects of strain rate and temperature.

Two different approaches have been adopted:

(i)   Development of phenomenological equations which do not necessarily have significance at a molecular level, but provide very accurate numerical representations which can be used for the modelling of polymer engineering processes.

(ii)   Development of equations which do have significance at a structural level either in terms of the deformation of a molecular network or classical plasticity.

These two approaches will now be discussed in some detail.

10.1.1 Phenomenological equations

10.1.1.1 Analogy with finite elastic behaviour

In some instances, a simple starting point is to assume that the stresses developed in a multidimensional deformation can be described by one of the constitutive equations of finite elasticity. It would be possible to do this by following the initial proposals of Rivlin [1], where the strain energy function U is considered to be a function of the strain invariants such that

$$U = f(I_1, I_2, I_3) \tag{1}$$

where $I_1 = \lambda_1^2 + \lambda_2^2 + \lambda_2^2,$ $I_2 = \dfrac{1}{\lambda_1^2} + \dfrac{1}{\lambda_2^2} + \dfrac{1}{\lambda_3^2}$

$I_3 = \lambda_1^2 \lambda_2^2 \lambda_3^2$ (= 1 for constant volume deformation) and $\lambda_1, \lambda_2, \lambda_3$ are the principal extension ratios.

As is now well documented (see for example references 2) although this is formally correct, it is equally correct and much more tractable, because it involves fewer fitting parameters, to obtain a satisfactory data fit to follow the approach of Ogden [3] (see also Valanis and Landel [4]).

Then

$$U = \sum_{i=1}^{n} A_i \left(\lambda_1^{\alpha_i} + \lambda_2^{\alpha_i} + \lambda_3^{\alpha_i} - 3 \right) \qquad (2)$$

is an n-term Ogden model where A_i and α_i (i = 1 to n) are material constants.

For a simple rubber where n = 1 and $\alpha_1 = \alpha = 2$ this gives the familiar equation:

$$U = A \left(\lambda_1^2 + \lambda_2^2 + \lambda_3^2 - 3 \right) \qquad (3)$$

with A = NkT, and N the number of cross-links/unit volume, k Boltzmann's constant and T absolute temperature.

The stresses are given by the first derivatives of U with respect to strain and assuming that the material is incompressible, we have for principal directions denoted by the index i

$$\sigma_{ii} = \lambda_i \frac{\partial U}{\partial \lambda_i} + p \qquad (4)$$

where p is a hydrostatic pressure and the true stress σ_{ii} (i = 1,2,3) is related to λ_i (principal extension ratio or draw ratio) by the equation.

$$\sigma_{ii} = A \alpha \lambda_i^{\alpha} + p \qquad (5)$$

for a one-term Ogden model. The extreme cases where $\alpha = 1$ and $\alpha = 2$ represent Varga[5] and neo-Hookeian behaviour respectively. Materials capable of being modelled by the one-term Ogden representation generally have a value of α between 1 (i.e. true stress) and 2 (ideal rubber like behaviour).

Ward and co-workers have used the Ogden representation to model the behaviour of PVC at temperatures near its glass transition. Two cases have been considered, namely (i) the biaxial stretching of PVC sheet [6] and (ii) an analysis of the mechanics for drawing of PVC tubes over an expanding mandrel [7].

In the first study, sheets of unplasticised PVC were drawn into three distinct states of finite strain using a custom-built high temperature biaxial drawing machine. Stress-strain curves were produced for uniaxial, planar (constant width) and equibiaxial extrusion over a range of constant octahedral shear strain rates. The latter were obtained by programming the governing algebraic relations into the computers which also control the motor.

Equations (2), (4) and (5) above give expressions for the stress-strain curves in terms of the material parameters A and α, which are allowed to depend on strain rate and temperature. Sweeney and Ward [7] showed that curve fitting to the constant width case, which is the most sensitive to an accurate fit because two independent stresses have to be fitted, gave good predictions for the uniaxial and equibiaxial data. A typical set of results is shown in Figures 10.1(a), (b) and (c). The values of A and α depend on strain rate and temperature with α in the range 1.1-1.4 and an increase in A is observed with increasing strain rate or decreasing temperature.

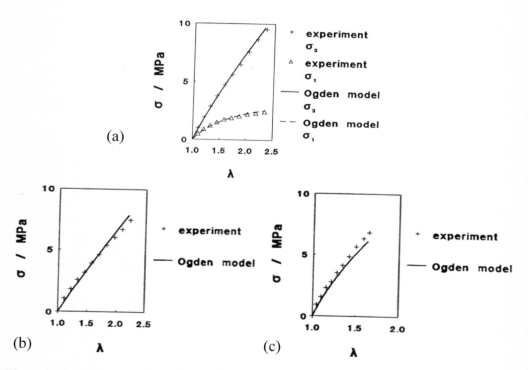

Figure 10.1 Stretching of PVC sheet at 90°C. (a) constant width (b) uniaxial (c) equibiaxial. +, Δ Experimental; full line is Ogden model with appropriate values of A and α.

10.1.1.2 Phenomenological equations based on true-stress strain curves

It has long been recognised that one viable approach to the plastic deformation of polymers is to construct a master flow curve, the true stress strain curve. This has been discussed in Chapter 1 where it was shown that this master curve can incorporate the effects of any previous deformation of the sample by a horizontal translation along the true strain axis (Figure 10.2). As discussed in Chapters 5 and 9 Ward and co-workers have used the concept of the true-stress strain curve to obtain quantitative analyses of the mechanics of the hydrostatic extrusion and die-drawing processes, with the additional ingredient that the curve depends on current strain rate and temperature.

This concept has been developed by G'Sell and co-workers [8-10] by showing that a constitutive relation for high density polyethylene can be obtained which accurately models the form of the true stress-strain curve and incorporates both strain hardening and strain rate sensitivity. For an initially isotropic sample (i.e. no previous deformation)

$$\sigma(\varepsilon, \dot{\varepsilon}) = K\left[1 - \exp tw\varepsilon\right]\exp(h\varepsilon^2)\dot{\varepsilon}^m \quad (6)$$

where the term in the square brackets describes the transient elastic behaviour, $\exp(h\varepsilon)$ the strain hardening and $\dot{\varepsilon}^m$ the strain-rate sensitivity.

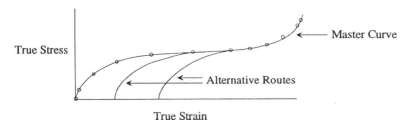

Figure 10.2 Schematic diagram illustrating concept of the true stress – true strain curve

For a high density polyethylene G'Sell and co-workers [10] showed that a good fit to experimental data for uniaxial deformation was obtained with K=46MPa, w=40, h=0.41 and m=0.075.

The strain rate coefficient m is small, corresponding to the comparatively low strain rate sensitivity. In terms of the Eyring activated rate theory

$$\dot{\varepsilon} = \dot{\varepsilon}_o \exp\left(-\frac{\Delta H - \sigma v}{kT}\right) \quad (7)$$

$$m = \frac{\partial(\log \sigma)}{\partial(\log \dot{\varepsilon})} = \frac{kT}{v} \quad (8)$$

where v is the activation volume, and is therefore comparatively large.

Accepting the concept of a true stress-strain curve equation (6) can be readily modified to include the effect of an initial deformation ε_p (pre-strain, as it is sometimes called). We have

$$\sigma\left(\varepsilon,\dot{\varepsilon},\varepsilon_p\right) = K\left\{1 - \exp\left[-w\left(\varepsilon_p\right)\varepsilon\right]\exp\left[h\left(\varepsilon+\varepsilon_p\right)^2\right]\dot{\varepsilon}^m\right\} \tag{9}$$

Clearly the strain rate sensitivity is unaltered and the strain hardening term is easily modified to $\exp\left[h\left(\varepsilon+\varepsilon_p\right)\right]$. The anelastic behaviour changes in a more complex manner, because it involves structural changes. G'Sell et al suggested that it was adequate to decrease the parameter $w\left(\varepsilon_p\right)$ with increasing strain to obtain a good fit to the data (Figure 10.3).

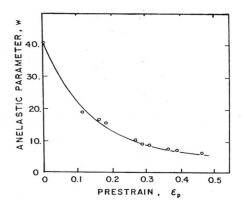

Figure 10.3 Influence of true-strain ε_p on the anelastic parameter w of constitutive relation for high density polyethylene prepared by G'Sell et al. Reproduced by permission of the publishers Elsevier Applied Science.

A similar form of constitutive equation has been adopted by Nazarenko et al.[11] in their study of neck development in tensile specimens of polycarbonate. They used a simplified version of equation (6) [8]:

$$\sigma\left(\varepsilon,\dot{\varepsilon}\right) = K\exp\left(\frac{\gamma_0}{2}\varepsilon\right)\dot{\varepsilon}^m \tag{10}$$

Values of the strain rate sensitivity parameter m and the strain hardening coefficient γ_0 were obtained from a series of experiments carried out at different temperatures and hydrostatic pressures. The experiments were in tension, using cylindrical specimens which necked. The two parameters m and γ_0 were obtained indirectly using the measured neck shapes. Numerical solutions to the differential equation governing the neck profile were generated as a function of m and γ_0, and comparison of measured and calculated neck shapes yielded values for the two parameters. The evaluations covered the temperature range –65 to 21°C and the pressure range from atmospheric to 210 MPa. γ_0 was found to be sensibly independent of pressure and temperature, with a value of approximately 3.3. m increased with pressure and decreased with temperature, with a lower limit of 0.28 levelling off at high pressures and low temperatures to a value of 0.035. The results were reinterpreted in terms of shear activation volume; as expected, the activation volume decreased with increasing pressure or decreasing temperature.

10.1.2 Constitutive relations incorporating the deformation of a molecular network

A common theme in the deformation of polymers is the stretching of a molecular network, and there are several indications of the importance of a stretched molecular network with regard to the physical properties of oriented polymers (e.g. the existence of internal stress effects, creep and thermal expansion behaviour). In an important early development, Haward and Thackray [12] proposed the representation shown in Figure 10.4. The key elements are a Hookean spring E, which is series with (a) a rubber network Langevin spring and (b) a thermally activated dashpot described by the Eyring equation (7 above).

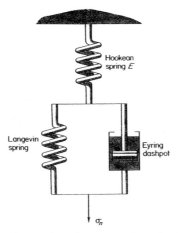

Figure 10.4 Schematic diagram illustrating the Haward and Thackray model for polymer deformation. Reproduced by permission of the publishers John Wiley & Sons.

Haward and Thackray related the total strain e and the plastic strain e_A from the activated dashpot to the nominal stress σ_n (load applied divided by initial cross-sectional area). This gives

$$e = \frac{\sigma_n(1+e)}{E} + e_A \qquad (11)$$

and
$$d\frac{[\ell n(1+e_A)]}{dt} = \dot{e}_A \exp\left(\frac{-\Delta H}{kT}\right) \sinh \frac{v(\sigma_n - \sigma_R)}{kT} \qquad (12)$$

where σ_R is the rubber like stress which describes the strain hardening and is determined from rubber elasticity theory [13] as

$$\sigma_R = \frac{1}{3}NkTn^{1/2}\left[\mathscr{L}^{-1}\left(\frac{1+e_A}{n^{1/2}}\right) - (1+e_A)^{-3/2}\mathscr{L}^{-1}\frac{1}{(1+e_A)^{1/2}n^{1/2}}\right] \qquad (13)$$

where $\mathscr{L}^{-1}$ is the inverse Langevin function, N is the number of chains between cross-link points per unit volume and n is the average number of random links per chain.

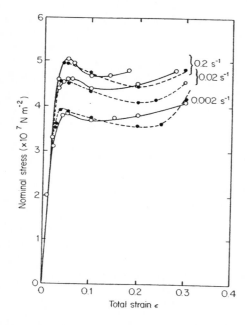

Figure 10.5   Stress – strain curves for cellulose nitrate at 23°C. Experimental curves (○) and calculated curves (●). Redrawn from Haward and Thackray, Proc.Roy.Soc. A302, 453 (1968) with permission of the Royal Society.

Equation (12) was integrated numerically, using equations (11) and (13) to describe the deformation of several polymers, including cellulose nitrate, for which results are shown in

Figure 10.5. It is important to note that the main features of the stress-strain curve are reproduced, and that this approach raises the realistic hope that links between phenomenology and structure can be established.

In a review article, Ward [14] emphasised the importance of Haward and Thackray's approach and showed how the deformation behaviour of three major polymers, polyethylene terephthalate, polymethylmethacrylate and high density polyethylene, could be considered in terms of two key factors, the stretching of a molecular network and the influence of thermal activated processes. Having satisfied ourselves that this is so, the main issue in how to incorporate these two key elements, the molecular network and the thermally activated viscosity (and the initial elastic response) into a consistent representation which can describe the three-dimensional deformation of polymers.

Much of the emphasis in the research of three groups of workers (Boyce et al.[15], Buckley et al. [16], Sweeney and Ward [17] and co-workers) has been to find an adequate description of the network, which then acts in parallel with the thermally activated viscosity. Sweeney [18] has explored the relationship between three theories of polymer network deformation; the eight-chain model of Arruda and Boyce [19], the full network model of Wu and Van der Giessen [20]; and the cross-link model of Edwards and Vilgis [21]. It was concluded that the Edwards-Vilgis model is capable of modelling a broader range of material behaviour than the other two models, owing to the additional feature of slip links, and that this model can replace the two other models to a good approximation provided that the finite chain extensibility limit is not approached too closely. As remarked previously by Sweeney [18] the choice of network model seems to be largely a matter of personal preference. Additionally, it is also usually necessary to add some additional ingredients to the phenomology to deal with temperature dependence and strain rate for multiaxial deformation, and accurate modelling of the strain hardening behaviour.

Recent attempts to model polymer behaviour have retained the essential structure of Haward and Thackray's model illustrated in Figure 10.3, albeit generalised to three dimensions. The theories differ in the choice of both the network and the viscous mechanism. Thus, the model of Boyce, Parks and Argon [22] makes use of the three-chain model of James and Guth [23] for the network, and the Argon model [24] of plastic flow for the viscous component. The three-chain model is constructed from three Langevin type chains (see equation (13)) arranged along the principal stretch directions. The eight-chain model [19] is a development of this, using the same model for the individual chain, and is used in place of the three-chain model in the theory of Arruda and Boyce [15].

In contrast, Buckley and Jones' theory [16] uses the Edwards-Vilgis network model combined with the Eyring model of viscosity. Sweeney and Ward's [17] theory also uses the Eyring process, but the network is that of Ball et al. [25], itself a special case of the Edwards-Vilgis model in that there is no finite chain extensibility. Wu and van der Giessen [26] use their own network [20] in combination with the Argon process. The position for all these theories of polymer deformation is summarised in Table 10.1.

Table 10.1

Theory	Viscous process		Network model				
	Eyring	Argon	3-chain	8-chain	Full network	Ball et al.	Edwards-Vilgis
Boyce, Parks & Argon		✓	✓				
Arruda & Boyce		✓		✓			
Buckley & Jones	✓						✓
Sweeney & Ward	✓					✓	
Wu & van der Giessen		✓		✓			

One of the simplest cases which has been considered, is the multiaxial drawing of PVC sheet at 90°C, which is above T_g. Sweeney and Ward [17] showed that the behaviour could be accurately described by the model of Ball et al. [25]. In this case the strain energy function is given by

$$\frac{F}{kT} = \frac{1}{2} N_c \sum \lambda_i^2 + \frac{1}{2} N_s \sum \left[\frac{(1+\eta)\lambda_i^2}{1+\eta\lambda_i^2} + \ln\left(1+\eta\lambda_i^2\right) \right] \quad (14)$$

where F is the energy, N_c and N_s are respectively the number of cross-links and the number of slip-links, and η is a parameter governing the slipperiness of the slip-links. A schematic model of the network is shown on Figure 10.6.

Rewriting equation (4), the stress in any stretching mode is given by

$$\sigma_{ii} = \lambda_i \frac{\partial F}{\partial \lambda_i} + p \quad (15)$$

where p is a hydrostatic pressure. Solving for p in each case gives a set of expressions for the stress in the three stretching modes (uniaxial, planar and equibiaxial extension).

Sweeney and Ward found that $N_c = 0$, irrespective of the strain-rate, which is physically reasonable for a non-crystalline polymer above its glass transition, and that a constant value of $\eta = 0.08$ also held. To bring together results for uniaxial drawing, equibiaxial drawing and drawing at constant width, in each case the octahedral shear stress was considered to be a function of the octahedral shear strain rate (obtained by transforming the strain rate tensor on to the axes set defined by the octahedral stresses). It was found that the stress was a linear function of the logarithm of the strain rate, which follows from the relation

$$N_s kT = A + B \ln(\dot{\gamma}/\dot{\gamma}_0) \qquad (16)$$

which applies when $\dot{\gamma} \geq \dot{\gamma}_0$.

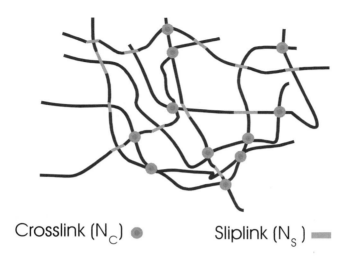

Crosslink (N_c) ●   Sliplink (N_s) ▬

Figure 10.6   Schematic diagram of the network of Ball et al

For multiaxial drawing of PVC at 84°C, which is in the glass transition region, a more complicated analysis was required, analogous to the Haward and Thackray model described above. We now require a viscous element, an Eyring thermally activated process in parallel with the Ball network, which is in series with an elastic response (see Figure 10.4). As shown in Figure 10.7, this representation describes the behaviour very well. It is important to note that the new ingredient compared with the earlier Haward and Thackray model is the introduction of a second thermally activated process to describe the strain rate and temperature dependence of the rubber network term (i.e. slip links rather than permanent cross-links).

Matthews et al [27] showed that the biaxial drawing of poly(ethylene terephthalate) (PET) at 85°C (i.e. above T_g) could be satisfactorily described by a similar analysis to that discussed for PVC at 84°C. The strain rate dependence was well described by equation 17:

$$N_s kT = 3.25 + 0.398 \ln(\dot{\gamma}/\dot{\gamma}_o) \qquad (17)$$

analogous to equation (16); in Equation 17 the units of $N_s kT$ are MPa. It was interesting that the validity of the network model was confirmed by photoelastic studies where the stress-optical coefficient agreed well with previous work, provided that the draw ratios did not reach a level where strain crystallisation occurred.

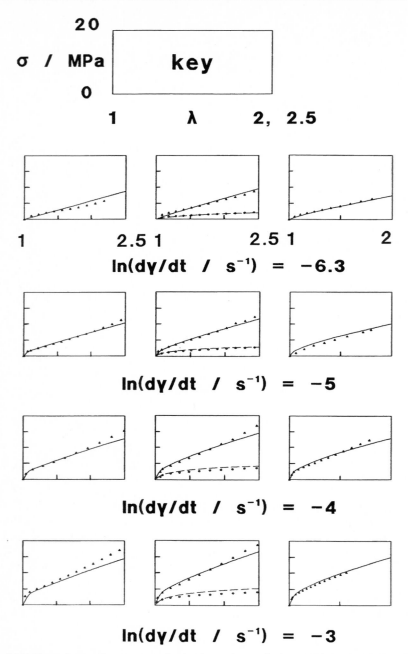

Figure 10.7 Experimental stresses (points) and predictions (lines) obtained with the adaptation of the Ball et al model used by Sweeney and Ward for biaxial deformation of PVC. (a) Uniaxial tension (b) Constant width (c) Equibiaxial tension. Reproduced by permission of Elsevier Applied Science.

In a contemporaneous series of publications, Buckley et al. [16,28,29] describe the modelling of drawing of PET in the temperature range 75-120°C at high strain rates, 1-16s⁻¹ comparable to those pertaining in an industrial film stretching process. The Edwards-Vilgis network model, which incorporates strain hardening, was used rather than the Ball et al model. Similar to other workers, Buckley and co-workers assumed that the viscous flow process followed an Eyring activated rate process, relating to the octahedral shear stress. There were some key similarities to that of other workers, and some differences. For example, the best fit for the Edwards-Vilgis term was obtained with $\eta=0$, but the number of entanglements N_e corresponded very well to that obtained by much simpler analysis of peak shrinkage stress data by Ward and co-workers. The major inadequacy of the model was that it did not include flow resulting from entanglement slippage, which is also modified by stress induced crystallisation, effects which become more important with increasing temperature.

Buckley et al. [28] attempted to give a physical interpretation to their results, in terms of commenting on the shear activation volume which is very large, suggesting a highly cooperative process and on the presence of a small pressure activation volume suggesting that the flow is accompanied by local dilation. As already remarked, the entanglement density appears to be similar to that observed by other workers and corresponds to an average chain length of 17 monomer units. The quantities of permanent chain slippage and crystallisation at high temperatures are considered, and must be taken into account in any satisfactory model.

10.2   Numerical Modelling of Forming Processes

Once the mechanical behaviour of a material is adequately defined in terms of its constitutive equation, then in principle we have the means to calculate the shapes and forces associated with any forming process. However, a significant degree of computational effort will be required when complex shapes or nonlinear mechanical behaviour is involved. The finite element method is now routinely used in many engineering fields, and is widely available in the form of commercial packages. The traditional linear elastic finite element formulation, which is often the basis of such packages, is clearly not applicable in the context of highly nonlinear material behaviours such as those defined by the variety of constitutive equations outlined above. In order to make use of complex constitutive equations, there are two alternatives: to write novel finite element code which is developed specially for a particular class of material behaviour; or to use a commercial code which is at a level of sophistication such as to enable its user to define the material behaviour in a general way. Both approaches have resulted in significant advances.

The importance of a molecular network in polymer deformation is now generally accepted, and elastic effects are observed at high temperatures. In some circumstances, it is possible to obtain useful modelling results on the assumption that the material behaviour is entirely elastic. The effects of temperature and strain rate are then taken into account implicitly, by obtaining the parameters characterising the model under appropriate experimental conditions. However, the material is in fact viscoelastic, with rate effects being readily observable. This limits the applicability of the purely elastic approach, as it is possible, particularly when there is the potential for necking instability, for rate effects to become crucially important such that they influence the deformed shape in a qualitative sense. In this survey, we divide the numerical models into two classes – elastic and rate-dependent.

10.2.1 Elastic constitutive behaviour

The Ogden model has been successfully used in the finite element modelling of polymers to large strains. Nied, Taylor and Delorenzi [30] have used this approach to model thermoforming, a process whereby flat sheets of polymer at high temperatures are forced by air pressure into a mould. They fitted one-term Ogden models (equation (2)) to uniaxial stress-strain curves for a modified polyphenylene oxide (PPO) material at various temperatures to good accuracy. They used their own finite element code to model the problem as a thin membrane being stretched in three dimensions. In these simulations, a two-term rather than a one-term Ogden model was used to aid numerical stability at large stretches. The only experimental observations were of the final product, so verification of intermediate shapes was not possible. Further, the external dimensions of the final shape are governed by the shape of the mould, so that the only verifiable prediction of the model is that of product thickness. Good predictions of thickness were made, and this is certainly a very important factor from the point of view of the manufacturer. Nied and Lorenzi [31] have used a similar approach to model blow-moulding and stretch blow-moulding, using the materials Acrylonitrile Butadiene Styrene (ABS) and an alloy of Polycarbonate (PC) and Poly(Butylene Terephthalate) (PBT) in addition to PPO. They have used Ogden models containing up to three terms, and point out that the thickness predictions are not greatly sensitive to the assumed material parameters. The attractiveness of the Ogden model lies partly in that it can be fitted easily to material data. . It seems likely that other constitutive models of incompressible material would give similar thickness predictions; the results demonstrate the usefulness of the Ogden model, but do not show it to be uniquely effective.

Another relevant application of the Ogden model in a finite element scheme is reported in the work of Sweeney and Ward [6], referred to above. They stretched PVC sheets multiaxially at high temperatures, and showed that, provided the polymer was above its glass transition temperature, the one-term Ogden model gave consistent representations of stress-strain behaviour in uniaxial, equibiaxial and planar extension. For testing temperatures of 95 and 100°C, values of the Ogden exponent were found to be in the range 0.9 – 1.4. The Ogden equation was incorporated into an in-house finite element scheme, so that non-uniform shapes could be modelled. Such shapes were generated experimentally by stretching square sheets along diagonals gripped at the corners; this is shown schematically in Figure 10.8. The finite element model of a stretched quarter specimen is shown in Figure 10.9, for a temperature of 100°C. Comparisons of the observed and modelled specimen shape, and of the observed and modelled drawing forces, are shown respectively in Figures 10.10 and 10.11. Both the shapes and drawing forces were predicted satisfactorily.

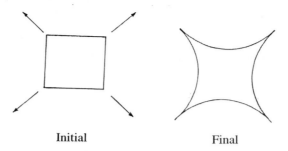

Initial   Final

Figure 10.8 Schematic diagram of the nonuniform drawing experiment in which the initially
square specimen is pulled from each corner. Reproduced by permission of the
publishers Institution of Chemical Engineers.

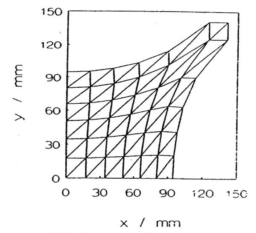

Figure 10.9 Finite element model of a quarter of the sheet specimen. The material is PVC
at 100°C. Reproduced by permission of the Institution of Chemical Engineers.

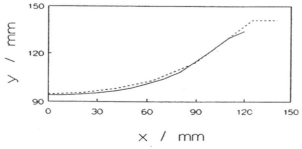

Figure 10.10   Observed and modelled boundary shapes for the nonuniform drawing
experiment of Figures 10.8 and 10.9. The full line is the observation and
the dashed line the model prediction. Reproduced by permission of the
Institution of Chemical Engineers.

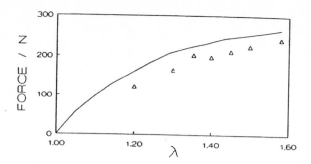

Figure 10.11   Observed and modelled drawing forces for the experiment of Figures 10.8
                      and 10.9. The full line is the observation and the triangles the model
                      prediction. Reproduced by permission of the publishers Institution of
                      Chemical Engineers.

In a study which further develops the application of the Ogden model to PVC, Kakadjian,
Craggs and Ward [32] showed that the representation provided straightforward constitutive
equations for the analysis of the mechanics of a simple engineering process, the expansion
of PVC tubes by drawing over a mandrel. The approach is analogous to the lower bound
Hoffman-Sachs analysis for hydrostatic extrusion through a conical die (see Chapter 9).

In Figure 10.12, the stresses acting in a general element n are described for both

    (a)   the axial and thickness direction, and

    (b)   the hoop direction

The analogous equation to equation (1) in Chapter 9 is

$$\frac{\delta}{\delta r}\left(\sigma_a\, tv\right) = \sigma'_h t + \frac{\mu\, Pr}{\sin\theta} \tag{18}$$

where σ'_h, the mean hoop stress, $= \sigma_h + \dfrac{\delta\sigma_n}{2}$

is given by $$\sigma'_n = \frac{Pr}{t\cot\theta} \tag{19}$$

combining (18) and (19) we have

$$\frac{\delta}{\delta r}\left(\sigma_a\, tr\right) = t\sigma^1_h\left(1 + \mu\cot\theta\right) \tag{20}$$

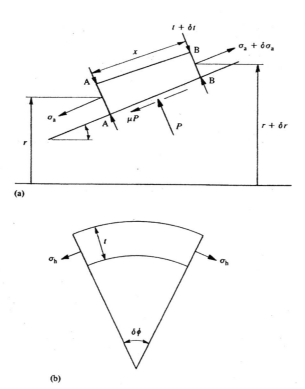

(a)

(b)

Figure 10.12 Stresses acting on a general element n in (a) the axial direction and thickness direction   (b) the hoop direction. Reproduced by the permission of the publishers Professional Engineering Publishing Ltd.

From the Ogden equations the axial and hoop stresses in an element are

$$\sigma_a = A\alpha\left(\lambda_a^\alpha - \lambda_t^\alpha\right)$$

$$\sigma_n = A\alpha\left(\lambda_n^\alpha - \lambda_t^\alpha\right) \tag{21}$$

The incremental increase in these stresses $\delta\sigma_a$ and $\delta\sigma_n$

can be related to their deformations by

$$\delta\sigma_a = A\alpha^2\left(\lambda_a^{\alpha-1}\delta\lambda_a - \lambda_t^{\alpha-1}\delta\lambda_t\right)$$
$$\delta\sigma_n = A\alpha^2\left(\lambda_n^{\alpha-1}\delta\lambda_n - \lambda_t^{\alpha-1}\delta\lambda_1\right) \tag{22}$$

For constant volume deformation $\quad \lambda_a \lambda_n \lambda_t = 1$

and
$$\frac{\delta\lambda_a}{\lambda_a} + \frac{\delta\lambda_n}{\lambda_n} + \frac{\delta\lambda_t}{\lambda_t} = 0 \qquad\qquad (23)$$

In any element

$$\lambda_a = \frac{x}{x_o}, \lambda_n = \frac{r}{r_o}, \lambda_t = \frac{t}{t_o}$$

so that
$$\frac{\delta\lambda_a}{\lambda_a} = \frac{\delta x}{x} \frac{dx_a}{\lambda_n} = \frac{\delta r}{r} \quad \text{and} \quad \frac{\delta\lambda_t}{\lambda_t} = \frac{\delta t}{t} \qquad (24)$$

Substituting equations (21) (22) and (23) into the equilibrium equation (18) gives

$$\delta\lambda_t = \frac{T_1\left(\dfrac{\delta\lambda_n}{\lambda_n}\right)\lambda_t - \sigma_a + T_2\left\{\sigma_h + \dfrac{A\alpha^2}{2}\left(\lambda_h^{\alpha-1}\delta h\right)\right\}}{-T_1\left(\lambda_h^\alpha \lambda_t^{2\alpha} + 1\right) + \dfrac{\sigma_a}{\lambda_t}\dfrac{\lambda_k}{\delta\lambda_h} + \dfrac{T_2 A\alpha^2}{2}\lambda_t^{\alpha-1}} \qquad (25)$$

where $\quad T_1 = \dfrac{A\alpha^2}{\left(\lambda_t \lambda_h\right)^2 \lambda_t}\left(\dfrac{r}{\delta r}\right)$

and $\quad T_2 = \left(1 + \mu\cot\theta\right)$

Equation (25) permits the mechanical change in the thickness in any element $\delta\lambda_t$ to be determined. Kakadjian et al. dealt with this by an iterative procedure, dividing the expanding tube into many elements, and calculating the incremental changes in deformation $\delta\lambda_a, \delta\lambda_h, \delta\lambda_t$ and the incremental stress changes $\delta\sigma_a$ and $\delta\sigma_n$ which occur over the length of each element.

It was assumed that $\alpha = 1.2$ and that the parameter A in each element depended on the equivalent strain rate $\dot{\varepsilon}$ (i.e. similar but not identical to the octahedral shear strain rate).

The isotropic billet is assumed to have an initial velocity when it arrives at the mandrel, with the downstream side of element being unstressed and retaining the initial isotropic billet dimensions r_o and t_o (Figure 10.12). At the section B, the value of $\delta\lambda_h$ can be calculated from the geometry for a given element length x_o and moulded semi-angle θ. The value of $\delta\lambda_t$ is found from equation (14) with the parameter A which has been obtained for uniaxial drawing of sheet, the equivalent strain rate being calculated from the time taken for element 1 to move from the undeformed state to that shown in Figure 10.12.

The hoop incremental increase $\delta\lambda_h$ is initially assumed to be the same for elements n and (n-1), $\delta\lambda_a$ can then be found from equation (23), and a new value of $\delta\lambda_h$ is obtained as

$$\delta\lambda_h = \frac{x\sin\theta}{r}\left(1 + \frac{1}{2}\frac{\delta\lambda_a}{\lambda_a}\right) \qquad (26)$$

A computer program was developed to evaluate the stresses and draw ratios, incorporating iterative procedures which take into account the changing shape of the element with increasing incremental stresses following the scheme shown by the flow diagram of Figure 10.13.

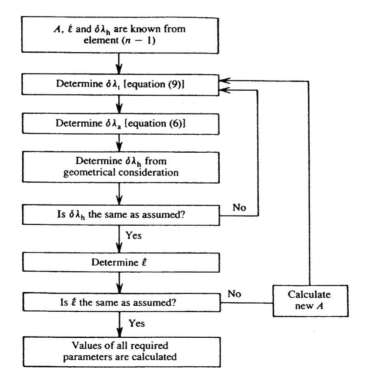

Figure 10.13 Flow diagram detailing solution for element n. Reproduced by permission of the publishers Professional Engineering Publishing Ltd.

This theoretical analysis was tested by drawing tubular PVC billets over a heated mandrel at 90°C. In each tube drawing test, the axial drawing load was measured, and by stopping the test, quickly cooling the drawn material it was possible to determine the axial, hoop and thickness draw ratios along the mandrel. In all cases any elastic recovery and relaxation of the polymer on stopping the test was small. The results of the analysis confirm the following.

(i)   The hoop stress (Figure 10.14(a)) dominates the behaviour during the early stages of drawing, causing a small lateral contraction in the axial and thickness direction. The axial stress and axial draw ratio (Figure 10.14(b) and 10.14(c)) become more prominent near the top of the mandrel rising sharply, with the axial stress becoming compatible with the draw load.

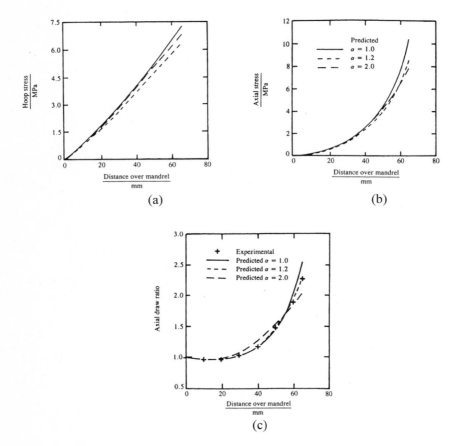

Figure 10.14   Predicted hoop (a) and axial (b) stresses and axial draw ratio (c) when drawing PVC tube over a mandrel of 67.5 mm final diameter and 15° semi-angle. Billet velocity is 0.28 mm/s and processing temperature 90°C. Reproduced by permission of Professional Engineering Publishing Ltd.

(ii)   Increasing mandrel angle produced a reduction in the frictional work, but this is counterbalanced by the increased strain rate, so that there is only a marginal decrease in the draw load.

(iii)   Increasing the final diameter increases the draw load, so that large diameter mandrels result in very thin walled tubes and very high axial stresses which can cause fracture of the product in some cases. Figures 10.15(a) and 10.15(b) show that these effects are accurately modelled by the analysis.

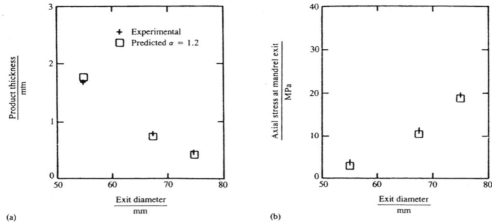

(a)   (b)

Figure 10.15   Experimental and predicted values of product thickness (a) and axial stress (b) for PVC tubes drawn over mandrels having 15° semi-angle and variable final diameter. Billet velocity is 0.28 mm/s and processing temperature 90°C. Reproduced by permission of Professional Engineering Publishing Ltd

The work described so far has only involved stable polymer deformations. In the case of semicrystalline polymers, necking instabilities routinely occur. For these to be modelled successfully, the associated constitutive equation must give rise to stress-strain curves of a particular form. When plotted in terms of engineering or nominal stress, the curve reaches a maximum which is a necessary condition for the onset of necking; this is shown in Figure 10.16, which corresponds to a specimen stretched in uniaxial tension.

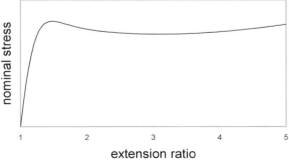

Figure 10.16   Nominal stress – extension ratio relation for a polymer necking in tension.

For material for which the stress is independent of the strain rate, this condition is also sufficient for necking to occur. The maximum is followed by a minimum, corresponding to the strain in the neck during stable necking. Given a suitable choice of material parameters, the model of Ball et al. [25] and the Edwards-Vilgis model [21] both produce stress-strain behaviour of this nature.

Sweeney and Ward [33] exploited this feature of the Ball et al. theory to model the behaviour of necking polypropylene sheet stretched at 150°C. They incorporated the Ball et al. strain energy function (equation (14)) into the same in-house finite element scheme as used in the Ogden modelling mentioned above. They were able to produce predictions of shapes and drawing forces for sheet specimens when subject to tensile stretching in uniaxial, equibiaxial and planar modes. Using parameter values for which $N_c / N_s = 0.15$, the predictions were sufficiently realistic to prove useful, but there were discrepancies in the detail. These arose from the incapability of the Ball et al. theory to model the onset of necking as occurring at extension ratios less than $\lambda \approx 1.8$; in the experiments on polypropylene, necking initiated at $\lambda \approx 1.3$.

In order to create a constitutive model that was more realistic and yet retained the desirable features of the Ball et al. model, Sweeney and Ward [34] introduced a modification to it. This took the form of replacing the constant sliplink number N_s with a function of strain, via the first invariant I_1. The sliplink number decreases as strain increases, and this has the effect of causing the maximum in nominal stress to occur at lower strains. The strain dependence was given the form

$$N_s(I_1) = \frac{N_{s0} - N_{s\infty}}{(I_1 - 2)^\beta} + N_{s\infty} \qquad (27)$$

where N_{s0} and $N_{s\infty}$ are constants corresponding respectively to the initial and ultimate sliplink numbers, and β is an additional fitting parameter. This theory is elastic, but the strain energy function no longer exists, so that it cannot be classified as hyperelastic; this has been discussed previously [34]. The strain dependence is introduced into the constitutive equations (15). These are obtained by differentiating equation (14) under the assumption of constant N_s; N_s is then replaced by the function $N_s(I_1)$ defined in (27) above.

The modified theory was shown to give an improved model of the deformation of polypropylene at 150°C [34], [35]. It was implemented within a finite element scheme via the use of the package ABAQUS. A user subroutine supplies the constitutive equation in the form of combinations of the derivatives $\dfrac{\partial F}{\partial I_1}$ and $\dfrac{\partial F}{\partial I_2}$, as well as higher-order derivatives of F which are required for the numerical solution. Since these derivatives must be in the form of functions of the invariants only, it is useful to re-express equation (14) as

$$\frac{F}{kT} = \frac{1}{2} N_c I_1 + \frac{1}{2} N_s \left[\frac{(1+\eta)(I_1 + 2\eta I_2 + 3\eta^2)}{1 + \eta I_1 + \eta^2 I_2 + \eta^3} + \ln(1 + \eta I_1 + \eta^2 I_2 + \eta^3) \right] \qquad (28).$$

Sheets were stretched both uniaxially and in planar extension, and formed inhomogeneous necked shapes. The planar extension tests are carried out using a biaxial testing machine in which all four specimen sides are gripped. Since the material has a tendency to neck locally at the grips, a thin circular inner gauge section is required to ensure that deformation takes place in the interior of the specimen; the specimen geometry is shown in Figure 10.17.

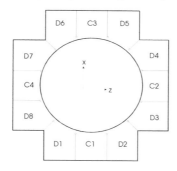

Figure 10.17 Specimen for planar extension experiments. Inner circular section is machined to half the thickness of the original (1.6 mm thick) sheet. Gripped areas D6, C3 and D5 and D1, C1 and D2 separate as the specimen is extended in the z direction. Reproduced by permission of Elsevier Applied Science

The values of the five parameters of the model are arrived at by fitting to experimental stress-strain curves using a process of trial and error. The set of parameters derived in this way for the polypropylene material at 150°C is given in Table 10.2. Uniaxial experiments provide insufficient data to determine uniquely the five parameters; as has been demonstrated [35], a set of values grossly different from that of Table 10.2 can generate a virtually identical uniaxial stress-strain curve. Such data sets are possible if η is allowed to vary freely – a value $\eta = 0.6$ was used in the alternative data set [35]. It is possible to fix the value of η via the use of additional experimental data, in the form of planar extension tests. η effectively controls the ratio of axial to lateral stress in these tests. Theoretical calculations for uniform deformations showed that, for $\eta = 0.6$, the nominal stress along the stretch direction was calculated to be less than that in the transverse direction, a clearly nonphysical prediction [35]. Practical planar extension tests were found to be particularly useful for establishing the value of η, despite their being associated with nonuniform deformation. The value of 0.2 was arrived at in this way, and is consistent with that recommended by Ball et al. [25] on theoretical grounds.

Table 10.2

$N_c kT$ / MPa	$N_{s0} kT$ / MPa	$N_{s\infty} kT$ / MPa	β	η
0.33	6.64	2.13	2.0	0.2

In order to make use of such a planar extension test, it must be modelled using finite elements. The stretched quarter model of the specimen of Figure 10.17 is shown in Figure

10.18. The drawing forces for the complete experiment are compared in Figure 10.19. The value of η was adjusted to give the good agreement between the observed and predicted ratios of the axial and transverse nominal stresses which is apparent in the figure. To illustrate the general effectiveness of the model, observed and modelled strain fields corresponding to the final state of deformation are compared in Figure 10.20. In the experiment, the shear strain rate was a constant $0.01s^{-1}$. The quality of these predictions is good. A similar quality of modelling was found for uniaxial stretching of cylindrical and plane specimens.

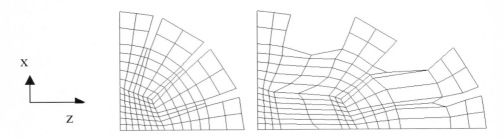

Figure 10.18 Finite element quarter model of the planar extension experiment, with the undeformed model above and the deformed model below. Reproduced by permission of Elsevier Applied Science.

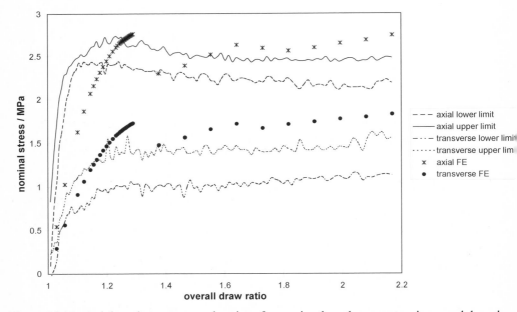

Figure 10.19 Axial and transverse drawing forces in the planar extension model and experiment. The experimental results (lines) span 90% confidence intervals. Reproduced by permission of the publishers Elsevier Applied Science.

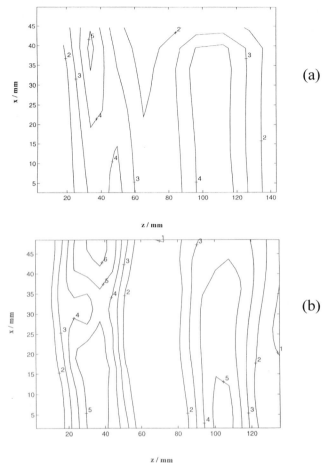

Figure 10.20 Contour plots of axial extension ratio λ_z for the (a) experimental results and (b) the finite element model. Reproduced by permission of Elsevier Applied Science.

An obvious limitation of all the elastic theories is their inability to model rate dependence of the stress. A set of model parameters is therefore associated with a particular rate of strain and is not generally applicable. More significantly, when there is the possibility of a necking instability, the rate dependence controls the speed with which the neck develops; if the rate dependence is sufficiently strong, necking is suppressed altogether. In contrast, for the elastic model, necks may develop at unrestricted speeds. This results in a too-sharp transition in strain between the necked and unnecked parts of the polymer body while the instability is at early stages of development. The good strain predictions in Figure 10.20 refer to a well-developed necked shape; predictions at intermediate stages are less good. A fully rate dependent theory is required to give a good representation of instabilities throughout the deformation history.

10.2.2 Rate dependent constitutive behaviour

The introduction of time dependence in material behaviour adds a level of complexity to any associated numerical process model. In principle, the stress at any time may depend on the entire strain history at any material point. In simpler cases, time is introduced via the dependence of the stress on the current strain rate, so that only the immediate past history of the strain need be known. A theory of this nature results when use is made of the Eyring process mentioned above, or alternatively of Argon's model [24].

The analysis of die drawing of Kukureka, Craggs and Ward [36] is an example of a numerical scheme that includes rate dependent material. In their finite difference analysis, rate dependence of the polymer was introduced in two ways: by the use of empirical data from tensile tests at a range of speeds; and via the Eyring process. The Eyring process gave the better representation of the material behaviour.

Their analysis was of an axisymmetric cylindrical rod being pulled through a conical converging die, as shown in Figure 10.21. It is assumed that the polymer rod maintains contact with the die wall, and the consequent change in geometry ensures that the material, which is hauled off from the die at a constant speed, is subject to an increasing strain rate. Furthermore, adiabatic heating as the material is compressed causes it to increase in temperature. Both these effects are included in the analysis.

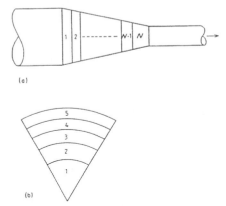

Figure 10.21 Elements used in the analysis of the die-drawing process. (a) axial elements. (b) radial division of each element. Reproduced by permission of Kluwer Academic Publishers.

As in the case of the mandrel-drawing analysis described above, a lower-bound Hoffman-Sachs approach was used for the mechanical analysis. The finite difference method involved dividing the polymer into thin cylindrical slices with axes along the axis of symmetry (see Figure 10.21); each slice is assumed to be in equilibrium. Flow stresses were determined either by the use of equation (7) or by interpolating between experimental tensile data obtained over a range of strain rate and temperature. Heat flow was assumed to be both axial and along the radius, and the temperatures were calculated using a two-dimensional finite difference scheme, which took account of both heat generation originating from mechanical work and of the change in thermal conductivity caused by increased orientation of the polymer.

In the model, the die was assumed to have a 15° semi-angle. A range of die lengths was used, such as to give final draw ratios of between 4 and 10. The material was assumed to be a polyethylene. For a range of haul-off speeds, all the relevant physical quantities were computed. The pressure between the polymer and the die wall was predicted to remain positive for all the conditions studied, implying that contact was maintained; this was in accordance with the observations with this particular polymer. As haul –off speed rises, the increased strain rate results in both higher drawing stress and a greater rise in temperature. The increase in temperature between die entry and exit was calculated to be as high as 10°C at the highest speed. To summarise the effectiveness of the model, observed and computed drawing stresses are compared in Figure 10.22 for two die lengths. The calculation of flow stress via equation (7) gives a more accurate prediction than does the use of the interpolation method.

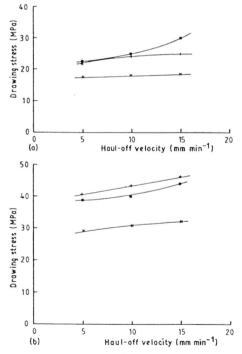

Figure 10.22 Comparison of (●) experimental and calculated values of drawing stress using (+) activated rate theory and (×) uniaxial tension data. (a) die with $R_N = 4$, (b) die with $R_N = 6$. Reproduced by permission of the Kluwer Academic Publishers.

Theories that feature rate dependent behaviour can also be implemented using the finite element method. The commercial code ABAQUS has the capability of including material behaviour defined by the user, with few *a priori* restrictions on the form of the constitutive model used. Buckley and his colleagues have shown that their theory [28], which features both rate dependence and large nonlinear deformations, can be incorporated into this package [37].

Another example of the use of ABAQUS in this way is that of Sweeney et al. [38]. This was an attempt to generalise the elastic modified Ball et al. model described above [34],[35] by introducing rate dependence in a simple way. The constitutive equations, in terms of the principal extension ratios, are derived by operating on the strain energy function (14) with equation (15). For convenience, we rewrite equation (14) as

$$\frac{F}{kT} = \frac{1}{2} N_c \sum_{i=1}^{3} \lambda_i^2 + N_s \sum_{i=1}^{3} b(\lambda_i) \qquad (29)$$

The theory has been implemented by assuming plane stress conditions in the 1-3 plane. Then, the use of equation (15) gives, once p has been eliminated, the constitutive equations for the principal stresses σ_{11} and σ_{33}

$$\sigma_{ii} = \frac{1}{2} N_c kT (\lambda_i^2 - \lambda_2^2) + N_s kT (\lambda_i b'(\lambda_i) - \lambda_2 b'(\lambda_2)) \qquad (i = 1,3) \qquad (30)$$

where the primes denote differentiation. To develop the rate dependent theory, the constant N_s in equation (30) is replaced by a function which is a generalisation of that defined in the relation (27):

$$N_s(I_1, \dot{\gamma}) = \frac{N_{s0}(\dot{\gamma}) - N_{s\infty}(\dot{\gamma})}{(I_1 - 2)^\beta} + N_{s\infty}(\dot{\gamma}) \qquad (31)$$

where $\dot{\gamma}$ is the octahedral shear strain rate and the functions N_{s0} and $N_{s\infty}$ are defined by

$$\left.\begin{array}{l} N_{s0}(\dot{\gamma}) = A_0 + B\ln(\dot{\gamma}/\dot{\gamma}_0) \\ N_{s\infty}(\dot{\gamma}) = A_\infty + B\ln(\dot{\gamma}/\dot{\gamma}_0) \end{array}\right\} \dot{\gamma} \geq \dot{\gamma}_0$$

$$\left.\begin{array}{l} N_{s0}(\dot{\gamma}) = A_0 \\ N_{s\infty}(\dot{\gamma}) = A_\infty \end{array}\right\} \dot{\gamma} < \dot{\gamma}_0 \qquad (32)$$

A_0, A_∞, B and $\dot{\gamma}_0$ are material constants defining the rate dependence. For the assumed two-dimensional plane stress conditions, the octahedral strain rate $\dot{\gamma}$ is related to the principal extension ratios λ_1 and λ_3 by

$$\dot{\gamma} = \frac{\dfrac{\dot{\lambda}_1}{\lambda_1} + \dfrac{\sigma_{33}}{\sigma_{11}} \dfrac{\dot{\lambda}_3}{\lambda_3}}{\sqrt{2} \left[1 + \left(\dfrac{\sigma_{33}}{\sigma_{11}} \right)^2 - \dfrac{\sigma_{33}}{\sigma_{11}} \right]^{\frac{1}{2}}} \qquad (33)$$

In the subroutine which defines the theory in the finite element package, nodal displacements are translated into principal stretches for each element, and the stresses calculated in principal directions using equation (30). Strain rates required in equation (33) are derived from the differences in strain between successive time steps. Principal stresses are transformed into stresses in the global axis set using the appropriate second-order transformation.

The theory thus defined is open to the objection that it predicts a discontinuity in stress when subject to a step change in strain rate. However, it is more realistic than the elastic theory described above [34], which permits strain to increase at an unlimited rate. The main motivation of this development was to eliminate this characteristic.

The values of the parameters A_0, A_∞, B and $\dot{\gamma}_0$ must be consistent with the observed rate dependence of the stress. Uniaxial experiments at different speeds were used to measure the peak stress at the onset of necking (see Figure 10.16), and these data helped to define the four parameters. Trial and error procedures similar to those used in the fitting of the elastic theory data of Table 10.2 were used to derive the remaining parameters. For the polypropylene behaviour at 150°C, the model parameters are given in Table 10.3.

Table 10.3

$N_c kT$ / MPa	$A_0\, kT$ / MPa	$A_\infty kT$ / MPa	BkT / MPa	$\dot{\gamma}_0$ / s^{-1}	η	β
0.37	6.2	2.0	0.2	0.0025	0.2	2.0

The present model was shown to give an overall improvement in the predictions of neck shapes, when compared with the modified Ball et al. elastic model described above. This is illustrated in Figures 10.21 and 10.22, which show predictions of the stretching of plane polypropylene tensile specimens. The tests are at a constant testing speed corresponding to an initial strain rate of 0.012 s^{-1}. Figure 10.23 compares quarter models of the specimens at an intermediate stage of necking. Results generated using the elastic theory and the material data of Table 10.2 are compared with those generated using the rate dependent model and the data of Table 10.3; it is clear that there is a more abrupt transition from necked to unnecked material in the elastic model. The two numerical results are compared with the experimental result at 150°C in Figure 10.24. The strain measurements were made using an image capture technique [39]. The comparison is made in terms of the axial extension ratios along the central axis of the specimen. The rate dependent model gives the more realistic strain prediction.

In essence, the model outlined above is an *ad hoc* improvement on an elastic theory, with a particular end in view – the simulation of neck development at a realistic rate. There are circumstances in which we would expect the model to give predictions which are qualitatively incorrect. In particular, a step strain history would give an instantaneous stress drop as the strain became constant. The theory [40] which is described next overcomes this difficulty and produces qualitatively correct responses for a more extensive set of conditions.

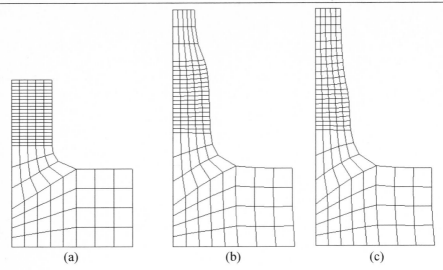

(a)   (b)   (c)

Figure 10.23 Quarter models of tensile specimens. (a) is undeformed, (b) the deformed elastic model, and (c) the deformed rate dependent model.

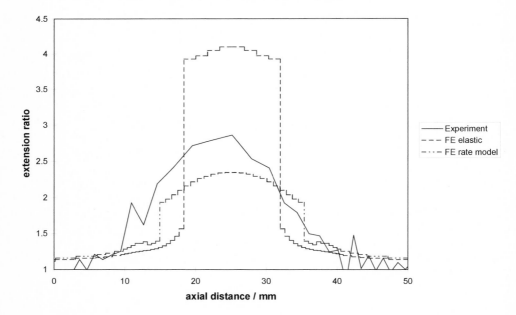

Figure 10.24 Comparisons of axial extension ratio profiles for the two models. The rate dependent theory gives the more realistic prediction.

There is some physical motivation behind this theory. It is recognised that semi-crystalline polymers are essentially two-phase materials, with a predominantly crystalline hard phase coexisting with a largely rubbery soft phase. Large deformations are accompanied by rearrangement of the microstructure. We model the rubbery phase using the Ball et al. network model, and simulate the effects of the hard phase by the device of a rigid sphere embedded in a representative volume of the network. The rigid inclusion is spherical in shape purely because of symmetry considerations. Its purpose is twofold: to introduce strain concentration, so that the Ball et al. network attains its stress peak at a realistic overall strain; and to introduce time dependence, by reducing in size at a rate determined by shear stress and thus lessening the strain concentrating effect. The latter process resembles the collapse of the microstructure, with the accompanying increase in the proportion of polymer chains participating in the deformation process.

A representative volume with its spherical inclusion is shown in Figure 10.25, in the form of a unit octant containing a sphere of radius R < 1. The effect of the sphere on the strain in the surrounding network is estimated using a simple geometrical argument. We consider lines of material along principal directions, and assume that on deformation they maintain the same orientation. In the undeformed state, consider such a line along the 1 direction joining the point (X_1,X_2,X_3) on the sphere surface to the point $(1,X_2,X_3)$ on the plane P of Figure 10.25.

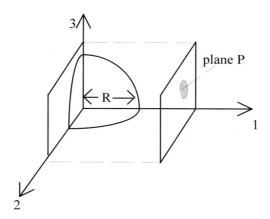

Figure 10.25 Schematic of sphere embedded in unit octant. Reproduced by permission of John Wiley & Sons.

On deformation to extension ratio λ_1, the point on plane P moves to (λ_1,X_2,X_3), so that the extension ratio locally is given by

$$\lambda_1' = \frac{\lambda_1 - X_1}{1 - X_1} \tag{34}$$

For parallel lines not touching the sphere, the local extension ratio remains λ_1. We obtain an effective extension ratio $\overline{\lambda}_1$ by averaging over the octant:

$$\overline{\lambda}_1 = \int_0^R \int_0^{\sqrt{R^2 - X_3^2}} \lambda_1'(X_2, X_3) dX_2 dX_3 + \lambda_1(1 - \frac{1}{4}\pi R^2) \quad (35)$$

Using the relation (34) and the equation for the sphere surface now gives

$$\overline{\lambda}_1 = \int_0^R \int_0^{\sqrt{R^2 - X_3^2}} \frac{\lambda_1 - \sqrt{R^2 - X_2^2 - X_3^2}}{1 - \sqrt{R^2 - X_2^2 - X_3^2}} dX_2 dX_3 + \lambda_1(1 - \frac{1}{4}\pi R^2) \quad (36)$$

The same argument applies for the other two principal directions, so that in general

$$\overline{\lambda}_i = \int_0^R \int_0^{\sqrt{R^2 - X_k^2}} \frac{\lambda_i - \sqrt{R^2 - X_j^2 - X_k^2}}{1 - \sqrt{R^2 - X_j^2 - X_k^2}} dX_j dX_k + \lambda_1(1 - \frac{1}{4}\pi R^2) \quad (37)$$

$$\{i, j, k\} = \{1, 2, 3\}$$

The theory consists of using the effective stretches defined in equation (37) in place of the macroscopic extension ratios in the Ball et al. model of equations (14) and (29). For the plane stress implementation used here, the constitutive equation is of the form of equation (30) with $\overline{\lambda}_i$ replacing λ_i.

Having introduced the strain concentration, the next step is to introduce time dependence by defining how the value of the radius R evolves throughout the deformation process. We assume that the time derivative of R is controlled by the current value of shear stress. We define a stress τ such that

$$\tau = \sigma_{max} - \sigma_{min} \quad (38)$$

where σ_{max} and σ_{min} are respectively the greatest and smallest principal stresses. Then, R is defined by

$$\frac{dR}{dt} = -(R - R_\infty)A[\exp(B\tau) - 1] \quad (39)$$

where t represents time and A, B and R_∞ are constants, the last defining a lower limit for R. This is an empirical relationship which has been shown to perform realistically. In particular, it provides a realistic relationship between peak nominal stress and strain rate when used to model tensile stretching. To complete the material specification, the initial value of R, R_0, must be specified.

The material model defined by equations (29), (37) and (39) has been programmed into ABAQUS as a user-defined material. The integral of equation (37) is evaluated numerically as described previously [40]. The solution is incremental; at each time step, the derivative of R is calculated for each element using equation (39), and R updated. Transformations in and out of principal directions are carried out in the same way as for the rate dependent theory described above.

Table 10.4

N_ckT / MPa	N_skT / MPa	η	A / s^{-1}	B / MPa	R_0	R_∞
0.257	1.70	0.23	5×10^{-5}	1.49	0.99	0.60

A set of material parameters which provide an effective model for polypropylene at 150°C is given in Table 10.4. It has been shown that these values result in a good overall performance, both for continuously increasing strains in uniaxial and planar tension experiments, and for stress relaxation in uniaxial tension. For uniaxial testing, we examine the predictions of both the shapes of the necking specimens and the associated drawing forces. The specimen is the same as that modelled in Figure 10.23, and is stretched at the same constant testing speed corresponding to a strain rate of 0.012 s^{-1}. The development of the axial strain profile is shown in Figure 10.26, as the specimen is stretched from soon after the onset of necking at an overall draw ratio of 2.1, to an overall draw ratio of 2.9.

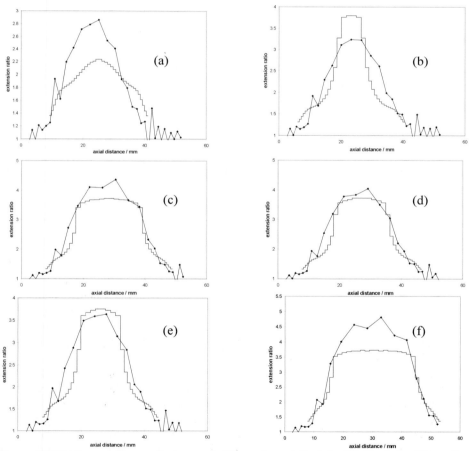

Figure 10.26 Comparisons of observed (♦) and modelled (solid line) profiles of axial extension ratio, as the tensile specimen is extended. Overall draw ratios are (a) 2.1, (b) 2.2, (c) 2.3, (d) 2.5, (e) 2.6 and (f) 2.9. Reproduced by permission of John Wiley & Sons.

The strains are modelled well during the intermediate stages, though there is some fall-off in accuracy at later stages of deformation. The model reaches a constant 'natural' draw ratio of around 3.6 in the neck, whereas in the experimental result the strain in the neck continues to slowly increase. The behaviour of the model at this point corresponds to the sphere radius having reached R_∞, at which point the material becomes elastic. The real material does not attain a true elastic state, though we would expect its rate dependence to decline as its molecular orientation develops.

The corresponding nominal stress-strain curve is shown in Figure 10.27. The strain is that corresponding to the neck centre; conditions are truly uniaxial here, though it should be borne in mind that that the rate of strain is not constant. The position of the nominal stress peak is modelled well. The only significant discrepancy is that, in the model, the strain does not increase beyond a limiting extension ratio of ~3.7. This corresponds to the elastic state of the neck referred to above; indeed, there are signs of elastic recovery which are a necessary accompaniment to relaxation of stress in the area outside the neck centre.

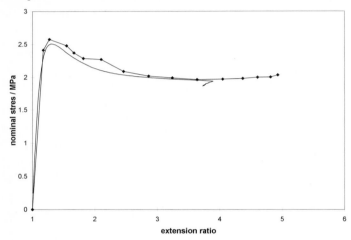

Figure 10.27 Observed (♦) and predicted (solid line) stress-strain curves for the tensile specimen. The observed strains are derived from captured images of the neck centre. Reproduced by permission of the publishers John Wiley & Sons.

Observed and predicted deformation fields for the planar extension test are shown in Figure 10.28. They correspond to the specimen geometry of Figure 10.17, stretched at a constant shear strain rate of $0.01s^{-1}$, and the finite element mesh of Figure 10.18. Necking instabilities occur between the separating grips as the deformation proceeds, and the resulting pattern of strains is well predicted. More validation of the model is provided by the drawing forces, shown in Figure 10.29. The boundary forces from by the finite element model are translated into nominal stresses, and compared with the measurements. Both lateral and transverse forces show good agreement with the predictions. This result verifies the choice of the η parameter in the Ball et al. model; as mentioned above in the discussion of the elastic modified Ball et al. model, the ratio of axial and transverse forces depends strongly on η.

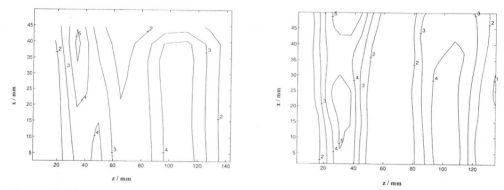

Figure 10.28 Contour plots of axial extension ratio λ_z for the experimental results (left) and the finite element model. Reproduced by permission of John Wiley & Sons

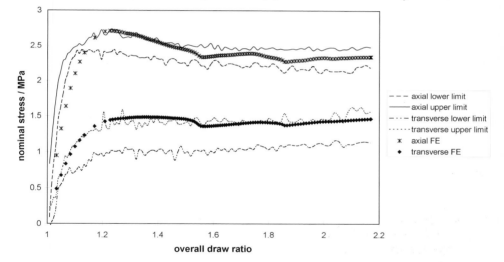

Figure 10.29 Axial and transverse drawing forces in the planar extension model and experiment. The experimental results span 90% confidence intervals. Reproduced by permission of John Wiley & Sons.

In this model, the presence of shear stress under conditions of constant strain causes the sphere radius to decay, a consequent lessening in the effective strain, and so a relaxation in stress. Further verification of the model is therefore possible by direct comparison with stress relaxation experiments. Tensile stress relaxation experiments were carried out using the same specimen geometry and finite element meshes (Figure 10.23) as for the tensile stress-strain experiments. The step-like deformation histories of the experiments were copied in the finite element model to produce the predictions, one of which is compared with experiment in Figure 10.30. The prediction is realistic up to the time of 900s.

In summary, the model has been subject to a broad range of experimental verification. It gives a good overall performance in the prediction of shapes, forces and rate-dependence.

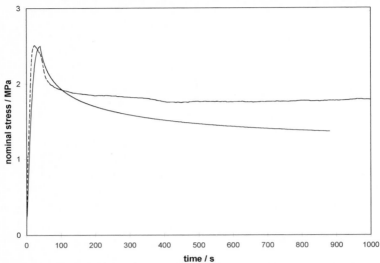

Figure 10.30 Stress relaxation of the tensile specimen at an overall draw ratio of 1.5. (- - - observed, ——— FE model) Reproduced by permission of John Wiley & Sons.

References

1   R.S Rivlin, Phil. Trans. Roy Soc., **A240**, 459, 491 (1948), **A241**, 379, (1949).
2   I.M. Ward, Mechanical Properties of Solid Polymers, Second Edition, John Wiley & Sons, Chichester, 1983, p55.
3   R.W. Ogden, P. Chadwick & E.W. Haddon, Quart. J. Mech. Appl. Math., **26**, 27 (1973).
4   K.C. Valanis & R.F. Landel, J. Appl. Phys., **38**, 2997 (1967).
5   O.H. Varga, Stress strain behaviour of elastic materials, John Wiley, New York, 1966.
6   J. Sweeney & I.M. Ward, Trans. Inst. Chem. Eng. A., **71**, 232, (1993).
7   J. Sweeney & I.M. Ward in IUTAM Symposium on Anisotropy, Inhomogeneity and Nonlinearity in Solid Mechanics, D. F. Parker and A. H. England (Eds.), Kluwer Academic Publishers (1995).
8   C. G'Sell & J.J. Jonas, J. Mater. Sci, **14**, 1979 (583).
9   C. G'Sell, J.H. Hiver, A. Dahoun & A. Sonahi, J. Mater. Sci., **27**, 5031, (1992).
10   C. G'Sell, N.A. Aly-Helal, S.L. Semiatin and J.J. Jonas, Polymer, **33**, 1244, (1992).
11   S. Nazarenko, S. Bensason, A. Hiltner & E Baer, Polymer, **35**, 3883 (1994).

12   R.N. Haward & G. Thackray, Proc. Roy. Soc., London, Ser. A., **302**, 453, (1968).
13   L.R.G. Treloar, The Physics of Rubber Elasticity, Third Edition, Clarendon Press, Oxford, 1975 p102.
14   I.M. Ward, Polym. Eng. Sci., **24**, 274 (1984).
15   E.M. Arruda & M.C. Boyce, International Journal of Plasticity, **9**, 697, (1993).
16   C.P. Buckley & D.L. Jones, Polymer, **36**, 3301 (1995).
17   J. Sweeney & I.M. Ward, Polymer, **36**, 299, (1995).
18   J. Sweeney, Comp. and Theoret. Poly. Sci., **9**, 27-33, (1999).
19   E M Arruda & M C Boyce, J. Mech. Phys. Solids, **41** ,389, (1993).
20   P.D. Wu & E. van der Geissen, Mech. Res. Comm., **19**, 427, (1992).
21   S.F. Edwards & T.A. Vilgis, Polymer, **27**, 48, (1986).
22   M.C. Boyce, D. M. Parks & A.S. Argon, Mech. Mater., **7**,15, (1988).
23   H.M. James & E.M. Guth, J. Chem. Phys., **11**, 455 (1943).
24   A.S. Argon, Phil. Mag., **28**, 839, 1973.
25   R.E. Ball, M. Doi, S.F. Edwards & M. Warner, Polymer, **22**, 1010, (1981).
26   P.D. Wu & E. van der Geissen, J. Mech. Phys. Solids, **41**, 427, (1993)
27   R.G. Matthews, R.A. Duckett, I.M. Ward & D.P. Jones, Polymer, **38**, 4795, (1997).
28   C.P. Buckley, D.C. Jones & D.P. Jones, Polymer, **37**, 2403 (1996).
29   A.M. Adams, C.P. Buckley & D.P. Jones, Mechanics of Plastics & Plastic Components, ASME, 1995, p365.
30   H.F. Nied, C.A. Taylor & H.G. Delorenzi, Polymer Engineering and Science, **30**, 1314 (1990)
31   H.G. Delorenzi & H.F. Nied in Modeling of Polymer Processing , A.I. Isayev (ed.) Hanser (1991).
32   S. Kakadjian, G. Craggs & I.M. Ward, Proc. Inst. Mech. Eng., **210**, 65, (1996).
33   J. Sweeney & I.M. Ward, J. Rheol., **39**, 861-872 (1995).
34   J. Sweeney & I.M. Ward, J. Mech. Phys. Solids, **44**, 1033 (1996).
35   J. Sweeney & I.M. Ward, Polymer, **38**, 5991-5999, (1997).
36   S.N. Kukureka, G. Craggs & I.M. Ward, J. Mat. Sci., **27**, 3379, (1992).
37   C. Gerlach, C.P. Buckley & D.P. Jones, Trans. I. Chem. E., **76**, 38, (1998).
38   J. Sweeney, T.L.D. Collins, A.P. Unwin & I.M. Ward, ASME Journal of Engineering Materials and Technology, **119**, 228, (1997).
39   A.R. Haynes, & P.D. Coates, , J. Mater. Sci., **31**, 1843, (1996).
40   J. Sweeney, T.L.D. Collins, P.D. Coates and R.A. Duckett, Journal of Applied Polymer Science, **72**, 563-575, (1999).

Index